Digital Electronics

Theory and Experiments

Digital Electronics

Theory and Experiments

(SECOND EDITION)

VIRENDRA KUMAR
MSc, Physics
(Indian Civil Service)

Publishing Globally

NEW AGE INTERNATIONAL (P) LIMITED, PUBLISHERS
New Delhi • Bangalore • Chennai • Cochin • Guwahati
Hyderabad • Kolkata • Lucknow • Mumbai
Visit us at **www.newagepublishers.com**

Published by New Age International (P) Ltd., Publishers
First Edition: 2002
Second Edition: 2015

BRANCHES

- **Bangalore** 37/10, 8th Cross (Near Hanuman Temple), Azad Nagar, Chamarajpet Bangalore-560 018, Tel.: (080) 26756823, Telefax: 26756820 **E-mail: bangalore@newagepublishers.com**
- **Chennai** 26, Damodaran Street, T. Nagar, Chennai-600 017, Tel.: (044) 24353401 Telefax: 24351463, **E-mail: chennai@newagepublishers.com**
- **Cochin** CC-39/1016, Carrier Station Road, Ernakulam South, Cochin-682 016 Tel.: (0484) 2377303, Telefax: 4051303 **E-mail: cochin@newagepublishers.com**
- **Guwahati** Hemsen Complex, Mohd. Shah Road, Paltan Bazar, Near Starline Hotel Guwahati-781 008 Tel.: (0361) 2513881, Telefax: 2543669 **E-mail: guwahati@newagepublishers.com**
- **Hyderabad** 105, 1st Floor, Madhiray Kaveri Tower, 3-2-19, Azam Jahi Road, Near Kumar Theater Nimboliadda, Kachiguda, Hyderabad-500 027, Tel.: (040) 24652456 Telefax: 24652457 **E-mail: hyderabad@newagepublishers.com**
- **Kolkata** RDB Chambers (Formerly Lotus Cinema) 106A, 1st Floor, S N Banerjee Road Kolkata-700 014 Tel.: (033) 22273773, Telefax: 22275247 **E-mail: kolkata@newagepublishers.com**
- **Lucknow** 16-A, Jopling Road, Lucknow-226 001, Tel.: (0522) 2209578, 4045297 Telefax: 2204098, **E-mail:lucknow@newagepublishers.com**
- **Mumbai** 142C, Victor House, Ground Floor, N.M. Joshi Marg, Lower Parel Mumbai-400 013, Tel.: (022) 24927869, Telefax: 24915415 **E-mail: mumbai@newagepublishers.com**
- **New Delhi** 22, Golden House, Daryaganj, New Delhi-110 002 Tel.: (011) 23262368, 23262370, Telefax: 43551305 **E-mail: sales@newagepublishers.com**

ISBN: 978-81-224-3892-5

₹ 299.00

C-14-11-8080

Printed in India at Kanishik Printing Press, New Delhi.
Typeset at Excellent Graphics, New Delhi.

PUBLISHING GLOBALLY
NEW AGE INTERNATIONAL (P) LIMITED, PUBLISHERS
7/30 A, Daryaganj, New Delhi - 110002
Visit us at **www.n wagepublishers.com**

To

DHARMVATI

My Wife

Preface to the Second Edition

After the good response of the earlier editions it is a pleasure for us to present the second edition. In this edition the style pattern of the book has been modified to give a new look. Necessary corrections are incorporated in response to the suggestions made by many teachers and students and all attempts are made to make the book error free. We would like to acknowledge the valuable advice and suggestions of many teachers and students who used the earlier edition. We would be grateful to our esteem readers for suggestions in the interest of the subject.

Publisher

Preface to the First Edition

This Laboratory Manual for experiments in digital electronics has been designed for students pursuing a course in digital technology to enable them to conduct experiments so that they may acquire a better understanding of digital circuits and digital logic, than is possible by merely following a text on digital technology. This Manual also serves the purpose of familiarizing students with the practical applications of digital devices, which should enable them to acquire adequate experience so that they develop the capability to devise their own solutions to problems which they come across in handling and designing practical devices.

This Manual is, therefore, intended to serve the purpose of providing hands-on experience to students, which is essential for a thorough grasp of digital logic, and for that purpose it is also meant to serve as a useful companion to a text on digital technology.

It was considered necessary to introduce students to the essential characteristics and operating features of logic devices before introducing them to experiments on digital devices. The first two chapters have, therefore, been devoted to the study of TTL and CMOS devices and their operating features. It was considered necessary that students should make themselves thoroughly familiar with the essential characteristics of logic devices so that, in time, they are in a better position to select the right device for the intended application and are also able to get the best out of these devices. To help the student in this area, lists of useful digital devices have been given in Appendices 6 and 7.

It is necessary to point out here that the symbols used in the Manual to designate digital devices are the conventional ones which, by and large, are normally used in digital texts. However, in recent years new standards for symbols have been laid down. In the long run it will help the student to make himself/herself familiar with these new standards, as some of these symbols are in use even today. These new standards are available as IEE Standard 91-1984.

Chapters 3 to 6 have been devoted to the study of various types of logic gates and include experiments which are designed to make the student familiar with the essential features and operating characteristics of these devices.

Chapter 7 deals with the problems associated with interfacing between logic devices belonging to different families. Many types of interfacing problems have been considered and practical solutions have been explored.

Chapters 8 to 17 deal with more complex logic devices, like multivibrators, shift registers, counters, decoders, encoders, multiplexers, demultiplexers, arithmetic logic circuits, memories, LED displays, Analog/Digital and Digital/Analog converters.

All the chapters begin with a discussion of the theory of the device being considered, its operating characteristics and the results that may be expected from the associated experiments. Each experiment is assigned a set of objectives which the student is expected to realize. This is followed by detailed step-by-step operating procedures for performing the experiment. The student is expected to record his experimental observations in the tables provided for the purpose and compare his/her observations with completed experimental tables given in Appendix 3. In some cases the expected results are provided in the theoretical discussion in the introductory section of each chapter.

The pin-out diagrams for all the integrated circuits used in experimental work in this Manual have been given in Appendix 1 and a complete list of equipment and parts required for experiments is given in Appendix 2. In the Appendix section we have also discussed the application of Timer and Operational amplifier (Appendices 5 and 11). Although these are linear devices, they still find considerable application in digital systems, and any discussion of digital systems would have been incomplete without reference to the applications of the Timer and the Op-amp. For the same reason a brief reference has been made to Boolean algebra in Appendix 10 as it is the backbone of all logic systems.

The experiments covered in this Manual can be performed on a logic trainer, which are commercially available or by assembling the various components and devices on a breadboard. To help students to extend the scope of functions normally available in logic trainers, circuit diagrams have been given in Appendix 12 for building logic monitors, pulser switches, clocks and power supplies for TTL devices as well as for Operational amplifiers. As Op-amps require a dual balanced supply, power supply splitters have been considered which will enable Op-amps to be operated on variable dual balanced power supplies.

It is considered desirable that the various experiments are performed in the numerical sequence followed in the Manual. This will enable students to progress from simpler devices to more complex logic devices and in this process they will enhance their own ability to perform experiments and interpret the results.

The Manual is entirely self-contained and it provides the necessary theoretical background for a proper understanding of the operation of the devices discussed. However any text on digital technology of an appropriate level and scope can be consulted if more information on digital devices is desired.

Virendra Kumar

Contents

CHAPTER 1

Logic Systems

1.1 INTRODUCTION

The binary number system can be visualized as one having two discrete states. The term binary can represent any two-state device. The two states of a binary system are designated by symbols 0 and 1. For instance, you can represent the OFF and ON states of an electrical circuit by these two symbols. If 1 represents the ON state; 0 can represent the OFF state. It really depends on the convention we choose to adopt. We can also adopt 0 for the ON state and 1 for the OFF state.

In digital electronics the binary system of representation is used mainly because the two states in the binary system are very specific and convenient to adopt. We can also represent voltage levels in the binary system. Consider the following representation of voltage levels in the binary system:

0 volt : Binary 0

and + 5 volt : Binary 1

or $\left\{ \begin{array}{l} 0 \text{ volt : Binary 1} \\ +5 \text{ volt : Binary 0} \end{array} \right.$

It is really a matter of convention that you choose to adopt. The binary designation of voltage levels can also be used to represent logic levels and in fact it is the common practice. For instance we can refer to + 5 V as logic 1 level and 0 V as logic 0 level. However, normally logic 1 level is considered to be the more positive of the two levels and the logic 0 level is assigned to the more negative of the two levels. This convention is referred to as positive logic. In negative logic the more negative of the two levels is designated as the logic 1 level. We will come back to the topic of logic polarity a little later.

1.2 LOGIC GATES

Logic gates are used in digital circuits for the purpose of logical decisions. There are the following basic types of gates for which symbols are given in Table 1.1.

NOT gate or Inverter
AND gate
NAND gate
OR gate
NOR gate

You can visualize a logic gate as a two-state logic element. In other words, you may say that its output may be either low or high, that is logic 0 or logic 1. The state of its output depends on its functional type, as well as on the present condition of its inputs. Logic gates do not possess memory and therefore, the output changes with the change of input levels. We will, at a later stage, consider flip-flops which are combinations of logic gates and possess memory.

Some other gates are derived from these basic gates. The NOT gate or Inverter is used to invert or complement the logic level. For instance, if the logic level is high or 1, it can complement it to 0 as the complement of 1 is 0. If the logic level is 0, it can complement it to 1, since 0 = 1 (it is read as Not $\overline{0}$ = 1). An example of complementation is seen in the NAND gate which is a contraction for NOT-AND. In other words, if the output of an AND gate is complemented by using an Inverter, it becomes a NAND gate. Similarly, the NOR gate is a contraction for NOT-OR. The OR gate functions as a NOR gate if its output is complemented.

The Inverter has a single input and a single output. However, the other gates may have two or more inputs and a single output. As there are only two logic levels in digital circuits, the output of a gate is either high (logic 1) or low (logic 0) in positive logic. The several inputs to a logic gate may all have different logic levels; but that depends on the functions being assigned to the various inputs.

Three methods are available to us for describing and predicting the input/output properties of gates and of combinations of gates. These are :

* Boolean algebra
* Truth tables, and
* Timing diagrams.

We will take up the Boolean algebra first and see how the input/output relationship can be expressed in an equation form.

1.3 BOOLEAN ALGEBRA

Boolean expressions consist of variables like A, B, C etc., which can have either of the two values 0 or 1, constants and operators. The result of a Boolean function is either 0 or 1. For instance, $A \cdot B$ (read as A and B) can either be 1 or 0. Boolean algebra uses symbols which represent operators. The symbols have been summarized below:

Table 1.1

	Device	*Abbreviation*	*Symbol*
1.	Inverter	$Y = \overline{A}$	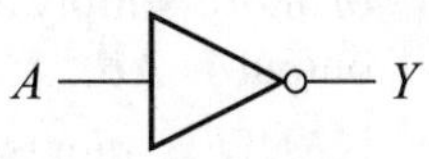
2.	Buffer	$Y = A$	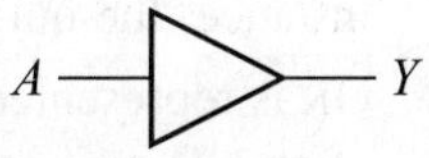
3.	AND gate	AND $Y = A \cdot B$	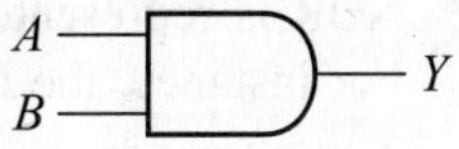
4.	NAND gate	NAND $Y = \overline{A \cdot B}$	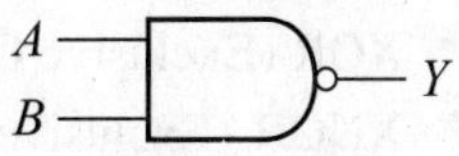
5.	OR gate	OR $Y = A + B$	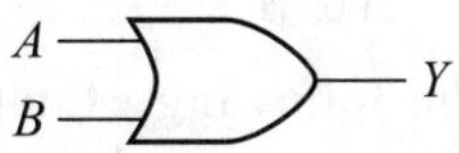
6.	NOR Gate (Not OR gate)	NOR $Y = \overline{A + B}$	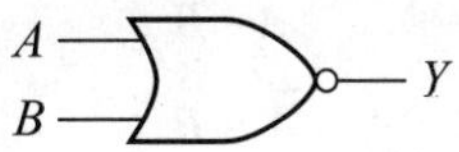
7.	Exclusive – OR gate	XOR $Y = A \oplus B$	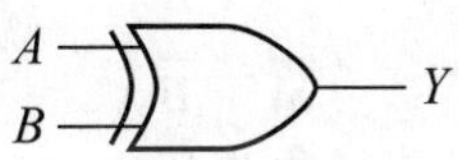
8.	Exclusive – NOR gate	XNOR $Y = \overline{A \oplus B}$	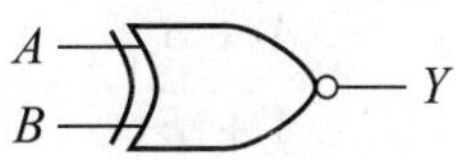
9.	Set-Reset flip-flop	FF	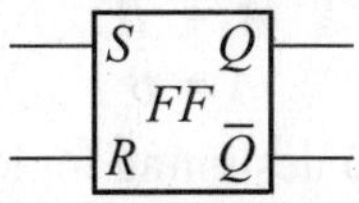
10.	J K flip-flop		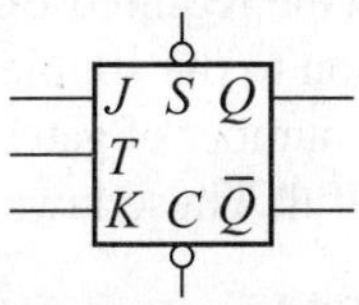
11.	Schmitt trigger		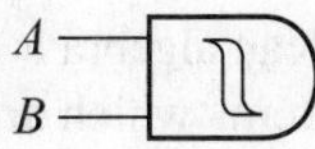

* INVERT (Complement) uses a bar over a variable.

  $\overline{A}$ is read as 'Not A'. It represents the complement of A.
* AND is represented by (·) or (×) between variables. It is written as $A \cdot B$ or more simply AB. If A and B are the two inputs of an AND gate, its output is AB.
* NAND is represented by a bar over the complemented variables. For instance, the output of a 2-input NAND gate can be stated as $\overline{AB}$.
* OR is represented by a + (plus) sign between variables. If the two inputs of an OR gate are A and B, its output will be $A + B$ (read as A or B).
* NOR is represented by a bar over both the variables which are ORed. For instance, the NOR output of a 2-input NOR gate can be written as $\overline{A+B}$.
* XOR (Exclusive OR) output of a 2-input XOR gate is written as $A \oplus B$.
* XNOR (Exclusive NOR) output of a 2-input XNOR gate is written as $\overline{A \oplus B}$.

The following examples will enable you to understand how Boolean expressions should be read :

$A \cdot B$	A and B
$\overline{A} \cdot B$	Not A and B
$A \cdot \overline{B}$	A and Not B
$\overline{A} \cdot \overline{B}$	Not A and Not B
$A + B$	A or B
$\overline{A} + B$	Not A or B
$A + \overline{B}$	A or Not B
$\overline{A} + \overline{B}$	Not A or Not B
$A = B$	A equals B

When designing logic circuits the normal practice is to draw up a truth table and then the required Boolean expression. It is necessary to reduce the Boolean expression to the simplest form so that the design can be implemented with the smallest number of gates. For this exercise the laws of Boolean algebra are found very useful. These laws have been summarized in Appendix 10.

Truth Table

Just as Boolean algebra is found useful in describing the input/output relationship of logic systems which are made up of combinations of gates; so also the input/output relationship of logic systems is summarized in a *Truth Table*. The truth table gives a summary in a tabular form of the outputs which result from the several combinations of the inputs to a logic gate or a combination of logic gates. For analyzing the performance of a logic circuit, it is useful to work out a truth table for the evaluation of output for the different input combinations.

Truth tables for both positive and negative logic for basic logic gates are given in Appendix 6. The implementation of logic gates using NAND and NOR logic has also been given in this Appendix for some logic gates.

1.4 LOGIC POLARITY

We have, earlier on, briefly referred to logic polarity. While mostly positive logic is normally employed in designing logic circuits, there are, however, some instances when negative logic is found more convenient. There are also some systems which use both positive and negative logic. By and large most of the systems use positive logic. This implies that + 5 V means logic 1 level and 0 V represents logic 0 level. In actual practice, however, it is not possible to achieve these voltage levels precisely. Logic circuits are therefore so designed that voltages exceeding 2.5 V are taken to be High (logic 1), and voltages less than 0.8 V are considered to be Low (logic 0).

Figure 1.1 shows the voltage level requirements for positive logic and Fig. 1.2 shows the voltage level requirements for negative logic.

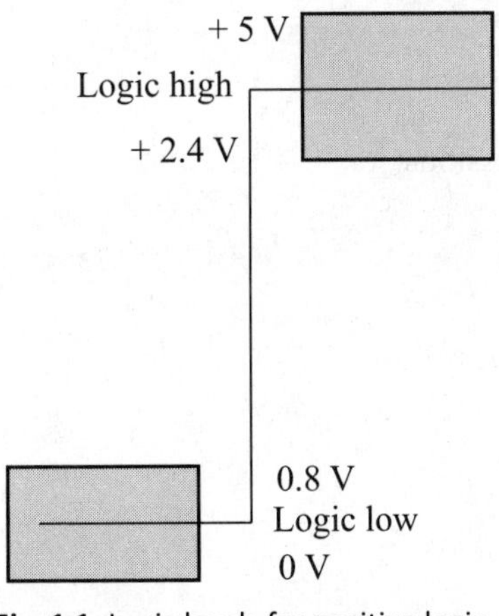

Fig. 1.1. Logic levels for positive logic.

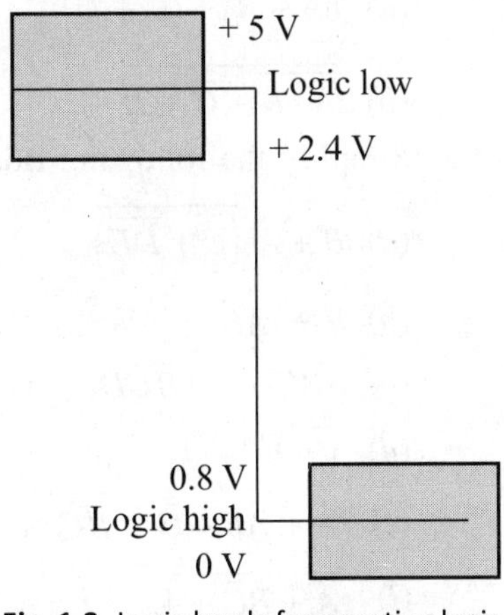

Fig. 1.2. Logic levels for negative logic.

1.5 LOGIC TRAINER

The experiments described in this manual can be performed on a Logic Trainer which has the following facilities:

1. Regulated Power Supplies
   (*a*) + 5 V D.C., better than 2%, 1 A.
   (*b*) + 12 V D.C., better than 1% at 100 mA.
   (*c*) – 12 V D.C., better than 1% at 100 mA.

2. One TTL Compatible Clock : Duty Cycle 45%
   Output frequencies : 1 Hz, 1 kHz, 10 kHz and 100 kHz
   5 V peak-to-peak, normal and complement outputs.
3. Four Logic Switches
   States : Logic High '1' (+5 V)
   Logic Low '0' (Ground).
4. Four Logic Level Monitors.
5. Two Bounceless Pushbutton Switches providing complementary pulser outputs of + 5 and + 0.2 V.

For experimenters who are interested in enlarging the scope of these facilities, some circuits have been provided in Appendix 12. The front panel layout of a typical logic trainer is given in Fig. AP-12.1 in Appendix 12. It should help students to follow the instructions given for carrying out experiments.

PROBLEMS

1.1. State DeMorgan's theorem and use it to simplify the following expressions:

(*a*) $\overline{\left(\overline{A+B}) + (\overline{C+D}\right)}$

(*b*) $\overline{\overline{\overline{A}\cdot\overline{\overline{B}}} + \overline{C} + D}$

1.2. Simplify the following Boolean expressions:

(*a*) $B + A\left(\overline{C + \overline{DE}}\right)$

(*b*) $B + A\overline{B}$

(*c*) $ABCD + A\overline{B}CD$

(*d*) $\overline{1 + \overline{\overline{B}}}$

(*e*) $\overline{A} + AB + \overline{C} + \overline{B}C$

(*f*) $\overline{\overline{A} + B\overline{\overline{C}} + 4}$

1.3. Prove that $\overline{\overline{A}\cdot\overline{B}} = A + B$

1.4. Prove that $\overline{\overline{A} + \overline{B}} = A \cdot B$

1.5. Illustrate the concept of logic polarity with examples.

1.6. From the following examples of logic polarity select those which stand for negative logic polarity:

(*a*) Logic 0 = + .1 V
   Logic 1 = + 3.5 V

(*b*) Logic 0 = – 6 V
Logic 1 = 0 V

(*c*) Logic 0 = + 15 V
Logic 1 = + 1 V

(*d*) Logic 0 = 0 V
Logic 1 = – 6 V

(*e*) Logic 0 = + 1 V
Logic 1 = + 15 V

(*f*) Logic 0 = + 3.5 V
Logic 1 = + .1 V

CHAPTER 2

Logic Families

2.1 INTRODUCTION

When we speak of a logic family we are referring to devices which have compatibility within the family in respect of the important parameters like supply voltage, input and output voltage levels, so that within the family different devices may be interconnected. These devices are fabricated using advanced techniques and are called *integrated circuits.* With the help of this technique integrated devices can be made containing a large number of transistors, diodes, resistors etc. They are broadly classified as follows :

Number of gates	*Integrated circuit category*	*Abbreviation*
1–12	Small Scale Integration	SSI
13–99	Medium Scale Integration	MSI
100–9999	Large Scale Integration	LSI
10,000–99,999	Very Large Scale Integration	VLSI
Over 100,000	Ultra High Scale Integration	UHSI

There are two main technologies which are used for the fabrication of integrated circuits. They are the bipolar and CMOS technologies. The most popular family in the bipolar technology is the TTL family. The following families also belong to the bipolar technology:

* Integrated injection logic (I^2L) and
* Emitter coupled logic

The following families belong to the MOS technology:

* CMOS logic
* PMOS logic and
* NMOS logic.

2.2 NUMBERING SYSTEM FOR DIGITAL ICs

Digital ICs carry a number code which is printed on top of the ICs to help in the identification of the devices.

The information code is usually as follows:

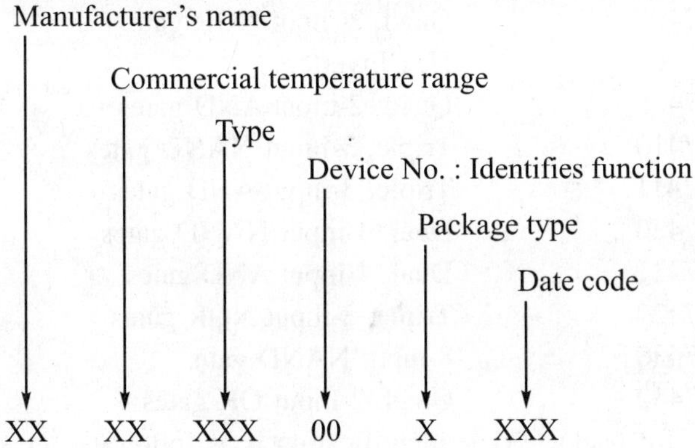

The codes of some of the manufacturers are as follows:

DM	National semiconductors
F	Fairchild
GD	Goldstar
HD	Hitachi
IM	Intersil
KS	Samsung
LR	Sharp
MC	Motorola
MN	Panasonic
N	Signetics
SN	Texas instruments
TC	Toshiba

The temperature range for TTL devices is as follows:

7400 series : 0-70°C : It is for commercial applications

5400 series : – 55°C to 125°C : For military applications

The type numbers for the TTL ICs are as follows:

000	Standard TTL
H00	High speed TTL
L00	Low power TTL
S00	Schottky TTL
AS00	Advanced Schottky TTL
LS00	Low power Schottky TTL
ALS00	Advanced low power Schottky TTL

The device numbers for some of the common TTL ICs are as follows:

7400	Quad, 2-input NAND gate
7402	Quad, 2-input NOR gate
7404	Hex Inverters
7408	Quad, 2-input AND gates
7410	Triple, 3-input NAND gates
7411	Triple, 3-input AND gates
7420	Dual, 4-input NAND gates
7421	Dual, 4-input AND gates
7427	Triple, 3-input NOR gates
7430	8-input NAND gate
7432	Quad, 2-input OR gates

The codes for package type identification is as follows:

N	Plastic DIP (Dual-in-line package)
J	Ceramic DIP
D	Glass/metal DIP
W	Flat pack

2.3 TTL CHARACTERISTICS

It is important to consider the limitations of these devices in order to get the best possible performance from devices which incorporate them. We will, therefore, consider the parameters which govern their performance.

Floating Inputs

If any gate in a TTL package is not used, it is not necessary to make any connections to it. This, however, is not true for CMOS devices. In a CMOS device any unused input must be connected to V_{CC} or ground.

An open TTL input, that is, one which is not connected is the equivalent of a high input; but it cannot be regarded as a reliable high because noise pulses can drive it low if it is left open. An open input can disable an entire logic system or make its performance erratic.

An unused input should be so connected that the output does not get stuck at any logic level. For instance consider an OR gate. If it has an open input which acquires a logic 1 level, the output will always be logic 1, no matter what you do with the other inputs. In this case, the unused input may be connected to other inputs or to ground.

Similarly, if an unused input in an AND gate acquires a logic 0 level its output will get stuck at logic 0 level. The simple rule to be followed is to so connect an unused input to a logic 0 or logic 1 level that the output does not get stuck.

Power Dissipation

Specification sheets give the average power dissipation at constant output; but in the active state when the output is constantly changing, power dissipation increases and will exceed 10 mW per gate of a Standard TTL.

Propagation Delay Time

The output state of a gate does not change immediately after a change in the input. The time that it takes to change is a measure of the speed at which a logic circuit can operate. Where the speed of operation is important, the propagation delay time has to be maintained within acceptable limits. Because of manufacturing tolerances the propagation delay time may also vary from the nominal indicated values. When a number of gates are cascaded the propagation delay time adds up.

Input/Output Logic Levels

Let us consider the valid input and output voltage levels for a TTL device. We will take up the case of an Inverter. According to manufacturer's specifications any input voltage between 0 V to 0.8 V will produce a valid high output voltage. It is listed in data sheets as follows:

$$V_{IL} \text{ max} = 0.8 \text{ V}.$$

If the low state input voltage exceeds this limit the output will be unpredictable.

Similarly, any voltage from 5.0 V down to 2.0 V can be considered to be a high input voltage for a TTL device. Any voltage within this range will produce a low voltage at the output of an Inverter. The worst case high input voltage is listed in data sheets as follows:

$$V_{IH} \text{ min} = 2.0 \text{ V}.$$

If the high state input voltage lies any where between 0.8 V and 2.0 V the output will be unpredictable. According to TTL specification sheets the worst case output voltages are as follows:

Max low output voltage V_{OL} max = 0.4 V
Min high output voltage V_{OH} min = 2.4 V

The high state output voltage generally has a value between 2.4 V and 3.9 V. It is typically 3.5 V.

The worst case input and output voltages have been listed below and they have also been indicated on the input and output profiles of TTL devices shown in Fig. 2.1.

Max low output voltage V_{OL} max = 0.4 V
Min high output voltage V_{OH} min = 2.4 V
Max low input voltage V_{IL} max = 0.8 V
Min high input voltage V_{IH} min = 2.0 V.

In Fig. 2.1 the output voltage range of a TTL driver has been shown alongside the input voltage range of a TTL load. You will notice from this diagram that the output voltage range from 2.4 V to 3.9 V falls within the permissible high input voltage range of 2.0 V to 5.0 V.

From the above discussion you will notice that the low output voltage range 0 V to 0.4 V lies within the low input voltage range 0 to 0.8 V of the

TTL load. Also the high output voltage range of 2.4 V to 5 V (normally 3.9 V) lies within the high input voltage range 2 V to 5 V of the TTL load. The driver output voltage is, therefore, fully compatible with the input voltage range of the TTL load.

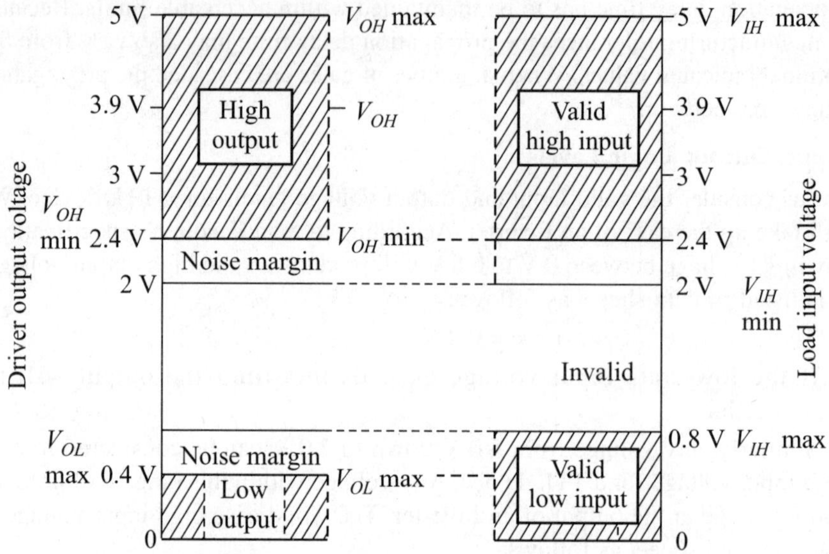

Fig. 2.1. (a) Output profile of a TTL driver.

Fig. 2.1. (b) Input profile of a TTL load.

The output of any TTL gate can be connected to the input of any TTL gate. However, it cannot be connected directly to the input of a CMOS gate. This can, however, be done with the help of interfacing devices.

Noise Margin

In the presence of a noise voltage not exceeding 0.4 V the minimum high level driver output will change to 2.4 – 0.4 V or 2 V in the negative direction and 2.4 + 0.4 or 2.8 V in the positive direction. These voltage levels are within the acceptable limits and this noise voltage will not cause any problem as long as the noise voltage is within 0.4 V. The noise margins have been shown in Fig. 2.1. However, if the noise voltage exceeds this figure it will lead to triggering of the TTL load.

Sourcing and Sinking of Current

When the TTL driver is connected to a load and the driver's output is low, current flows from the load to the driver. From a unit (single emitter) the maximum current will not exceed 1.6 mA. The driver transistor is said to sink a current of 1.6 mA.

When the driver output is high a current of 40 μA flows from the driver to the load. The driver now acts as a current source. Figure 2.2 shows the magnitude and direction of the current.

A standard TTL device is capable of sinking a current of 16 mA when its output is low and can source a current of upto 400 μA when its output is high.

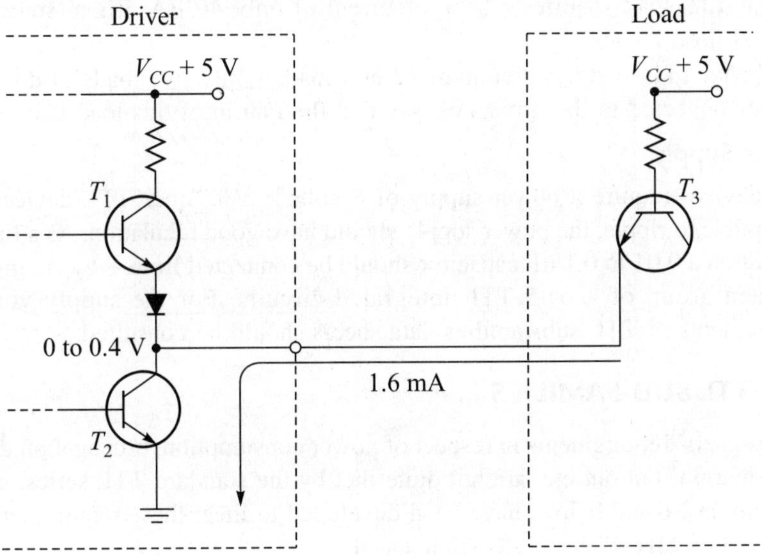

Fig. 2.2. (a) Shows direction and magnitude of current flowing from TTL load to TTL driver when the driver output is low (0-0.4 V).

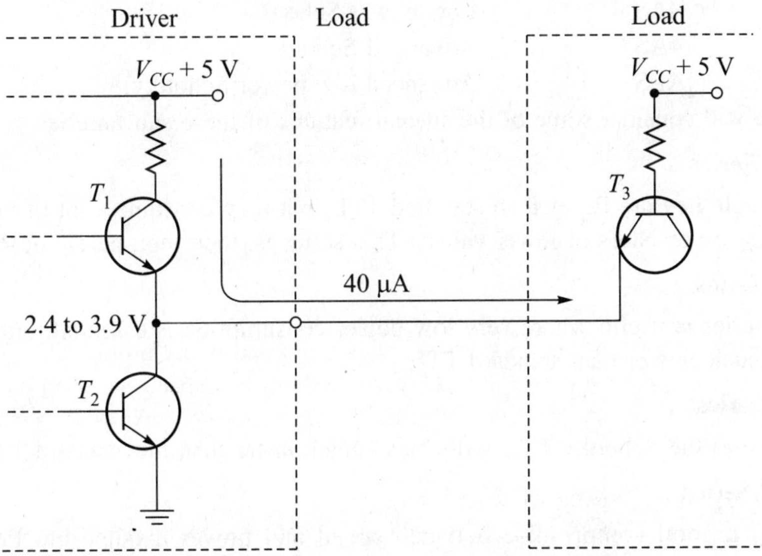

Fig. 2.2. (b) Shows direction and magnitude of current flowing from TTL driver to TTL load when the driver output is high (2.4–3.9 V).

Fan-in and Fan-out

Since a TTL driver can sink up to 16 mA of current when its output is low and a unit TTL load needs to sink only 1.6 mA of current; 10 emitters (unit loads) can be connected to the output of a TTL driver. It means that the Fan-out is ten. Also when the driver output is high it can source a current of 400 μA and since a TTL load requires a source current of only 40 μA, it can source 10 emitters (unit loads).

When a TTL load has 2 emitters (2 unit loads), only five loads of this type can be connected to the driver. We say that the Fan-in of this load is two.

Power Supply

TTL devices require a power supply of 5 volts ± 5%. Since TTL devices are susceptible to ripple, the power supply should have good regulation. As a further precaution a 0.01 to 0.1 μF capacitor should be connected from + V_{CC} to ground for each group of 4 or 5 TTL integrated circuits. For the supply voltage requirements of TTL sub-families data sheets should be consulted.

2.4 TTL SUB-FAMILIES

As designers' requirements in respect of power consumption, propagation delay, noise margin, fan-out etc. are not quite met by the standard TTL series, many sub-families, listed below, have been developed to meet these requirements:

74H	High speed
74L	Low power
74S	Schottky
74LS	Low power Schottky
74AS	Advanced Schottky
74ALS	Advanced low power Schottky

We will consider some of the special features of these sub-families.

74H Series

Although they are faster than standard TTL; but they consume a lot of power as they use resistors of lower values. This series is, therefore, nearly obsolete.

74L Series

This series is useful where very low power consumption is essential; but it is also much slower than standard TTL.

74S Series

Known as the Schottky TTL series, it is much faster than the standard TTL.

74LS Series

Offers a good compromise between speed and power dissipation. Power consumption is about one-fifth of the standard TTL series and the low power Schottky series. This series is, therefore, used more than other series in the TTL range.

74AS Series

This series is of special value where speed is a critical requirement and power consumption is not so important.

74ALS Series

These devices are the most advanced in the TTL family. Power dissipation per gate is only 1 mW and propagation delay is only 4 ns.

Summary of TTL Family Specifications

Parameters specified by manufacturers to define the operating characteristics of ICs are given in Table 2.1. Characteristics of TTL, IC families which we have discussed so far are given in Table 2.2.

Table 2.1
Glossary of Symbols used to Specify the Operating Characteristics of ICs

V_{CC}	Supply Voltage.
V_{1H}	High Level Input Voltage: This is the minimum input voltage which is recognized as logic '1' by the gate.
V_{1L}	Low Level Input Voltage: This is the maximum input voltage which is recognized as logic '0' by the gate.
V_{0H}	High Level Output Voltage: This is the minimum voltage at the output corresponding to logic '1'.
V_{0L}	Low Level Output Voltage: This is the maximum voltage at the output which will correspond to logic '0'.
I_{CC}	Supply Current: The current into the V_{CC} supply terminal of an IC.
I_{1CCH}	High Level Supply Current: This is the supply current when the output of the gate is at logic '1'.
I_{CCL}	Low Level Supply Current: This is the supply current when the output of the gate is at logic '0'.
I_{1H}	High Level Input Current: This is the current into the input when a high level voltage is applied to that input.
I_{1L}	Low Level Input Current: This is the current into an input when a low level voltage is applied to that input.
I_{0H}	High Level Output Current: This is the maximum current which the gate can source in logic '1' level at the output.
I_{0L}	Low Level Output Current: This is the maximum current which the gate can sink in logic '0' level at the output.
t_{PHL}	Propagation Delay Time: High to Low Level Output: This is the delay time when the output goes from high to low level at the output.
t_{PLH}	Propagation Delay Time: Low to High Level Output: This is the delay time when the output goes from low to high level at the output.
t_{pd}	Propagation Delay Time: This is the time between reference points on the input and output voltage waveforms with the output changing from one level (high or low) to the other level ($t_{pd} = t_{PHL}$ or t_{PLH}).

Table 2.2

Characteristics of TTL IC Family

				TTL Family				
Parameter	*7400*	*5400*	*74H00*	*54L00*	*74S00*	*74AS00*	*74LS00*	74ALS00
V_{CC} (V) min	4.75	4.5	4.75	4.5	4.75	4.5	4.75	4.5
V_{CC} (V) max	5.25	5.5	5.25	5.5	5.25	5.5	5.25	5.5
V_{IH} (V) min	2.0	2.0	2.0	2.0	2.0	2.0	2.0	2.0
V_{IL} (V) max	0.8	0.8	0.8	0.7	0.8	0.8	0.8	0.8
V_{OH} (V) min	2.4	2.4	2.4	2.4	2.7	$V_{CC}-2$	2.7	$V_{CC}-2$
V_{OL} (V) max	0.4	0.4	0.4	0.3	0.5	0.5	0.5	0.4
I_{IL} (mA) max	– 1.6	– 1.6	– 2.0	– 0.18	– 2.0	– 0.5	– 0.4	– 0.1
I_{IH} (mA) max	0.04	0.04	0.05	0.01	0.05	0.02	0.02	0.02
I_{OL} (mA) max	16	16	20	2	20	20	8	8
I_{OH} (mA) max	– 0.4	– 0.4	– 0.5	– 0.1	– 1.0	– 2.0	– 0.4	– 0.4
t_{phl} (ns)	7	7	6	31	5	4	10	8
t_{plh} (ns)	11	11	6	35	5	4.5	9	11
P_d (per gate) mW	10	10	22	1	19	8	2	1.2

2.5 MOS FAMILY CHARACTERISTICS

In the MOS family the P-channel devices have almost gone out of use and the N-channel devices are used mostly in large scale ICs. This leaves the CMOS family which uses both N-channel and P-channel MOSFETS. We will, therefore, confine our discussion to the characteristics of the CMOS devices.

Power Supply

CMOS devices can work over a voltage range extending from about 3 to 15 V. The disadvantage with a high supply voltage is that dissipation increases and it may even cause excessive heating. However, the advantage with a high operating voltage is decreased susceptibility to noise pick up. Operation at much lower voltages is accompanied by some adverse effects. It increases propagation delay and decreases noise immunity. In circumstances where the determining factors are propagation delay and noise immunity, the operating voltage should preferably be higher than 9 V.

Power Dissipation

The static power dissipation of a CMOS gate is about 10 nW. However, CMOS gates consume more power during a change of state, from high to low or low to high. Power dissipation also depends on the switching speed and at 1 MHz the dissipation increases to 1 mW.

Floating Inputs

With a floating input, as it is capacitative, and the input has very high impedance, the time constant may be very high, resulting in erratic and unpredictable behaviour. If the input capacitance acquires sufficient charge, it may even bias the input transistor to operate in the linear operating region, which will totally upset the system operation and the current consumption may become excessive. CMOS inputs should not, therefore, be left floating. Methods suggested earlier may be adopted.

Sourcing and Sinking of Current

In CMOS devices the currents are very much smaller than in TTL devices. Figure 2.3 (*a*) shows that when the driver output is low the driver can sink a current of 1 μA. Figure 2.3 (*b*) shows that when the driver output is high it can source a current of l μA. In the first case the current has a negative sign as the current is flowing out of the load when the driver output is low.

$$I_{OL} \text{ max} = -1 \ \mu\text{A}$$

In the second case the current has a positive sign as it is flowing into the load

$$I_{OH} \text{ max} = 1 \ \mu\text{A}$$

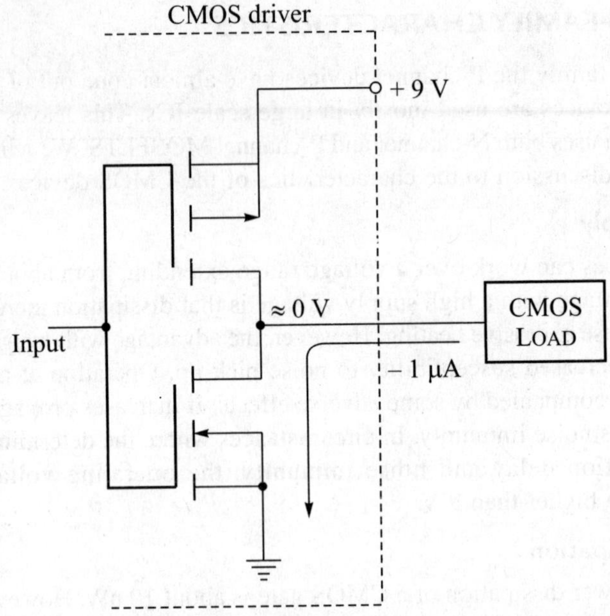

Fig. 2.3. (a) Shows the direction and magnitude of current flowing from a CMOS load to a CMOS driver when the driver output is low (≈ 0 V).

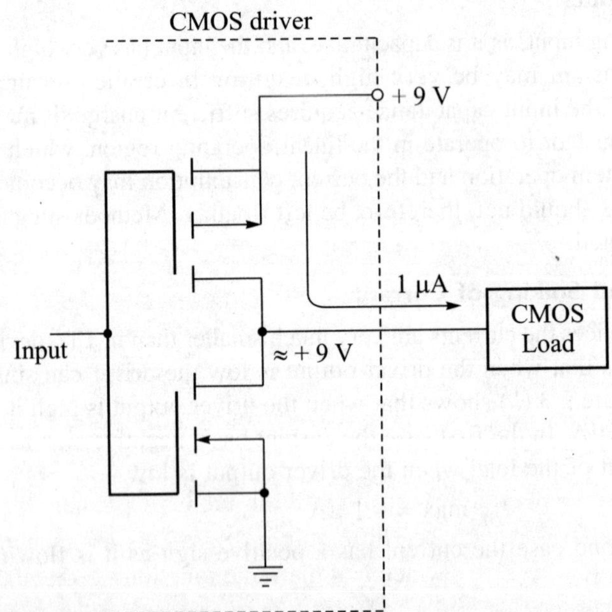

Fig. 2.3. (b) Shows the direction and magnitude of current flowing from a CMOS driver to a CMOS load when the driver output is high (≈ + 9 V).

Input/Output Logic Levels

The input/output logic levels for voltages and currents for CMOS IC families are given in Table 2.3. We are reproducing below the input/output voltage levels for CMOS devices in the 4000A and 4000B series when operated from a 5 V supply.

The worst case voltages are as follows:

V_{IH} min 3.5 V
V_{IL} max 1.5 V
V_{OH} min 4.9 V
V_{OL} max 0.01 V

The input/output profiles for these CMOS sub-family are shown in Fig. 2.4. You will observe that the input and output logic levels for the CMOS driver of the 4000 A/B series fall within the acceptable high and low input voltage levels of the CMOS load. There is, therefore, no interfacing problem between the driver and the load and they can be directly connected. However, there may be interfacing problem between a CMOS driver and a TTL load or *vice-versa*. This problem will be discussed later in one of the chapters.

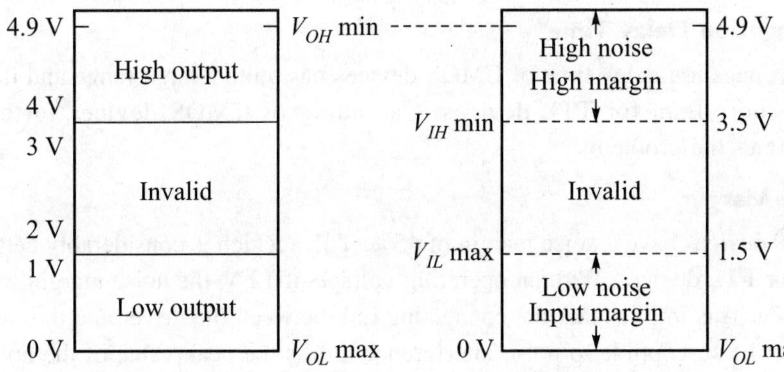

Fig. 2.4. (a) Showing the output profile and noise margins for a CMOS driver.

Fig. 2.4 (b) Showing the input profile for a CMOS load.

2.6 CMOS SUB-FAMILIES

The 4000 A series was the first sub-family in the CMOS category; but it was not very popular as it had very little output drive. This led manufacturers to introduce the 4000 B series which had the same functions as the 4000 A series and was in many ways similar to it. The main problem of compatibility with TTL devices remained unsolved.

These problems led to the introduction of the 54C/74C series of CMOS devices. This new series performs the same functions and has the same pin connections as the 54/74 TTL family. The 54C series has a temperature range of – 55°C to + 125°C. For the 74C series the temperature range is – 40°C to + 85°C. Both these sub-families have a supply voltage range of 3 to 15 V. The specifications for 74C sub-family are given in Table 2.3. Quite a large variety of CMOS ICs in the 74C sub-family are available. A list of the currently available ICs in this category is given in Appendix 8.

A high speed series of CMOS devices, 74HC is also now available. They have the same logic functions and pin numbers as the corresponding TTL devices. The 74HC series has a shorter propagation delay than the 74C series. The output low (I_{OL}) and output high (I_{OH}) current of 4 mA in 74HC devices is sufficient to drive more than one TTL device directly.

74HCT is another sub-family in the CMOS category. This sub-family has better compatibility with TTL devices than the 74HC series. If you refer to Table 2.3 you will observe that all the CMOS sub-families which we have discussed can be operated from a power supply of + 5 V.

Propagation Delay Time

The propagation delay time of CMOS devices has quite a large range and it is much more than for TTL devices. Cascading of CMOS devices further aggravates the problem.

Noise Margin

CMOS devices have a noise margin of 45% of V_{CC} which is considerably better than for TTL devices. With an operating voltage of 12 V, the noise margin will be 5.4 V. This implies that the connecting link between the driver and the load will not be susceptible to noise interference unless the peak value of the noise exceeds 5.4 V. Figure 2.4 (*b*) shows the noise margins when the operating voltage is 5 V.

2.7 INTEGRATED CIRCUIT PACKAGING

There are many types of packages for ICs; but conventional logic gates are often supplied in 14-pin or 16-pin DIL (Dual-in-line) packages. Figure 2.5 shows a 14-pin and a 16-pin DIL package. It should be noted that the diagrams show the devices as viewed from the top, not from the underside. In both the cases the IC pins are numbered sequentially, starting from the notch (or dot) and moving in the anti-clockwise direction.

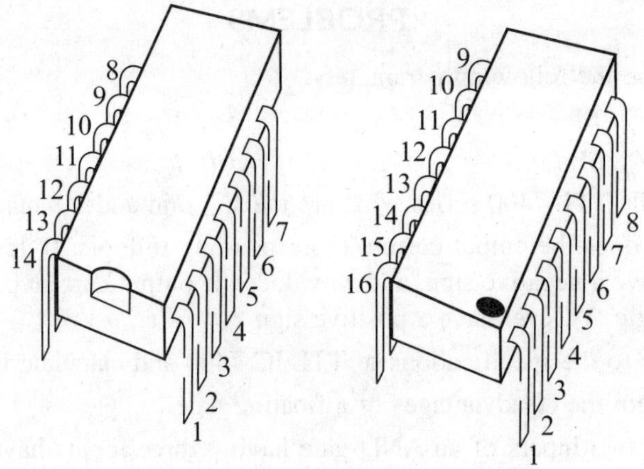

Fig. 2.5. Showing 14-pin and 16-pin DIL package pin numbering as viewed from the top of the package.

Table 2.3

Characteristics of CMOS IC Families

Parameter	*4000A*	*4000B*	*74C00*	*74HC00*	*74HCT00*
V_{CC} (V) min V_{CC} (V) max	3 15	3 18	3 15	2 6	4.5 5.5
Oper. @ (V)	5	5	5	4.5	4.5
V_{IH} (V) min	3.5	3.5	3.5	3.15	2
V_{IL} (V) max	1.5	1.5	1.5	0.9	0.8
V_{OH} (V) min	4.5	4.5	4.5	4.4	4
V_{OL} (V) max	0.01	± 0.01	0.5	0.1	0.33
I_{IL} (μA) max	10 pA	± 1	1	– 1	0.1
I_{IH} (μA) max	10 pA	1	– 1	1	1
I_{OL} (mA) min	0.12	0.4	1.75	4	4
I_{OH} (mA) min	0.12	1.6	– 1.75	– 4	4
t_{PHL} (ns)	180	160–320	90	23	20
t_{PLH} (ns)	125	210–420	90	23	20

Note: Specifications for 74C, 74HC and 74HCT families are for specific devices.

PROBLEMS

2.1. Define the following parameters:

(*a*) V_{OL} Max (*b*) V_{OH}

(*c*) V_{IH} Min (*d*) I_{IH}.

2.2. For the TTL 7400 series what are the V_{IH} min and V_{IL} max voltages ?

2.3. Why does the output current corresponding to logic '0' level of a TTL IC have a negative sign, and why does the output current corresponding to logic '1' level have a positive sign ?

2.4. Refer to the specifications of TTL IC 7400 and calculate its fan-out.

2.5. Explain the disadvantages of a floating gate.

2.6. Only two inputs of an AND gate having three inputs have been used. What happens if it is left floating?

2.7. In the above problem, how will you handle the third input to avoid problems?

2.8. In an OR gate having three inputs only two inputs are being used. Will you connect the third input to ground or to V_{CC}?

2.9. What is an acceptable low level input voltage for a TTL IC? What happens if it is more or less than V_{IL}?

2.10. If the minimum high input voltage to a TTL gate is less than V_{IH} min what effect will it have on its performance?

2.11. What is the highest valid low input voltage for a TTL gate?

CHAPTER 3

Inverters : Buffers

3.1 INTRODUCTION

In this chapter, we will study the functioning and characteristics of **Inverters** and **Buffers.** Both these electronic devices have a single input and a single output. However, there is a basic difference in their functioning. Whereas Inverters complement any input applied to them; the Buffers allow the input signal to pass through without any change. Inverters are used to complement a signal and the Buffers are used to isolate the input from the output.

The symbols for an Inverter are shown in Fig. 3.1 (*a*) and Fig. 3.1 (*b*). Both these symbols are in use. The triangle in these symbols indicate an amplifier and the bubbles signify inversion or complementation. The basic function of an Inverter is given in truth, Table 3.1.

(*a*) (*b*)

Fig. 3.1. Inverter symbols.

Table 3.1

Inverter Truth Table

Input	*Output*
A	$\overline{A}$
1	0
0	1

The following are some of the important points to be remembered about an Inverter:

* A LOW input produces a HIGH output and a HIGH input produces a LOW output.
* For an Inverter, the output is always the opposite of the input. If the input is logic 0, the output is logic 1 and if the input is logic 1, the output is logic 0.

* In other words, the output is always the complement of the input. The Inverter is sometimes called a NOT gate as the output is opposite of the input.

Buffer

The symbol for a Buffer is given in Fig. 3.2. If the input to a Buffer is *A*, the output is also *A*. If two Inverters are connected in cascade, the circuit behaves like a non-inverting amplifier. In other words, for a logic 1 input the output is also logic 1 and for a logic 0 input the output is also logic 0. During the course of the following experiment you will observe that an even number of Inverters in tandem act as a non-inverter (buffer) and an odd number in tandem invert the input signal.

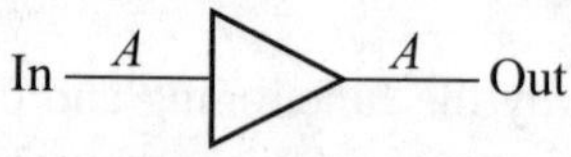

Fig. 3.2. Buffer symbol.

A Buffer is used as a non-inverting amplifier to isolate two other devices. Buffers have a high input and a low output impedance. This means a low input current and a high output current. A basic gate does not provide adequate isolation of two devices as the input current of a TTL gate is only about one-tenth of the output current. IC 7437 is a quad 2-input NAND Buffer in which the output current can be as much as thirty times the input current.

EXPERIMENT 3.1 : INVERTERS

Objective

To study the operation and characteristics of Inverters.

Materials Required

Logic trainer

Volt-ohm-milliammeter (VOM)

TTL Hex Inverter IC 7404

Resistor 470 ohm

Procedure

1. Connect the circuit shown in Fig. 3.3. The Hex Inverter used in this experiment contains six identical and independent Inverters, only four of which are used in this experiment.
2. The input for the Inverter is obtained through logic switch Sw_1, which can connect the input either to + 5 V for a high logic input or to ground for a low input.
3. The output of each of the Inverters is connected to Logic monitors (LEDs). The LEDs light up on a high Inverter output and are off when the Inverter output is low.

4. Connect pin 14 of the IC to V_{CC} (+ 5 V dc) and pin 7 to ground (Gnd).
5. You will measure the input and output voltage of Inverters with a Voltmeter and also observe the output on the Logic monitors.
6. Now measure the input (V_i) and output (V_o) voltages with both low and high input positions and logic switch Sw_1. You will record the input voltage and the output of each of the Inverters and enter your observations in Table 3.2.

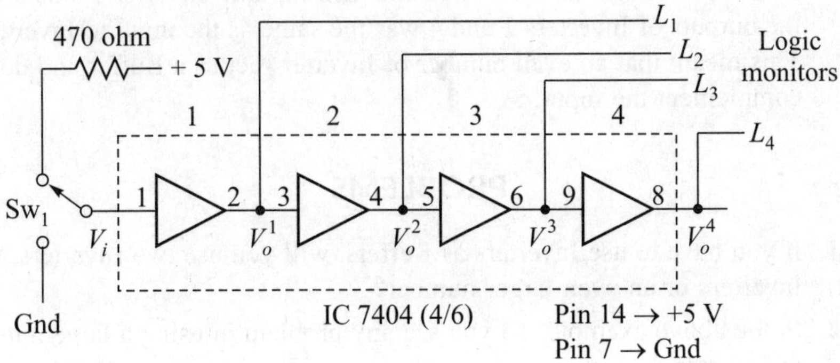

Fig. 3.3

Table 3.2

	Input V_i volt	*Output V_o^1 volt*	*Output V_o^2 volt*	*Output V_o^3 volt*	*Output V_o^4 volt*
1					
2					

Now using positive logic assignments convert the voltage levels in Table 3.2 into logic 1s and 0s and write your result in Table 3.3.

Table 3.3

	V_i	V_o^1	V_o^2	V_o^3	V_o^4
1					
2					

Discussion of Data

1. You applied logic 0 and logic 1 data at the input of Inverter 1 and measured the resultant outputs of Inverters 1, 2, 3 and 4.
2. Your observations as recorded in Table 3.2 showed that when input, V_i, to Inverter 1 was logic 0 or 1 the outputs of Inverters 1 and 3 showed complementation of the input V_i. This leads to the conclusion that an odd number of Inverters in tandem complement the input.
3. Your observation also showed that when input to Inverter 1 was 0 or I, the outputs of Inverters 2 and 4 was the same as the input to Inverter 1. This meant that an even number of Inverters act as a Buffer and do not complement the input.

PROBLEMS

3.1. If you have to use Inverters as Buffers, will you use two Inverters, four Inverters or an even larger number?

3.2. In the above example, do you see any problem in using a large number of Inverters?

3.3. List as many applications of Inverters as you can think of.

CHAPTER 4

AND : OR Gates

4.1 INTRODUCTION

Logic circuits are basically of two types—decision-making circuits and memory circuits. Their functioning depends on the binary inputs they receive and produce binary outputs which are a function of the inputs as well as of the characteristics of the logic circuit they employ.

After you have become familiar with the basic logic elements, common configurations of these devices and their characteristics, you will be in a position to evaluate the operation of digital equipments in common use and also acquire capability for designing digital devices.

Logic gates are the basic decision-making elements. Logic gates have a single output and they may have two or more inputs. For specific decision-making functions there are several types of gates. Gates are supplied binary inputs and produce an output, which is a function of their characteristics and the nature of the inputs. The output shows the decision made by gates. In this chapter, we will discuss the operation and characteristics of AND and OR gates.

4.2 AND GATES

The AND gate has two or more inputs and a single output. The logic symbol for an AND gate is given in Fig. 4.1.

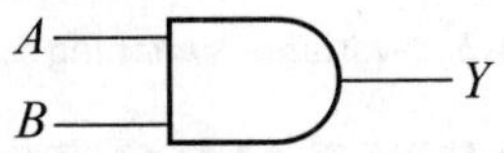

Fig. 4.1. Logic symbol for 2-input AND gate.

The distinctive characteristics for an AND gate are as follows:

* The output of the AND gate is logic 1, only if all the inputs are logic 1.
* If any one or more inputs are logic 0, the output will be logic 0.

The first characteristic of the AND gate cited above makes it an ideal control element. A single input will control the output irrespective of the state of the other input. The truth table for a 2-input AND gate is given in Table 4.1. The truth table indicates how it will perform under the different conditions of the states of the inputs *A* and *B*. Since there are two inputs, the total number of input combinations will be 2^2 or 4.

Table 4.1
Truth Table for a 2-Input AND Gate

Inputs		*Output*
A	*B*	*Y*
0	0	0
0	1	0
1	0	0
1	1	1

You will notice from the truth table that the output is logic 0 when either of the two or both inputs are logic 0. The output is logic 1 only when both the inputs are logic 1.

The logical operation performed by the AND gate is expressed as follows in Boolean algebra.

$$Y = A \cdot B = AB$$

The AND function is designated by the dot $(\cdot)$ between the input variables *A* and *B*. It is read as *Y* equals *A* AND *B*.

The 2-input AND gate can be compared to a switching circuit shown in Fig. 4.2. This circuit will produce a high output which will light up the lamp only when both the switches are closed. When any one or both the switches are open the lamp will not light up. This circuit will have the same truth table as the one shown in Table 4.1 for a 2-input AND gate.

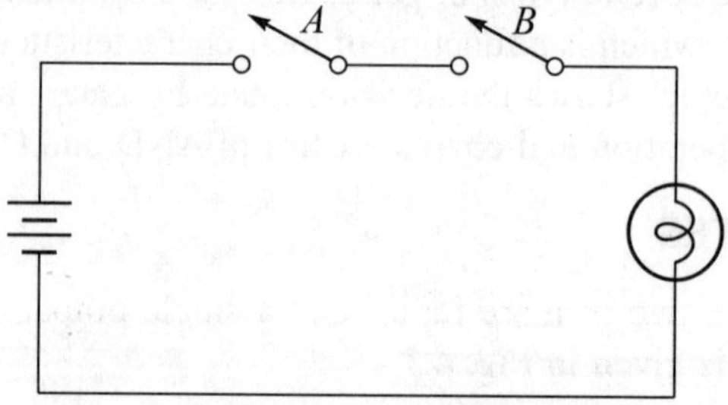

Fig. 4.2. 2-variable switching circuit.

EXPERIMENT 4.1: TWO-INPUT AND GATE

Objective

To verify the operation and characteristics of the 2-input AND gate.

Materials Required

Logic trainer

Volt-ohm-milliammeter (VOM)

TTL IC 7408 : Quad 2-input AND gates

Resistor 470 ohm

The circuit for the experiment is given in Fig. 4.3.

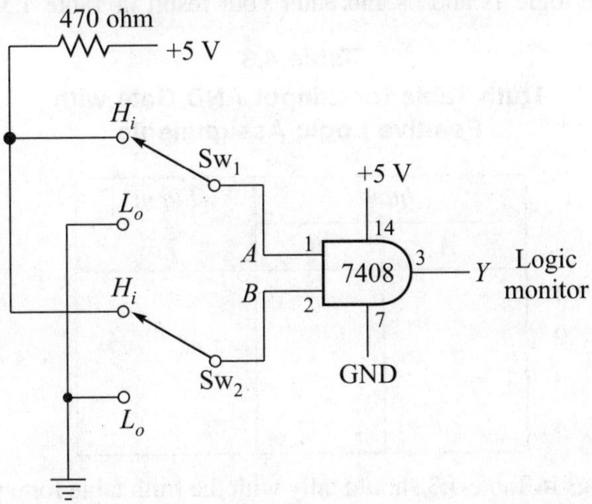

Fig. 4.3. Circuit for AND gate Experiment 4.1.

Procedure

1. Connect the circuit shown in Fig. 4.3. The IC used in this experiment contains four identical and independent AND gates only one of which has been used in this experiment.
2. The input for the AND gate is obtained through logic switches Sw_1 and Sw_2 which can connect the input either to + 5 V for a high logic input or to ground for a low logic input.
3. The output of the AND gate is connected to a Logic monitor (LED) which lights up on a high output and is off when the output is low.
4. Connect pin 14 of the IC to V_{CC} (+ 5 V) and pin 7 to ground (Gnd).
5. Now, with all possible combinations of Sw_1 and Sw_2 (high and low input states), record the output voltage in truth table [Table 4.2 and also observe the state of the Logic monitor (LED)].

Table 4.2
Truth Table for a 2-Input AND Gate

	Inputs		*Output*
	A	*B*	*Y*
1	0 V	0 V	
2	0 V	+ 5 V	
3	+ 5 V	0 V	
4	+ 5 V	+ 5 V	

6. Now using positive logic assignments convert the voltage levels in Table 4.2 into logic 1s and 0s and enter your result in Table 4.3.

Table 4.3
Truth Table for 2-Input AND Gate with Positive Logic Assignments

	Inputs		*Output*
	A	*B*	*Y*
1			
2			
3			
4			

Your readings in Table 4.3 should tally with the truth table for a positive logic 2-input AND gate in Appendix 6.

7. Now using negative logic assignments convert the voltage levels in Table 4.2 into logic 1s and 0s and enter your result in Table 4.4.

Table 4.4
Truth Table for 2-Input AND Gate with Negative Logic Assignments

	Inputs		*Output*
	A	*B*	*Y*
1			
2			
3			
4			

Your observations in Table 4.4 should tally with the truth table for negative logic of a 2-input AND gate as given in Appendix 6. If you look at the truth table for a 2-input OR gate at S. No. 7, column 4 of Appendix 6 using positive logic you will find that the two tables are identical. This shows that an AND gate using negative logic functions as an OR gate with positive logic. This shows the dual nature of logic gates.

The symbol for an AND gate using negative logic is given in Fig. 4.4.

Fig. 4.4. Symbol for AND gate using negative logic. (Functions as positive OR)

Three-Input AND Gate

The mechanical equivalent of a 3-input AND gate is like a circuit with three switches wired in series as shown in Fig. 4.5 (*a*) and its electronic symbol is given in Fig. 4.5 (*b*). In this arrangement the bulb will light up when all the switches are switched on. If a single switch is off the bulb will not light up. This feature is reflected in the truth table for the 3-input AND gate in Table 4.5.

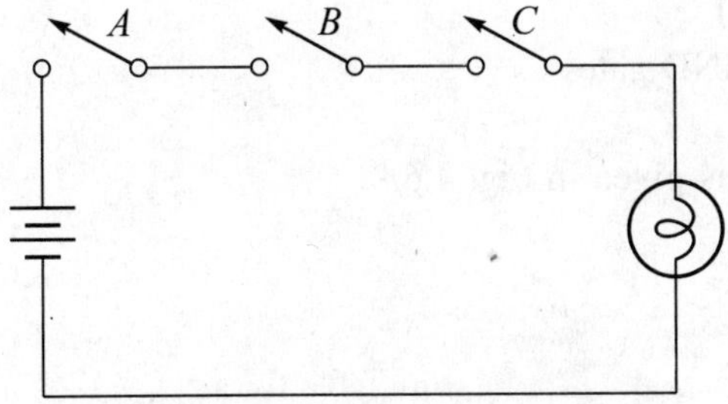

Fig. 4.5. (a) Mechanical equivalent of a 3-input AND gate.

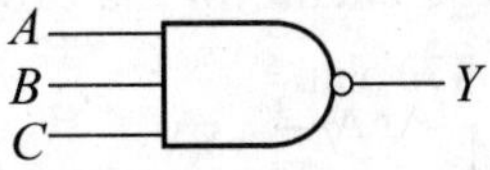

Fig. 4.5. (b) Symbol for 3-input AND gate.

Table 4.5

Truth Table for a 3-Input AND Gate

Inputs			*Output*
A	*B*	*C*	*Y*
0	0	0	0
0	0	1	0
0	1	0	0
0	1	1	0
1	0	0	0
1	0	1	0
1	1	0	0
1	1	1	1

A 3-input AND gate will have 2^3 or 8 states. In the truth table these 8 states are shown by the binary equivalents of the 8 states (0 through 7 in decimal form).

You will notice from the truth table that the output is high only when all the inputs are high and the output is low when any one or more inputs are low.

The operation of a 3-input AND gate can be stated as follows in algebraic form.

$$Y = A \cdot B \cdot C$$

EXPERIMENT 4.2 : THREE-INPUT AND GATE

Objective

* To build a 3-input AND gate using two 2-input AND gates.

* Draw up truth tables using positive and negative logic.

Materials Required

Logic trainer

Volt-ohm-milliammeter (VOM)

TTL IC 7408 : Quad 2-input AND gates

Resistor 470 ohm

The circuit for the experiment is given in Fig. 4.6.

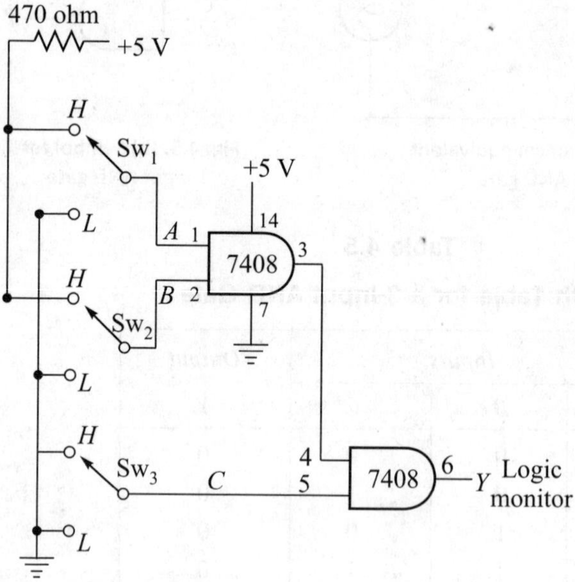

Fig. 4.6. 3-input AND gate assembled using 2-input AND gates.

Fig. 4.7. Symbol for 3-input AND gate.

Procedure

1. Connect the circuit shown in Fig. 4.6. Only two AND gates from the IC have been used.
2. The AND gate inputs can be connected through switches to high (H) and low (L) inputs.
3. The output of the lower AND gate is connected to the Logic monitor.
4. Connect pin 14 of the IC to V_{CC} (+ 5 V) and pin 7 to ground.
5. Now connect switches Sw_1, Sw_2 and Sw_3 to the logic levels shown in the eight sets of inputs in Truth Table, 4.5 and record the result of each input combination in Table 4.6 and compare this table with Table 4.5. Since both the tables use positive logic assignment, they should be identical.

Table 4.6

Truth Table for 3-Input AND Gate with Positive Logic Assignment

Inputs			*Output*
A	*B*	*C*	*Y*
0	0	0	
0	0	1	
0	1	0	
0	1	1	
1	0	0	
1	0	1	
1	1	0	
1	1	1	

Table 4.7

Truth Table for 3-Input AND Gate with Negative Logic Assignment

Inputs			*Output*
A	*B*	*C*	*Y*

6. Now draw up a truth table for the 3-input NAND gate using negative logic assignment. To do this substitute 1s for 0s and 0s for 1s in Table 4.6 and enter the result in Table 4.7. From this truth table you can draw the following conclusions:

   * The output is 0 only when all the inputs are also 0.
   * The output is 1 when at least one of the inputs is 1.

When you come to the subject of OR gates you will observe that these two properties are the characteristic features of an OR gate with positive logic assignment. You can, therefore, conclude that a 3-input AND gate with negative logic assignment functions as an OR gate with positive logic assignment.

The logic symbol for this 3-input AND gate with negative logic assignment is as given in Fig. 4.8.

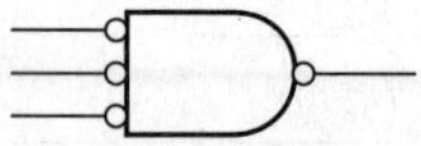

Fig. 4.8. Symbol for 3-input AND gate using negative logic assignment. (Functions as positive OR)

4.3 OR GATES

The OR gate has two or more inputs and a single output as shown in Fig. 4.9.

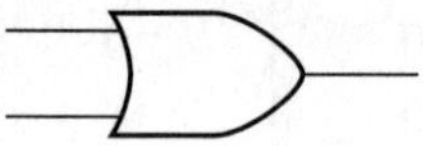

Fig. 4.9. Logic symbol for 2-input OR gate.

The distinctive characteristics for an OR gate are as follows:

* The output of an OR gate is logic 0 only when all inputs are logic 0.
* The output of an OR gate is logic 1 when any one or more inputs are logic 1.

The mechanical equivalent of a 2-input OR gate is like a circuit with two switches connected in parallel as shown in Fig. 4.10. In this arrangement the bulb will light up when any one or both the switches are turned on. If both the switches are off the bulb will not light up. This feature is reflected in the truth table, Table 4.8. As the circuit has two switches there will be 2^2 or 4 input combinations for the switches. The truth table, therefore, shows four input combinations.

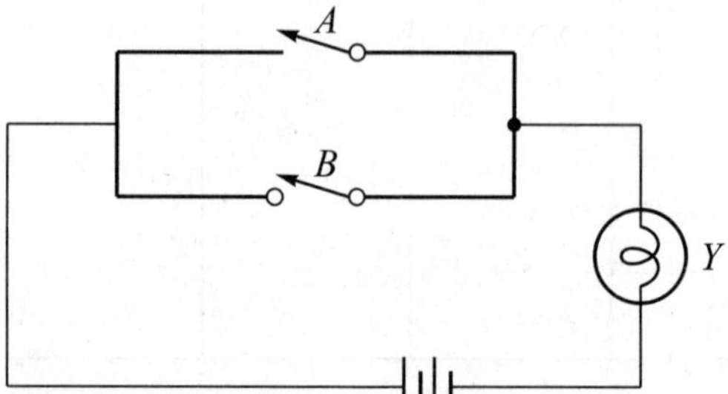

Fig. 4.10. Mechanical equivalent of a 2-input OR gate.

Table 4.8

Truth Table for a 2-Input OR Gate

Inputs		*Output*
A	*B*	*Y*
0	0	0
0	1	1
1	0	1
1	1	1

You will notice from this truth table that the output is 0 when both the inputs are 0. You will notice that in the remaining three cases at least one of the inputs is 1 and in these cases the output is also 1. The output of the OR gate is a function of both the inputs *A* and *B*. In equation form it is written as follows:

$$Y = A + B$$

The plus sign is an indication, in logic terms, that the logical sum operation (called disjunction of terms so connected) has taken place. The terms so connected are said to be **ORed.** The symbol + is not the same as the algebraic symbol for addition.

We have considered a 2-input OR gate; but like the AND gate the OR gate may also have more than two inputs. Truth Table 4.9 shows the truth table for a 3-input OR gate.

Table 4.9

Truth Table for a 3-Input OR Gate

Inputs			*Output*
A	*B*	*C*	*Y*
0	0	0	0
0	0	1	1
0	1	0	1
0	1	1	1
1	0	0	1
1	0	1	1
1	1	0	1
1	1	1	1

You will notice from this truth table that the output is 0 only in one case and that is when all inputs are 0. In the remaining cases the output is 1 since at least one of the inputs is 1.

EXPERIMENT 4.3 : TWO-INPUT OR GATE

Objective

To draw up truth tables for a 2-input OR gate using positive and negative logic.

Materials Required

Logic trainer

Volt-ohm-milliammeter (VOM)

TTL IC 7432 : Quad 2-input OR gates

Resistor 470 ohm

The circuit for the OR gate Experiment 4.3 is given in Fig. 4.11.

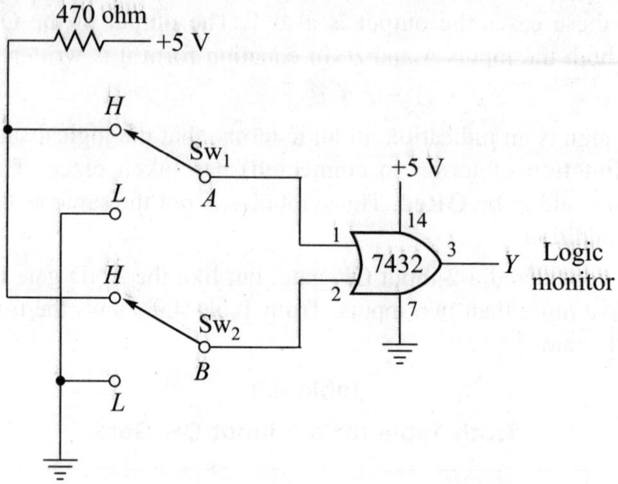

Fig. 4.11. Circuit for OR gate Experiment 4.3.

Procedure

1. Connect the circuit given in Fig. 4.11. Only two OR gates from the IC have been used.
2. Now connect pin 14 of IC to + 5 V and pin 7 to ground.
3. Follow the input combinations of Table 4.8 and set switches Sw_1 and Sw_2 to each combination of the inputs.
4. Observe the output as shown by the logic monitor and record the result for each input combination in Table 4.10.
5. As your experiment was intended for positive logic assignment your observations should tally with Table 4.8 which was also for positive logic assignment.

Table 4.10

Truth Table for 2-Input OR Gate Using Positive Logic Assignment

Inputs		*Output*
A	*B*	*Y*

6. Now draw up a truth table for the 2-input OR gate using negative logic assignment. To do this substitute 1s for 0s and 0s for 1s in Table 4.10 which is for positive logic assignment. Enter your result in Table 4.11.

Table 4.11

Truth Table for 2-Input OR Gate Using Negative Logic Assignment

Inputs		*Output*
A	*B*	*Y*

From Table 4.11, which is the OR gate truth table using negative logic assignment, you can draw the following conclusions:

* The output is 1 when both the inputs are 1.
* If any one or both inputs are 0, the output is 0.

If you refer to the characteristic features for the 2-input AND gate you will notice that the OR gate with negative logic assignment fulfils the criteria for the AND gate with positive logic assignment. Thus, the OR gate using negative logic assignment clearly functions as an AND gate using positive logic.

The OR gate symbol using negative logic is given in Fig. 4.12.

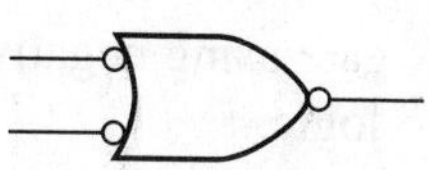

Fig. 4.12. Symbol for OR gate using negative logic. (Functions as an AND gate)

PROBLEMS

4.1. How many different input combinations will be possible for an AND gate having

(*a*) four inputs? (*b*) six inputs?

4.2. Write the truth table for an AND gate having four inputs.

4.3. Write the output equation for the AND gate given in Fig. P-4.3.

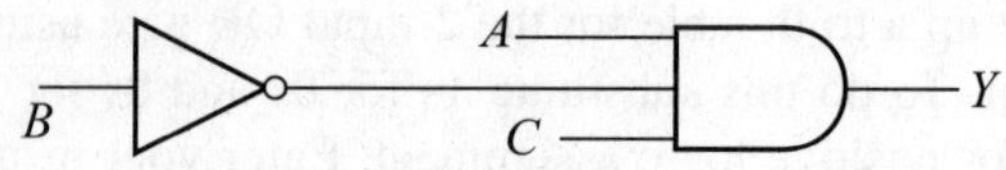

Fig. P-4.3

4.4. Prove using Boolean algebra that a 2-input AND gate using negative logic functions as an OR gate.

4.5. What is the typical low input voltage for a TTL gate?

4.6. State the highest valid low input voltage for a TTL gate.

4.7. Would you consider 0.9 V as the highest valid low input voltage for a TTL gate?

4.8. Would you consider 2.0 V as the lowest valid high input voltage for a TTL gate?

4.9. Write the output equation for the OR gate given in Fig. P-4.9.

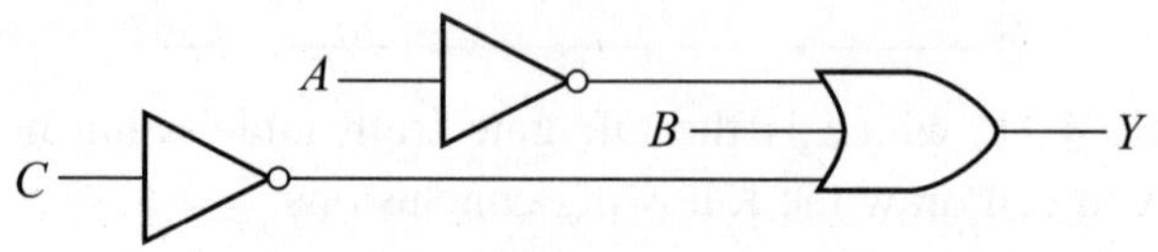

Fig. P-4.9

4.10. When is the output of an OR gate?

(*a*) Logic 0

(*b*) Logic 1

4.11. Draw a logic diagram to implement a 4-input OR gate using 2-input OR gates and give its logic symbol and truth table.

4.12. Write the equation for the output for the AND gate given in Fig. 4.8.

4.13. Prove that a 2-input OR gate using negative logic functions as a 2-input AND gate with positive logic.

CHAPTER 5

NAND : NOR Gates

5.1 INTRODUCTION

The NAND and NOR gates are the most widely used of all the digital elements as digital devices are more easily implemented with NAND and NOR gates than with AND, OR gates. As has been stated earlier NAND is a contraction for NOT-AND. In other words, an AND gate followed by an Inverter functions as a NAND gate. Similarly, NOR is a contraction for NOT-OR. An OR gate followed by an Inverter functions as a NOR gate.

In this chapter, we will consider these two gates and also discuss how these gates can be used to implement the functions of other gates.

5.2 NAND GATE

A NAND gate may have two or more inputs and a single output. The symbol for a 2-input NAND gate is given in Fig. 5.1 (*a*). It is equivalent to an AND gate followed by an Inverter as shown in Fig. 5.1 (*b*).

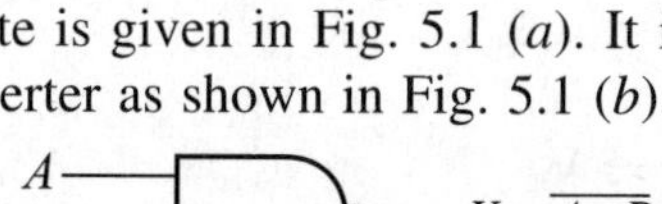

Fig. 5.1. (a) Symbol for 2-input NAND gate.

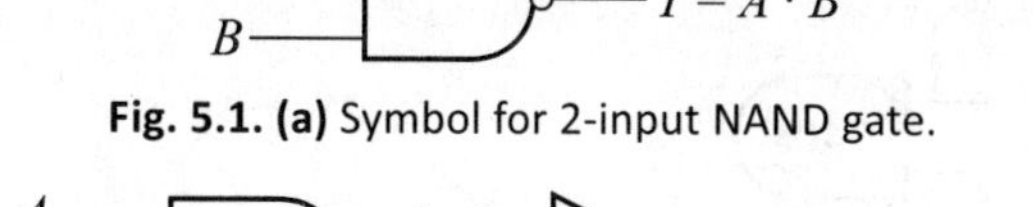

Fig. 5.1. (b) NAND function using AND gate plus an Inverter.

Knowing the implementation of the NAND gate given in Fig. 5.1 (*b*) you can work out its truth table from Table 4.1 by substituting 1s for 0s and 0s for 1s in the output column as shown in Table 5.1.

Table 5.1

Truth Table for a 2-Input NAND Gate

Inputs		*Output*
A	*B*	*Y*
0	0	1
0	1	1
1	0	1
1	1	0

From Table 5.1 you will notice that the unique feature of the NAND gate is that the output is 0 when both the inputs are 1.

EXPERIMENT 5.1 : THREE-INPUT NAND GATE

Objective

* To draw a truth table for a 3-input NAND gate,
* To consider the conversion of NAND gates to
  (*a*) inverters
  (*b*) AND gates
  (*c*) OR gates and
  (*d*) NOR gates.

Materials Required

Logic trainer
Volt-ohm-milliammeter (VOM)
TTL IC 7400 : Quad, 2-input NAND gates
TTL IC 7408 : Quad, 2-input AND gates
Resistor 470 ohm

We will first consider drawing up a truth table for a 3-input NAND gate. The circuit diagram for this experiment is given in Fig. 5.2 (*a*).

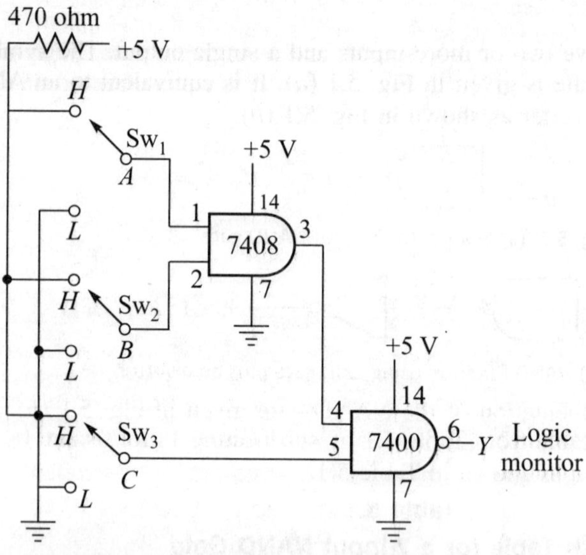

Fig. 5.2. (*a*) Circuit diagram for NAND gate experiment 5.1.

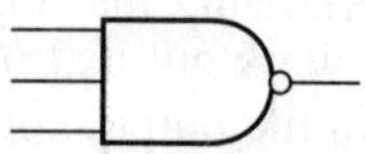

Fig. 5.2. (*b*) Logic symbol for 3-input NAND gate.

Procedure : 3-Input NAND Gate Truth Table

1. Connect the circuit shown in Fig. 5.2 (*a*).
2. Connect pins 14 of the ICs to + 5 V and pins 7 to ground.

3. Follow the input combinations given in Table 5.2 and set switches Sw_1, Sw_2 and Sw_3 to each combination of the inputs.
   Observe the output as shown by the logic monitor and record the result of each input combination in Table 5.2.
4. Your observations should tally with the truth table given in Table 5.2 in Appendix 3.

Table 5.2

Truth Table for 3-Input NAND Gate

Inputs			*Output*
A	*B*	*C*	*Y*
0	0	0	
0	0	1	
0	1	0	
0	1	1	
1	0	0	
1	0	1	
1	1	0	
1	1	1	

You will notice from the truth table for the 3-input NAND gate that, as with the 2-input NAND gate truth table, there is a unique state when the output is low when all the inputs are high.

Procedure : NAND Gate as an Inverter

1. Assemble the circuit given in Fig. 5.3 (*a*). Notice that both the inputs are connected together and then connected to logic switch.
2. Connect the power supply. Apply logic inputs 0 and 1, and record the output in Table 5.3.

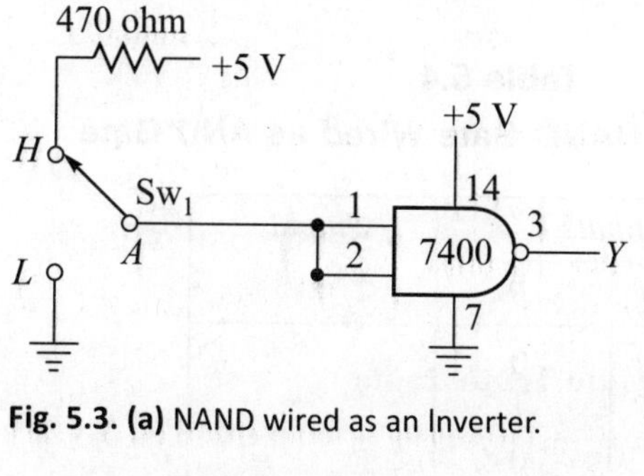

Fig. 5.3. (a) NAND wired as an Inverter.

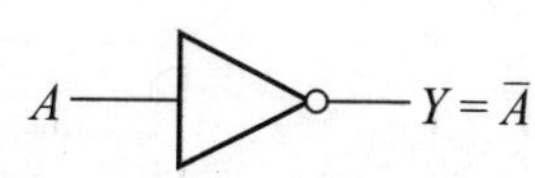

Fig. 5.3. (b) Equivalent Inverter.

Table 5.3

Truth Table for NAND Gate Wired as an Inverter

Input A	*Output* Y
0	
1	

Your observations should tally with Table 3.1.

Procedure : NAND Gates Wired as an AND Gate

1. Assemble the circuit given in Fig. 5.4 (*a*). Notice that a NAND gate wired as an Inverter has been connected to the output of a NAND gate.
2. Proceed as in AND gate Experiment 4.1.
3. Record your readings in Table 5.4. Your observations should tally with Table 4.1.

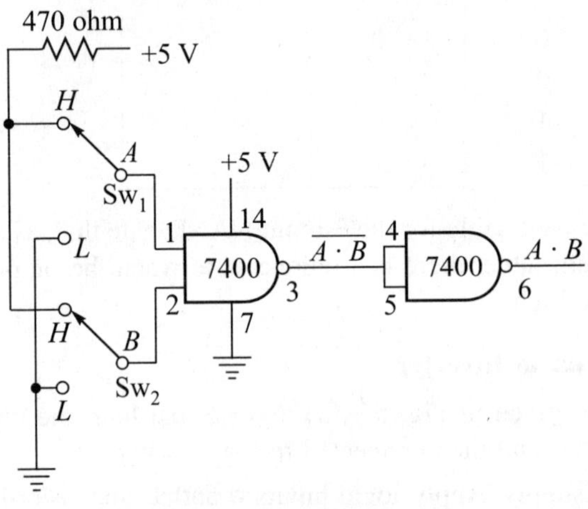

Fig. 5.4. (a) NAND gates wired as AND gate.

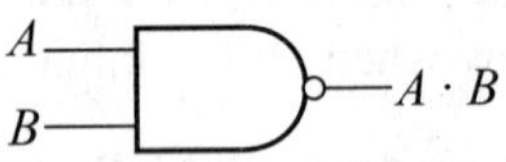

Fig. 5.4. (b) Equivalent AND gate.

Table 5.4

Truth Table for NAND Gate Wired as AND Gate

Inputs		*Output*
A	B	Y
0	0	
0	1	
1	0	
1	1	

Procedure : NAND Gates Wired as an OR Gate

1. Assemble the circuit given in Fig. 5.5 (*a*). Notice that NAND gates used as Inverters are connected at the inputs of a NAND gate.
2. Proceed as in the OR gate Experiment 4.3.
3. Record your observations in Table 5.5. Your observations should tally with Table 4.8.

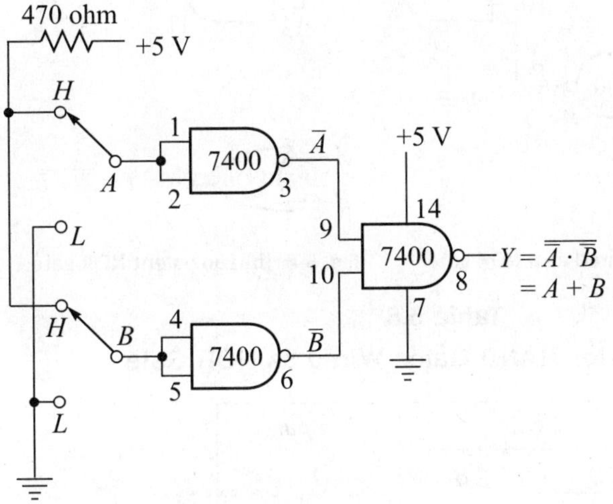

Fig. 5.5. (a) NAND gates wired as an OR gate.

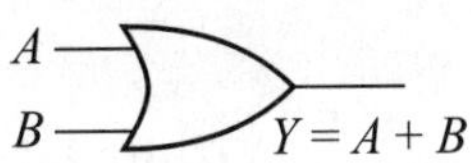

Fig. 5.5. (b) Equivalent OR gate.

Table 5.5

Truth Table for NAND Gates Wired as OR Gate

Inputs		*Output*
A	*B*	*Y*
0	0	
0	1	
1	0	
1	1	

Procedure : NAND Gates Wired as NOR Gate

1. Assemble the circuit given in Fig. 5.6 (*a*). Notice that three NAND gates have been used as Inverters.
2. Apply the logic inputs as listed in Table 5.6 and record your observations. Your readings should tally with Table 5.7 in Appendix 3.

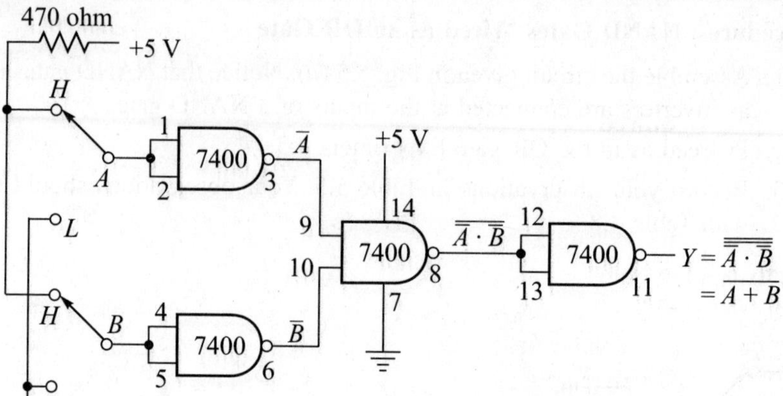

Fig. 5.6. (a) NAND gates wired as a NOR gate.

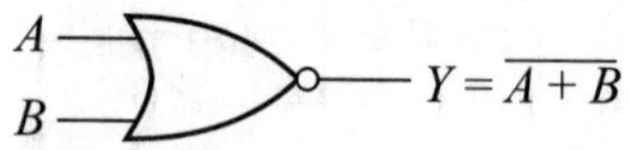

Fig. 5.6. (b) Equivalent NOR gate.

Table 5.6

Truth Table for NAND Gates Wired as NOR Gate

Inputs		*Output*
A	*B*	*Y*
0	0	
0	1	
1	0	
1	1	

5.3 NOR GATES

NOR gates have two or more inputs and one output. The symbol for a 2-input NOR gate is given in Fig. 5.7 (*a*). It is equivalent to an OR gate followed by an Inverter as shown in Fig. 5.7 (*b*).

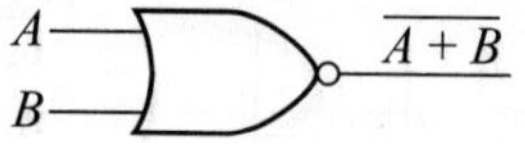

Fig. 5.7. (a) Symbol for 2-input NOR gate.

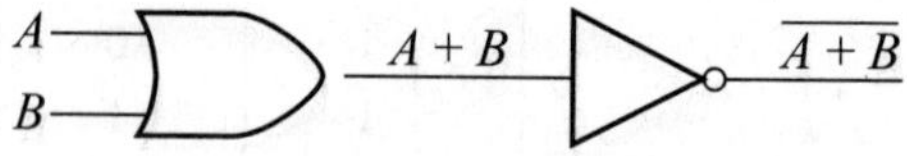

Fig. 5.7. (b) NOR function using OR gate plus an Inverter.

EXPERIMENT 5.2 : TWO-INPUT NOR GATE

Objective

* To verify the truth table for a 2-input NOR gate.
* To consider the operation of NOR gates as

(*a*) Inverters (*b*) OR gates
(*c*) AND gates and (*d*) NAND gates.

Materials Required

Logic trainer
Volt-ohm-milliammeter (VOM)
TTL IC 7402 : Quad, 2-input NOR gates
Resistor 470 ohm

We will first consider drawing up a truth table for a 2-input NOR gate. The circuit diagram for this experiment is given in Fig. 5.8.

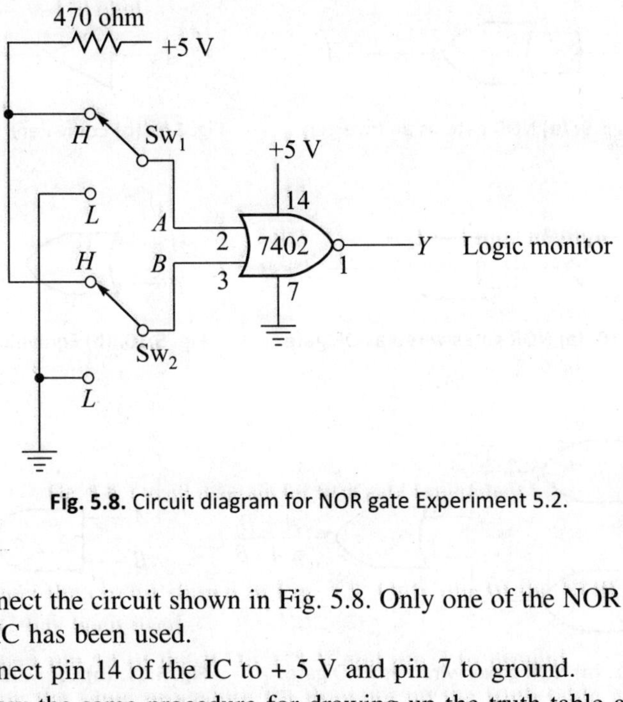

Fig. 5.8. Circuit diagram for NOR gate Experiment 5.2.

Procedure

1. Connect the circuit shown in Fig. 5.8. Only one of the NOR gates from the IC has been used.
2. Connect pin 14 of the IC to + 5 V and pin 7 to ground.
3. Follow the same procedure for drawing up the truth table as in earlier experiments and enter your result in Table 5.7. Your observations should tally with Table 5.7 in Appendix 3.

Table 5.7
Truth Table for 2-Input NOR Gate

Inputs		*Output*
A	*B*	*Y*
0	0	
0	1	
1	0	
1	1	

The distinctive feature of the NOR gate is that the output is high when all inputs are low.

5.4 NOR LOGIC

Like the NAND gate the NOR gate can also be used to implement the following logic gates. Some examples follow:

(*a*) Inverters (*b*) OR gates

(*c*) AND gates (*d*) NAND gates

Inverter

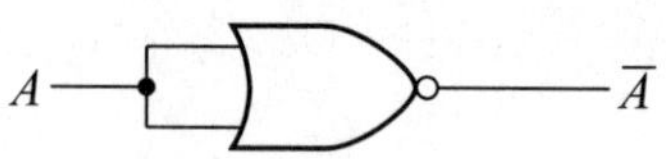

Fig. 5.9. (a) NOR gate as an Inverter.

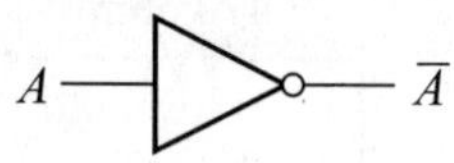

Fig. 5.9. (b) Equivalent inverter.

OR Gate

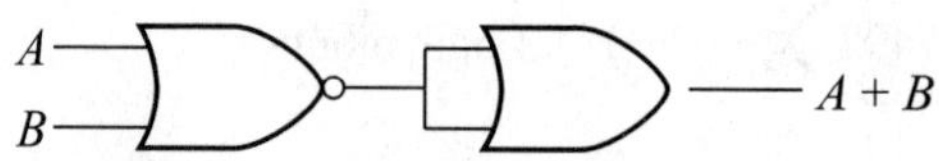

Fig. 5.10. (a) NOR gates wired as OR gate.

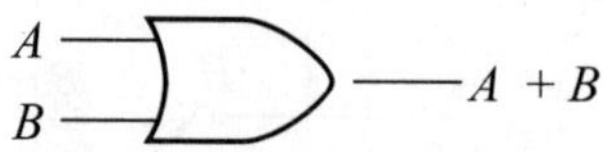

Fig. 5.10. (b) Equivalent OR gate.

AND Gate

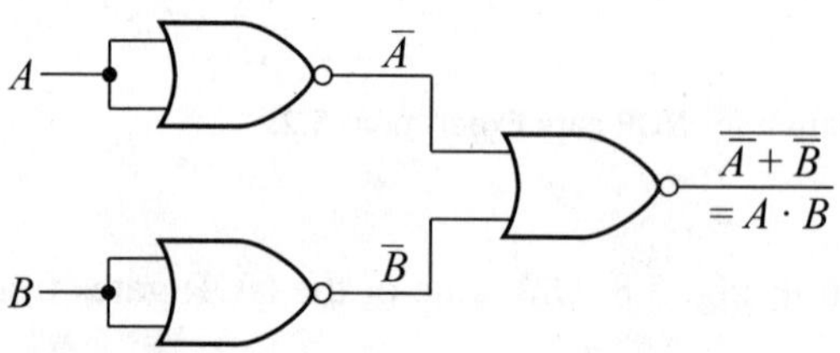

Fig. 5.11. (a) NOR gates wired as AND gate.

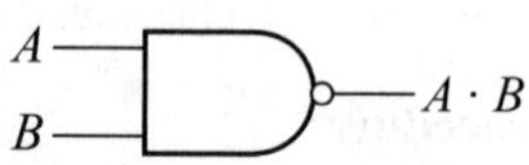

Fig. 5.11. (b) Equivalent AND gate.

NAND Gate

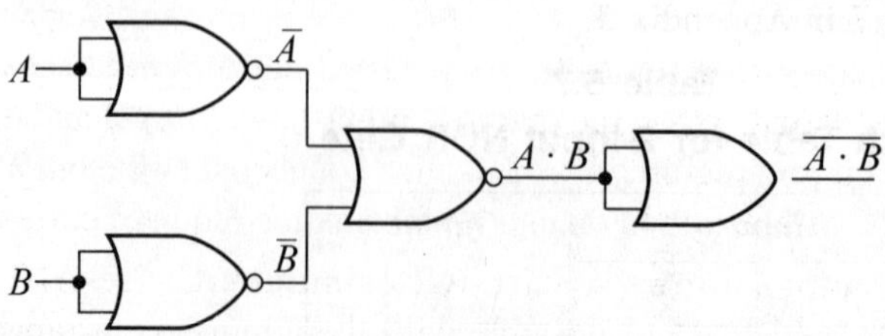

Fig. 5.12. (a) NOR gates as NAND gate.

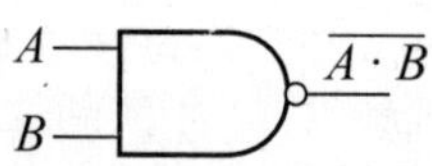

Fig. 5.12. (b) Equivalent NAND gate.

5.5 TTL OUTPUT CONFIGURATIONS

TTL devices are by and large available in totem-pole output types. Some of them are also available in open-collector and tri-state output configurations. We will briefly consider their advantages and applications.

Totem-Pole Output

Figure 5.13 shows a TTL NAND gate with totem-pole output. When the inputs are low, transistors T_1 and T_3 are turned off and the output goes high. T_2 can now source a current from the V_{CC} to the input of the following gate. When the inputs are high, T_1 and T_3 will be on and the output will be low. T_3 can now sink a current from the input of the following gate.

There are advantages and disadvantages in this configuration. The advantage is that in both the low and the high states, t_{PHL} and t_{PLH} are almost equal as a transistor drives the output in both the states. The main disadvantage is that the outputs cannot be connected together to feed a common line. If they are connected together the output transistors may get damaged.

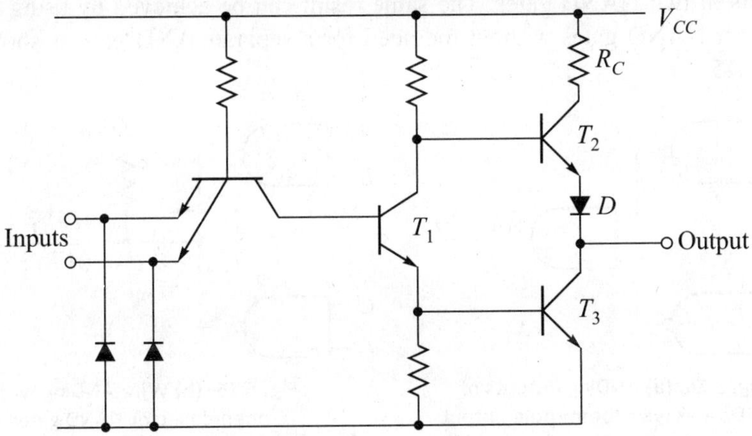

Fig. 5.13. TTL NAND gate with totem-pole output.

Open-Collector Output

Figure 5.14 (*a*) shows the open-collector configuration. This is done by slightly modifying the totem-pole circuit of Fig. 5.13. You will notice that resistor R_C, diode *D* and transistor T_2 have been removed. An external pull-up resistor *R* is connected from V_{CC} to the output. When the output is high, T_3 will be off and the resistor *R* will source the current required by the device connected to the output. When the output is low, T_3 will be turned on and it will sink the required current.

The open-collector output allows a wire-OR (similar to wire-AND) connection of the outputs, which is not possible with the totem-pole arrangement. The advantage of a wire-AND connection is that the output of a number of gates can be combined by tying the outputs together without the need for a separate AND gate. In this arrangement either of the outputs, when it goes low, can pull the common connection low.

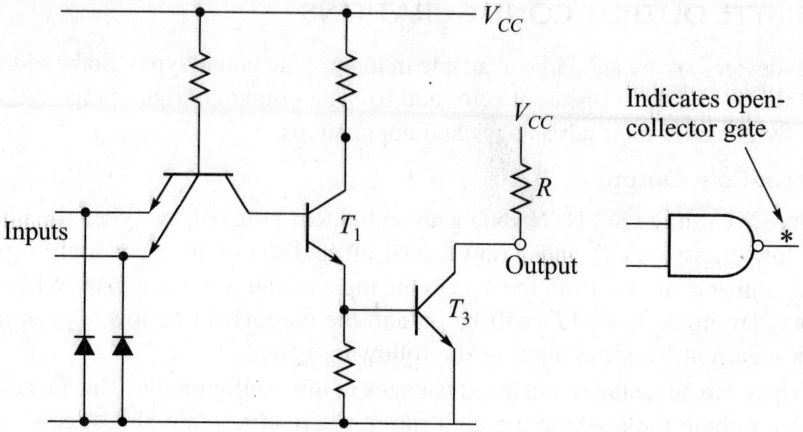

Fig. 5.14. **(a)** Open-collector TTL NAND gate. **Fig. 5.14.** **(b)** Logic symbol.

Consider the circuit given in Fig. 5.15 (*a*), which shows the ANDing of the outputs of two NAND gates. The same result can be achieved by using open-collector NAND gates without the need for a separate AND gate as shown in Fig. 5.15 (*b*).

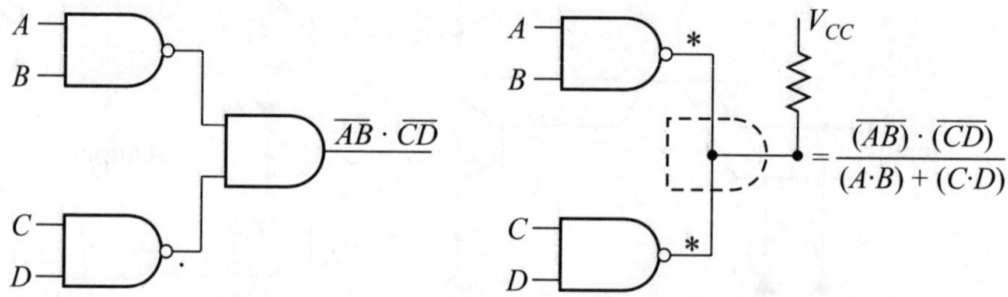

Fig. 5.15. **(a)** ANDing outputs of NAND gates with totem-pole outputs.

Fig. 5.15. **(b)** Wire-ANDing with open-collector NAND gates.

You will notice from these diagrams that the output in both the cases in the same, which will lead you to the conclusion that wire-ANDing of the outputs of open-collector gates produces an AND connection of their inputs. The output can also be expressed as

$$\overline{(A \cdot B) + (C \cdot D)}$$

by applying DeMorgan's theorem. This expression looks like the output of a NOR gate. You will see this more clearly when you perform the following experiments.

EXPERIMENT 5.3 : OPEN-COLLECTOR INVERTERS

Objective

* To study the logic function performed by open-collector Inverters using the wired-OR connection.

Materials Required

Logic trainer
Volt-ohm-milliammeter
TTL IC 7403 : Quad, 2-input, open-collector NAND gates
Resistor 1 K ohm
Resistor 470 ohm
The circuit diagram for this experiment is given in Fig. 5.16.

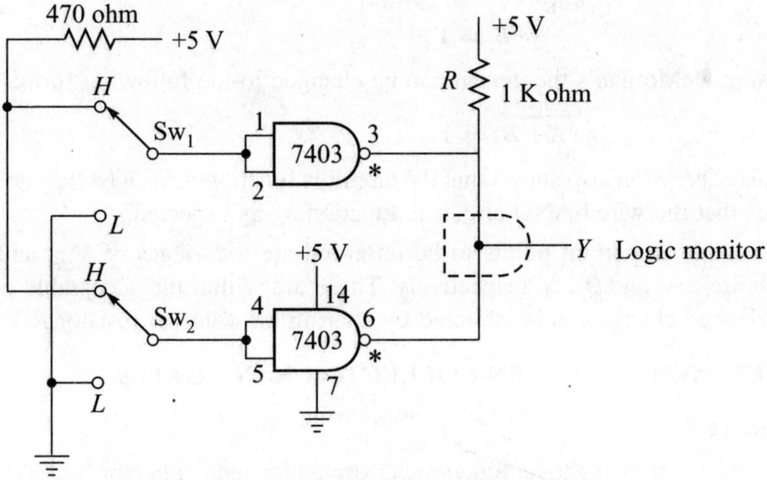

Fig. 5.16. Circuit diagram for Experiment 5.3.

Procedure

1. Connect the circuit given in Fig. 5.16. Notice that only NAND gates have been used and they have been connected as Inverters.
2. Connect pin 14 of the IC to + 5 V and pin 7 to ground.
3. Apply the inputs shown in Table 5.8 and follow the same procedure for drawing up the truth table as in earlier experiments. Enter your observations in Table 5.8.
4. Measure the voltage with a voltmeter at pin 3 or pin 6 of the IC when both logic switches Sw_1 and Sw_2 are in the high position and also when both are in the low position. Enter your voltage readings in Table 5.8.
5. Your observations must tally with Table 5.8 in Appendix 3.

Table 5.8

Inputs		*Output*	*Voltage at pin 3*
A	*B*	*Y*	
0	0		
0	1		
1	0		
1	1		

If you look at the last row of Table 5.8 you will notice that when both the inputs are 1, the output is low. If you prepare a Boolean expression from this observation it will be as follows:

$$\overline{(A + B)} = Y$$

This shows clearly that the circuit is functioning as a NOR gate. There is another way of analyzing the circuit. You can obtain a Boolean expression for the output from the first row of Table 5.8. It will be as follows:

$$\overline{A} \cdot \overline{B} = Y$$

Using DeMorgan's theorem it can be changed to the following form:

$$\overline{(A + B)} = Y$$

This expression also shows that the circuit is functioning as a NOR gate. This proves that the wired-AND circuit is functioning as expected.

The other important points to be observed are the values of V_{OH} and V_{OL} which are 5 V and 0.1 V respectively. These are within the acceptable limits. Any desired change can be effected by altering the value of resistor *R*.

EXPERIMENT 5.4 : OPEN-COLLECTOR NAND GATES

Objective

* To study the logic function performed by open-collector NAND gates.

Materials Required

Logic trainer

Volt-ohm-milliammeter

TTL IC 7403 : Quad, 2-input open-collector NAND gates

Resistor 470 ohm

Resistor 1 K ohm

The circuit for this experiment is given in Fig. 5.17.

Procedure

1. Connect the circuit given in Fig. 5.17. Notice that only two NAND gates have been used.
2. Connect pin 14 of the IC to + 5 V and pin 7 to ground.
3. Apply the inputs shown in Table 5.9 and follow the same procedure for drawing up the truth table as in Experiment 5.3.
4. Measure the voltage with a voltmeter at pin 3 or pin 6 of the IC when both the logic switches Sw_1 and Sw_2 are in the high position and also when they are in the low position.
5. Your observations should tally with Table 5.9 in Appendix 3.

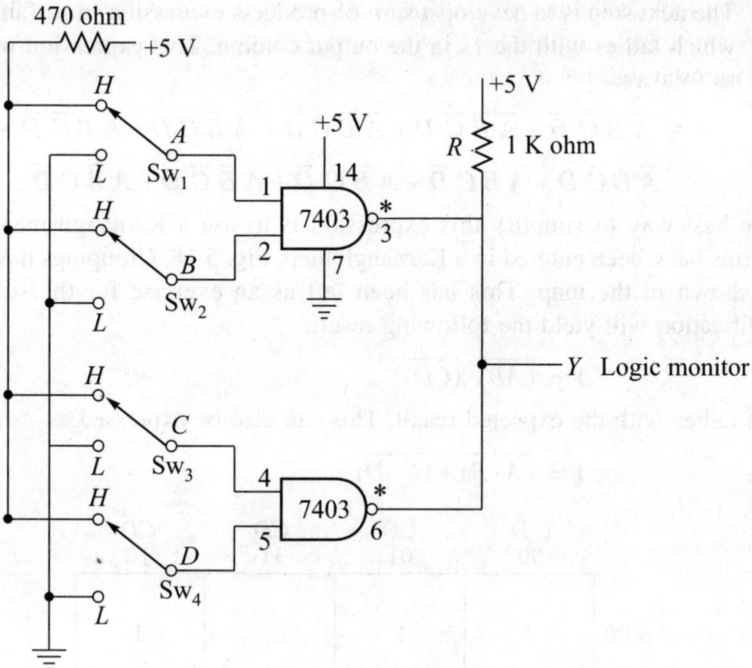

Fig. 5.17. Circuit diagram for Experiment 5.4.

Table 5.9

Inputs				Output	Voltage
A	*B*	*C*	*D*	*Y*	*at pin 3*
0	0	0	0		
0	0	0	1		
0	0	1	0		
0	0	1	1		
0	1	0	0		
0	1	0	1		
0	1	1	0		
0	1	1	1		
1	0	0	0		
1	0	0	1		
1	0	1	0		
1	0	1	1		
1	1	0	0		
1	1	0	1		
1	1	1	0		
1	1	1	1		

6. The next step is to develop a sum-of-products expression from Table 5.9 which tallies with the 1s in the output column. This expression will be as follows:

$$Y = \bar{A}\,\bar{B}\,\bar{C}\,\bar{D} + \bar{A}\,\bar{B}\,\bar{C}\,D + \bar{A}\,\bar{B}\,C\,\bar{D} + \bar{A}\,\bar{B}\,C\,D + \bar{A}\,B\,\bar{C}\,\bar{D} +$$
$$\bar{A}\,B\,\bar{C}\,D + \bar{A}\,B\,C\,\bar{D} + A\,\bar{B}\,\bar{C}\,\bar{D} + A\,\bar{B}\,\bar{C}\,D + A\,\bar{B}\,C\,\bar{D}$$

The best way to simplify this expression is to use a Karnaugh map. The minterms have been entered in a Karnaugh map, Fig. 5.18. Groupings have not been shown in the map. This has been left as an exercise for the student. Simplification will yield the following result:

$$Y = \overline{(AB)} \cdot \overline{(CD)}$$

which tallies with the expected result. This can also be expressed as

$$Y = \overline{(A \cdot B) + (C \cdot D)}$$

	$\bar{C}\,\bar{D}$ 00	$\bar{C}D$ 01	CD 11	$C\bar{D}$ 10
$\bar{A}\,\bar{B}$ 00	1 0	1 1	3	1 2
$\bar{A}\,\bar{B}$ 01	1 4	1 5	7	1 6
$A\,B$ 11	12	13	15	14
$A\,\bar{B}$ 10	1 8	1 9	11	1 10

Fig. 5.18

Tri-State TTL Gates

Another output configuration in TTL gates is known as Tri-state output. You are already familiar with high and low logic levels. There is a third variation known as the high impedance state. When a number of high and low outputs are connected to a common bus, interaction causes some problems. To overcome this a high impedance state has been conceived. These gates are known as 3-state gates.

A TTL Inverter with 3-state output is given in Fig. 5.19. Notice that there is a control input which was not there in 2-state gates. When the control input is low the output is in the high impedance state and when the control input is high, the output will be high or low depending on the state of the input.

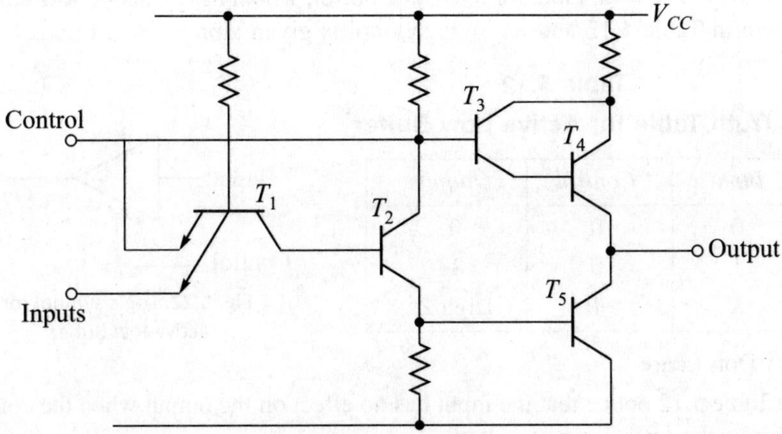

Fig. 5.19. Tri-state Inverter.

The truth table for this Tri-state Inverter, based on what we have discussed is given in Table 5.10 and the symbol for this Inverter is shown alongside in Fig. 5.20. Notice that this Inverter is active High.

Table 5.10

Truth Table for Active High Inverter

Input	*Control*	*Output*
0	0	High *Z*
1	0	High *Z*
0	1	1
1	1	0

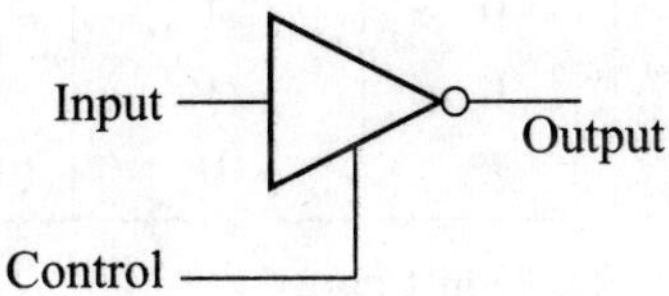

Fig. 5.20. Logic symbol for Tri-state Inverter.

This Inverter can also be designed so that it is active low. The truth table for an active low Inverter is given in Table 5.11 and its logic symbol is given alongside in Fig. 5.21.

Table 5.11

Truth Table for Active Low Inverter

Input	*Control*	*Output*
0	1	High *Z*
1	1	High *Z*
0	0	1
1	0	0

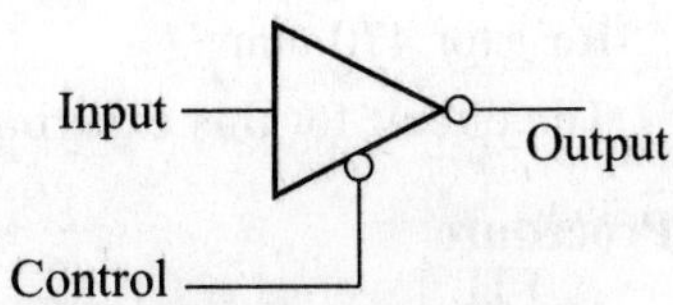

Fig. 5.21. Logic symbol for Tri-state Inverter.

Tri-State Buffer

A tri-state buffer can be made on the same basis as the tri-state Inverter shown in Fig. 5.19. The truth table for a tri-state buffer, which has an active low control is given in Table 5.12 and its logic symbol is given alongside in Fig. 5.22.

Table 5.12

Truth Table for Active Low Buffer

Input	*Control*	*Output*
0	0	0
1	0	1
X	1	High *Z*

X : Don't care

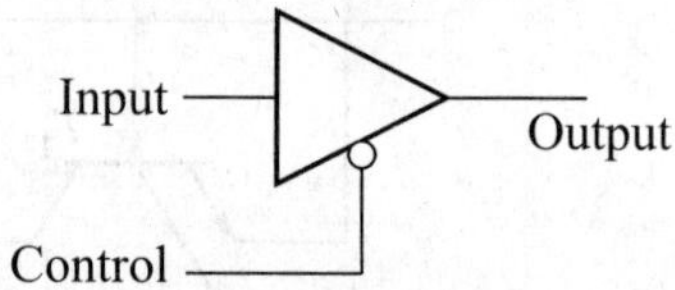

Fig. 5.22. Logic symbol for active low Buffer.

In Table 5.12 notice that the input has no effect on the output when the control is high as the buffer is active low.

The truth table for a tri-state buffer which is active high is given in Table 5.13 and its logic symbol is given in Fig. 5.23.

Table 5.13

Truth Table for Active High Buffer

Input	*Control*	*Output*
0	1	1
0	1	0
X	0	High *Z*

X : Don't care

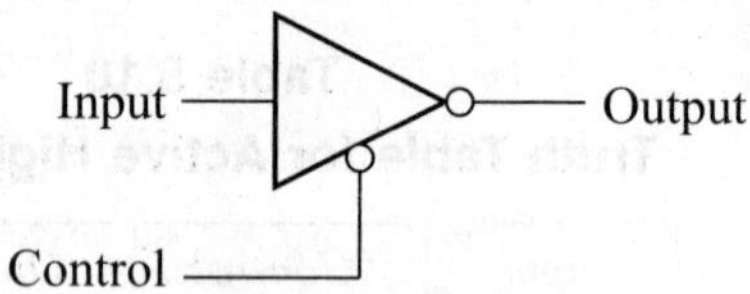

Fig. 5.23. Logic symbol for active high Buffer.

EXPERIMENT 5.5 : TRI-STATE BUFFER

Objective

* To study the operation of a Tri-state Buffer.

Materials Required

Logic trainer

TTL IC 74126 Tri-state Quad Buffers

Resistor 470 ohm

The circuit for this experiment is given in Fig. 5.24.

Procedure

1. Connect the circuit given in Fig. 5.24. Connect pin 14 of the IC to + 5 V and pin 7 to ground.

2. Apply high and low logic inputs at point *A* of the buffer while holding the control input high. Observe the output and record your observations.

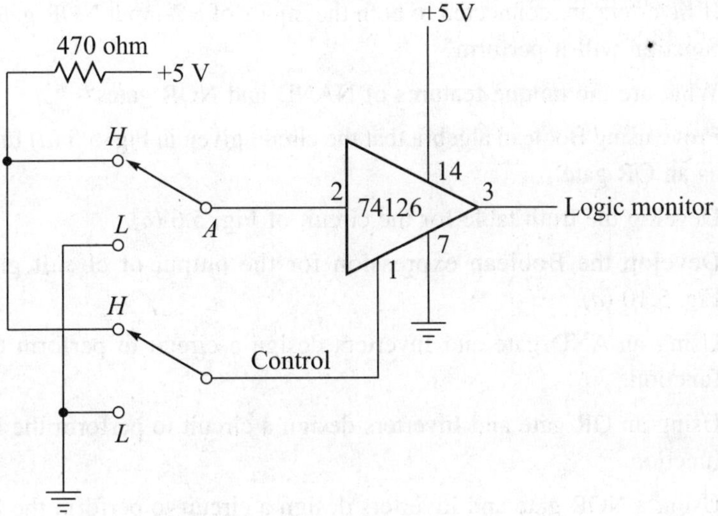

Fig. 5.24. Circuit for Experiment 5.5.

3. Now hold the control input at low and high logic levels at input *A*. Record your observations in Table 5.14.
4. Your observations should tally with Table 5.13.

Table 5.14

Truth Table for Tri-state Buffer (Active High Control)

Input	*Control*	*Output*
0	1	
1	1	
0	0	
1	0	

PROBLEMS

5.1. From the following select the output equations for NAND and NOR gates :

(*a*) $\overline{A + B}$ (*b*) $A \cdot B$

(*c*) $A + B$ (*d*) $\overline{A \cdot B}$

5.2. If an Inverter is connected to the output of a NAND gate what function will it perform?

5.3. If Inverters are connected to both the inputs of a 2-input NOR gate, what function will it perform?

5.4. What are the unique features of NAND and NOR gates?

5.5. Prove using Boolean algebra that the circuit given in Fig. 5.5 (*a*) functions as an OR gate.

5.6. Develop the truth table for the circuit of Fig. 5.6 (*a*).

5.7. Develop the Boolean expression for the output of circuit given in Fig. 5.10 (*a*).

5.8. Using an AND gate and Inverters design a circuit to perform the OR function.

5.9. Using an OR gate and Inverters design a circuit to perform the NAND function.

5.10. Using a NOR gate and Inverters design a circuit to perform the NAND function.

5.11. Draw a truth table for a 2-input NOR gate using negative logic. Does it resemble the truth table for any gate known to you?

5.12. Implement a NOR gate using only NAND gates.

5.13. Write the output equation for the circuit given in Fig. P-5.13.

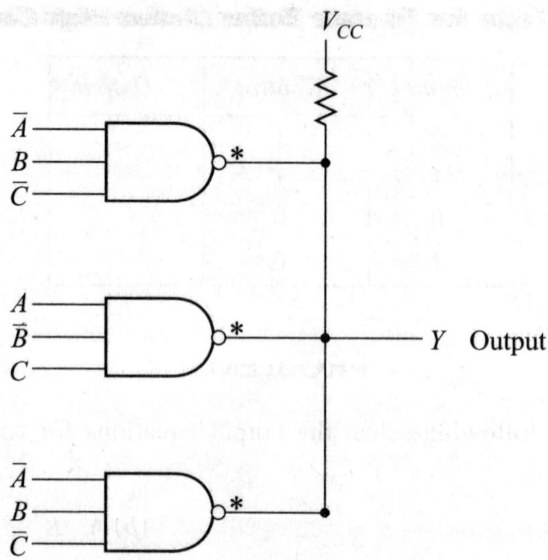

Fig. P-5.13

5.14. Perform an experiment on the circuit given in Fig. P-5.14 to determine its truth table and output.

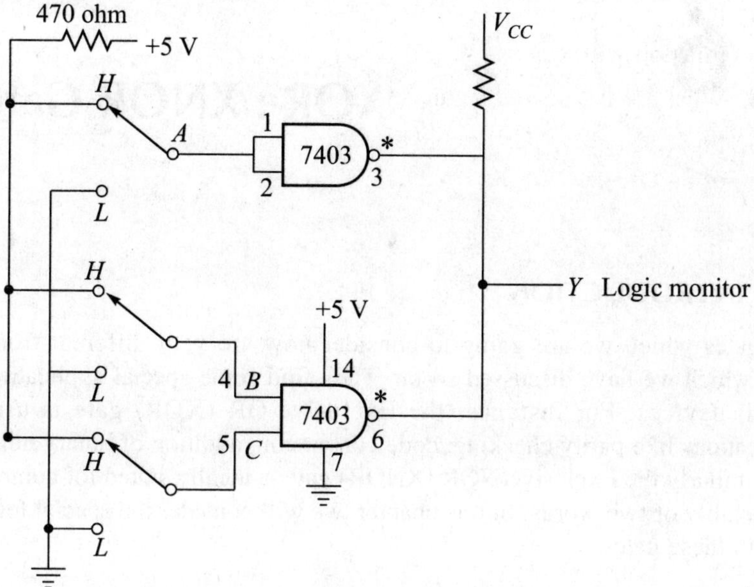

Fig. P-5.14

CHAPTER 6

XOR : XNOR Gates

6.1 INTRODUCTION

The gates which we are going to consider now are very different from the gates which we have discussed so far. They find some special applications in digital devices. For instance, the Exclusive-OR (XOR) gate is used in applications like parity checking, code conversion, addition of binary numbers etc. Similarly the Exclusive-NOR (XNOR) gate is ideally suited for comparing the equality of two words. In this chapter, we will consider the special features of both these gates.

6.2 EXCLUSIVE-OR (XOR) GATE

We refer to the OR gate, with which you are already familiar as the Inclusive-OR gate. The Exclusive-OR (XOR) gate is a special case of the general operation. If you refer to the truth tables of the OR gate, Table 6.1 and XOR gate, Table 6.2, given side-by-side below, you will be able to notice their distinctive features.

Table 6.1
Truth Table for OR Gate (2-Inputs)

Inputs		*Output*
A	*B*	*Y*
0	0	0
0	1	1
1	0	1
1	1	1

Table 6.2
Truth Table for XOR (2-Inputs)

Inputs		*Output*
A	*B*	*Y*
0	0	0
0	1	1
1	0	1
1	1	0

You will notice from the OR gate table, Table 6.1, that either input *A* or *B* or both must be true (binary 1) in order that the OR function is true. However, in the XOR operation, either *A* or *B*, but not both, must be true (binary 1) for the function to be true, as is evident from Table 6.2 for the XOR operation. If both *A* and *B* are true at the same time the function will be false (binary 0).

On comparing the two truth tables, Tables 6.1 and 6.2 for the 2-input OR and XOR gates you will find no difference in the outputs in the first three rows. However, when both the inputs are 1, as in the last row, the output of the XOR gate is 0, whereas it is 1 for the OR gate.

We can now draw very important conclusions from these observations:

* The XOR gate is enabled with an odd number of 1s at the input, as in rows 2 and 3.
* The XOR gate is disabled by an even number of 1s (0 and 2), as in rows 1 and 4 and a 0 appears at the output.
* The XOR gate can thus be used as an odd-bits checker. We can summarize the rules for XOR operation as follows:

$$0 \oplus 0 = 0$$
$$0 \oplus 1 = 1$$
$$1 \oplus 0 = 1$$
$$1 \oplus 1 = 0$$

EXPERIMENT 6.1 : TWO-INPUT XOR GATE

Objective

* To prepare the truth table.
* To write the minterm equation from the truth table.
* To draw a logic diagram from the minterm equation.

Materials Required

Logic trainer

TTL IC 7486 : Quad Exclusive-OR gates

Resistor 470 ohm

The circuit diagram for the experiment is given in Fig. 6.1 (*a*). The logic symbol for the XOR gate is given in Fig. 6.1 (*b*).

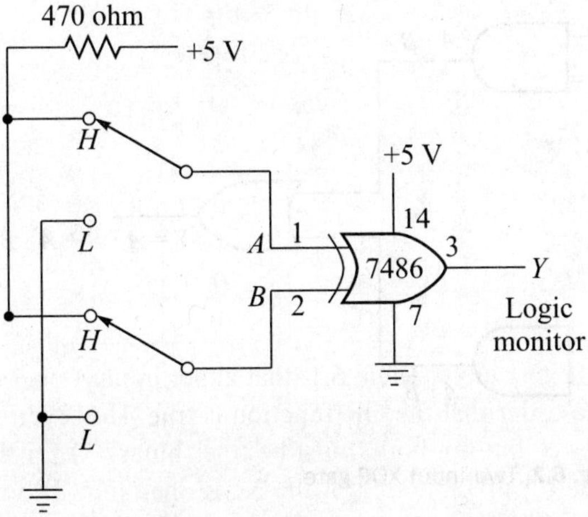

Fig. 6.1. (a) Circuit for Experiment 6.1.

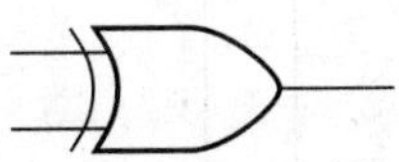

Fig. 6.1. (b) Logic symbol for 2-input XOR gate.

Procedure

1. Connect the circuit given in Fig. 6.1 (*a*). Connect pin 14 of the IC to + 5 V and pin 7 to ground. Notice that only one of the gates is being used for this experiment.
2. Apply the inputs shown in Table 6.3 and follow the same procedure for drawing up the truth table as in earlier experiments.
3. Record your observations in Table 6.3. Your observations should tally with Table 6.2.
4. You can write the minterm equation by following the inputs and the outputs in the 2nd and 3rd rows of Table 6.3. The equation should be as follows:

$$y = (\overline{A} \cdot B) + (A \,.\, \overline{B})$$

5. You can now draw the logic diagram with the help of the minterm equation. The diagram should be as in Fig. 6.2.

Table 6.3

Truth Table for 2-Input XOR Gate

Inputs		*Output*
A	*B*	*Y*
0	0	
0	1	
1	0	
1	1	

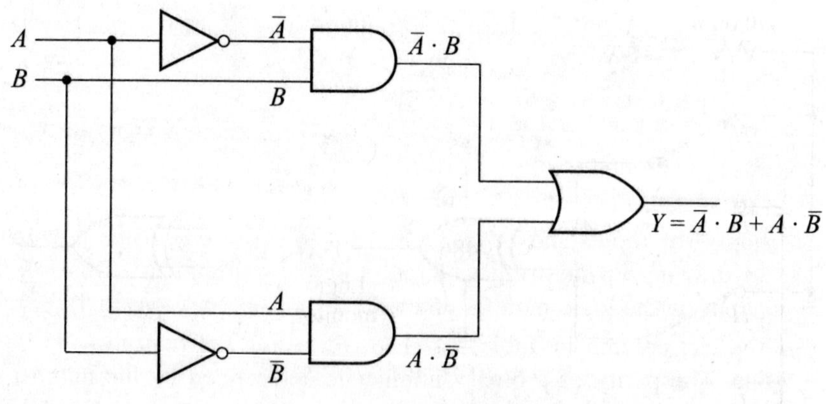

Fig. 6.2. Two-input XOR gate.

EXPERIMENT 6.2 : THREE-INPUT XOR GATE

Objective

* To draw up a truth table for a 3-input XOR gate.
* To determine the parity of the input bits.

Materials Required

Logic trainer

TTL IC 7486 : Quad Exclusive-OR gates

Resistor 470 ohm.

The circuit diagram for the experiment is given in Fig. 6.3 (*a*).

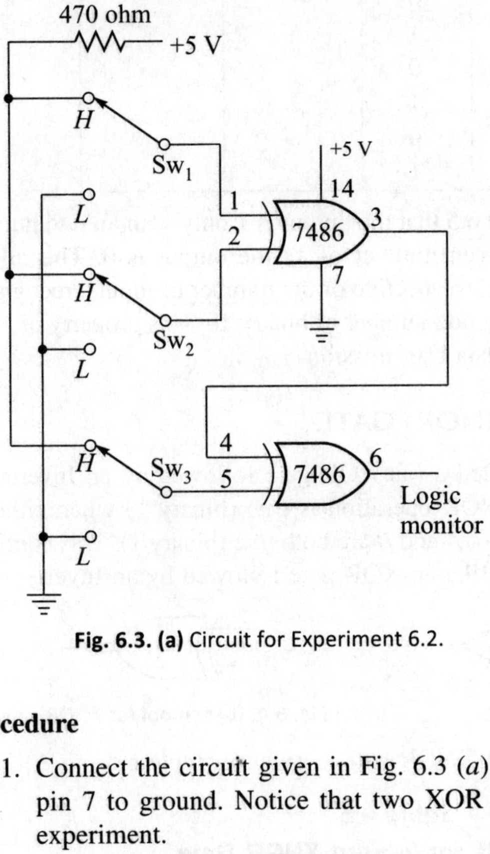

Fig. 6.3. (a) Circuit for Experiment 6.2.

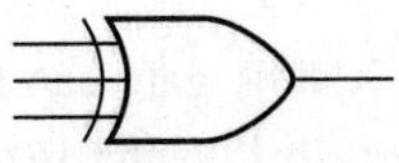

Fig. 6.3. (b) Logic symbol for 3-input XOR gate.

Procedure

1. Connect the circuit given in Fig. 6.3 (*a*). Connect pin 14 to + 5 V and pin 7 to ground. Notice that two XOR gates have been used for this experiment.
2. Apply the inputs shown in Table 6.4 and follow the same procedure for drawing up the truth table as in earlier experiments. Observe the output on the logic monitor and record your observations in Table 6.4.
3. The last column in Table 6.4 is for recording the parity of the input bits. The parity of a binary number is determined by the number of 1s in the binary number. If the number of 1s is even, the binary number is said to have even parity. If the number of 1s is odd, the binary number is said to have odd parity.

4. Record the parity of the input bits in the parity column of Table 6.4.
5. Your observations should tally with Table 6.4 (Appendix 3).

Table 6.4

Truth Table for 3-Input XOR Gate

Inputs			*Output*	*Parity*
A	*B*	*C*	*Y*	
0	0	0		
0	0	1		
0	1	0		
0	1	1		
1	0	0		
1	0	1		
1	1	0		
1	1	1		

You will observe from Table 6.5 that the output is 1 only with an odd number of binary 1 inputs. With an even number of 1s the output is 0. The rule to remember is that an XOR gate, irrespective of the number of inputs, recognizes only those words which have an odd number of binary 1s. This property of XOR gates is found very useful in data transmission.

6.3 EXCLUSIVE-NOR (XNOR) GATE

The XNOR gate can be regarded as an XOR gate followed by an Inverter as shown in Fig. 6.4 (*a*). The XNOR operation is true (binary 1) when either *A* and *B* are both false (binary 0), or *A* and *B* are both true (binary 1). This confirms that the XNOR gate functions like an XOR gate followed by an Inverter.

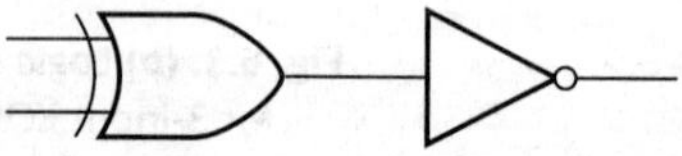

Fig. 6.4. (a) XNOR gate.

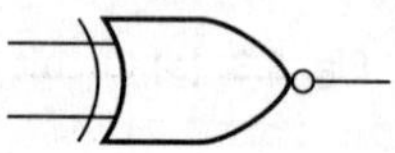

Fig. 6.4. (b) Symbol for XNOR gate.

The truth table for a 2-input XNOR gate is given in Table 6.5.

Table 6.5

Truth Table for 2-Input XNOR Gate

Inputs		*Output*
A	*B*	*Y*
0	0	1
0	1	0
1	0	0
1	1	1

You will notice from this table that the output of the XNOR gate is the complement of the output of the XOR gate.

The equation for the output of the XNOR gate can be written as follows:

$$Y = A \text{ XNOR } B$$

or as

$$Y = \overline{A \oplus B}$$

The rules for the XNOR operation can be stated as follows:

$$\overline{0 \oplus 0} = 1$$

$$\overline{0 \oplus 1} = 0$$

$$\overline{1 \oplus 0} = 0$$

$$\overline{1 \oplus 1} = 1$$

You will notice that the output is binary 1 only when both inputs are either binary 0 or binary 1. This is exactly the opposite of the rules for the XOR gate. With this property of the XNOR gate it will function as an equality detector for one bit of data. In some applications it is necessary to detect equalities involving several bits of data. This is accomplished by establishing the equality of each bit individually with separate XNOR circuits and then determining if all equalities exist at the same time by connecting all the XNOR outputs to the inputs of an AND gate.

Table 6.6

Truth Table for 3-Input XNOR Gate

Inputs			*Output*	*Parity*
A	*B*	*C*	*Y*	
0	0	0	1	Even
0	0	1	0	Odd
0	1	0	0	Odd
0	1	1	1	Even
1	0	0	0	Odd
1	0	1	1	Even
1	1	0	1	Even
1	1	1	0	Odd

The truth tables for 2-input and 3-input XNOR gates show that with an even number of 1s the output is 1 and with an odd number of 1s the output is 0. This holds good for XNOR gates with any number of inputs.

PROBLEMS

6.1. Determine the function of the circuit given in Fig. P-6.1.

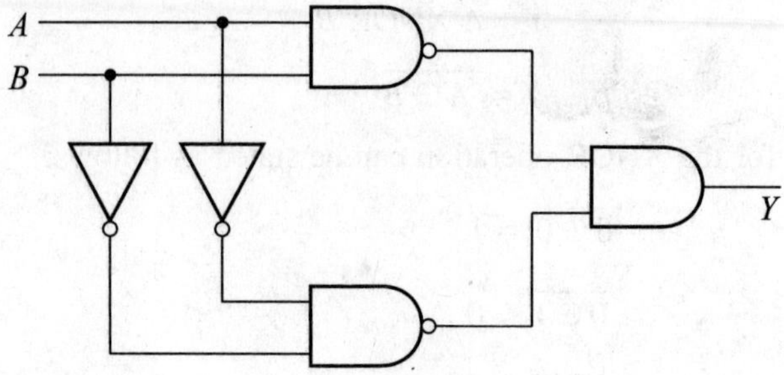

Fig. P-6.1

6.2. State the parities of the following binary numbers:

(*a*) 1101001 (*b*) 101011110

(*c*) 1000010110 (*d*) 10101001

6.3. Draw a circuit to determine the parity of a 4-bit binary number.

6.4. Draw a circuit to determine whether two 4-bit binary numbers are equal.

6.5. Draw a circuit diagram to implement a 4-input XOR gate.

6.6. Determine the function of the following circuit (Fig. P-6.6):

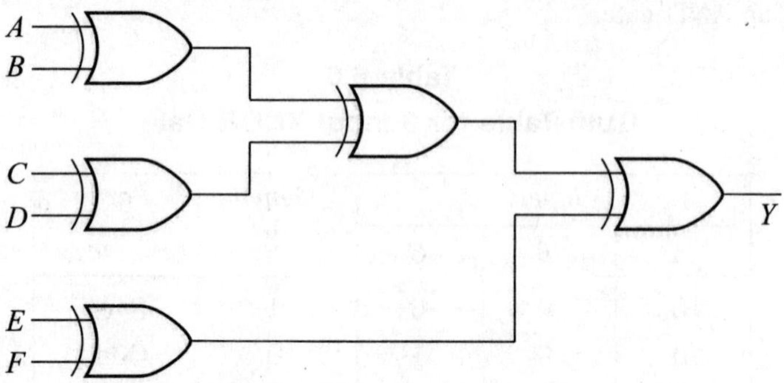

Fig. P-6.6

CHAPTER 7

Interfacing TTL : CMOS Gates

7.1 INTRODUCTION

It is sometimes necessary to use logic devices of different families in the same system. When that happens there is bound to be an interfacing problem as the input and output characteristics of two different families are likely to be different. The output of the driving gate has to match the input of the load being driven. When they do not match some external circuitry can always be employed to match the logic characteristics of the driver to the characteristics of the load.

In this chapter, we will consider how TTL and CMOS drivers can be interfaced with CMOS and TTL loads. There is no interfacing problem between devices which belong to the same family. Therefore, any TTL or CMOS device can be interfaced with any device of the same family.

7.2 CMOS TO CMOS INTERFACE

The input and output profiles of CMOS devices are shown in Fig. 2.4 (*a*) and Fig. 2.4 (*b*). You will notice from these profiles that the high and low output logic levels lie within the acceptable high and low input voltage levels of a CMOS load. A CMOS driver can therefore be directly connected to a CMOS load.

The minimum and maximum input voltage levels for CMOS devices are listed below

CMOS driver	**CMOS load**
V_{OH} min : 4.9 V	V_{IH} min : 3.5 V
V_{OL} max : 0.1 V	V_{IL} max : 1.5 V

The input and output currents for CMOS devices in the 74C00 series are as follows:

74C00 drivers	**74C00 loads**
I_{OL} max : 10 μA	I_{IL} max : – 1 μA
I_{OH} max : – 10 μA	I_{IH} max : 1 μA

Since the output current of the driver is ten times the input load current, a CMOS driver can drive 10 unit loads. Therefore a fan-out of 10 will be possible.

7.3 TTL TO TTL INTERFACE

The input and output profiles of TTL devices are shown in Fig. 2.1 (*a*) and Fig. 2.1 (*b*). You will notice from these diagrams that the output logic levels of a TTL driver are fully compatible with the input logic levels of a TTL load. There is, therefore, no interfacing problem within the TTL family.

7.4 CMOS TO TTL INTERFACE

For the CMOS driver of the 74C00 series and the TTL load of the 7400 series, the output and input voltages and currents are given below:

CMOS 74C00 driver	**7400 load**
V_{OH} min : 4.9 V	V_{IH} min : 2.0 V
V_{OL} max : 0.1 V	V_{IL} max : 0.8 V
I_{OH} max : – 360 μA	I_{IH} max : 40 μA
I_{OL} max : 360 μA	I_{IL} max : – 1.6 mA

The high and low output logic levels of the CMOS driver lie within the permissible range of the high and low input levels of a TTL input, and as such there is no interfacing problem as far as the voltage levels are concerned. However, there is a problem where the levels of current are concerned. When the output is high the CMOS driver can source a current of 360 μA which a TTL load will accept as the maximum current required is 40 μA. However, when the output of the CMOS driver is low, it can sink a current of only 360 μA which is inadequate for a TTL load, since the minimum current required is 1.6 mA. We will now consider some solutions to solve this problem.

* Using a low power Schottky TTL
* Using a CMOS driver of 74H series
* Using CMOS Buffers

We will first of all consider using a low power Schottky TTL device. The voltage and current requirements of a TTL load (74LS TTL series) are as follows:

I_{IH} max : 20 μA	V_{IH} min : 2.0 V
I_{IL} max : – 360 μA	V_{IL} max : 0.8 V

The logic voltage levels present no problem as they are the same as for the standard TTL series. If both the devices are operated from a 5V supply the current requirements will be within the permissible limits of CMOS drivers in the 74C00 series. For this series of CMOS devices the low level and high level output currents are listed as follows:

I_{OL} max : 360 μA

I_{OH} max : – 360 μA

This shows that the 74C00 driver can sink 360 μA which is just right to sink the input current of a low power Schottky device. As the high state current is only 20 μA, the driver has the capability to source a current of 360 μA which is more than adequate for the purpose. However, the fan-out is limited to one.

One of the alternatives is to use a CMOS driver of the 74H series to drive a standard TTL load. The output currents and voltages for this CMOS series are as follows:

I_{OH} max : 4.0 mA	V_{OH} min : 4.9 V
I_{OL} max : 4.0 mA	V_{OL} max : 0.1 V

The CMOS driver can source a current of 4 mA when its output is high, which is adequate for a standard TTL load, as it requires a current of only 40 μA. In the low state the CMOS driver can sink a current of 4 mA which is sufficient for a TTL load, as it requires a current of only 1.6 mA. Thus, a fan-out of 2 is possible with this arrangement if both the devices are operated from a 5V supply.

Yet another possibility is to use CMOS inverting or non-inverting Buffers. CMOS Buffers have an available output current of up to 6 mA. Some of the available Buffers are listed below.

CD 4049	Inverting Buffer
CD 4050	Non-inverting Buffer
74C901	Hex inverting Buffer
74C902	Hex non-inverting Buffer

A CMOS Buffer can drive up to two standard TTL inputs. A typical setup is shown in Fig. 7.1.

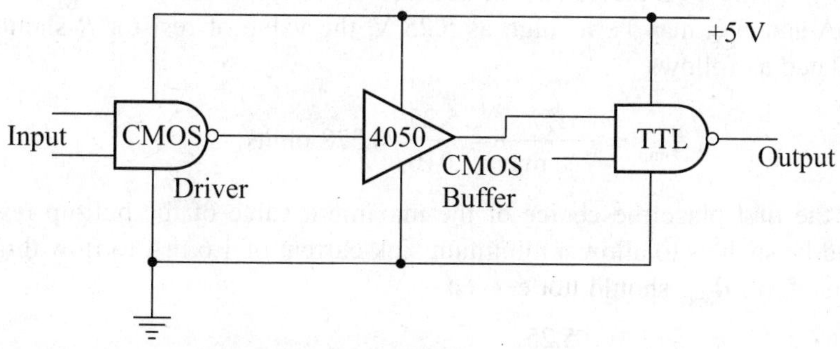

Fig. 7.1

7.5 TTL TO CMOS INTERFACE

The worst case output voltages for a TTL driver and the required input voltages for a CMOS load, when both are operated from a 5V supply, are as follows:

TTL driver	**CMOS load**
V_{OH} min : 2.4 V	V_{IH} min : 3.5 V
V_{OL} max : 0.4 V	V_{IL} max : 1.5 V

The low state output voltage of a TTL driver which is 0.4 V is suitable for a CMOS load, as it will accept any voltage up to 1.5 V as a logic 0 input. However, when the TTL driver output is high, the minimum output voltage is 2.4 V, whereas the voltage required for a CMOS load is 3.5 V. The high state output voltage of a TTL driver is, therefore, not acceptable for a CMOS load.

The simplest solution to the problem is the use of a pull-up resistor as shown in Fig. 7.2:

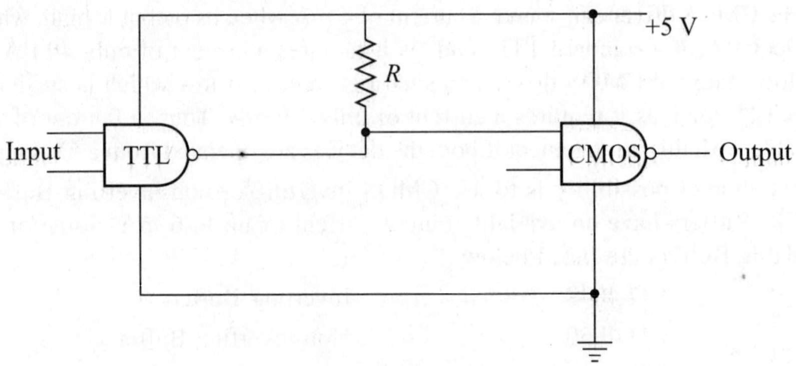

Fig. 7.2. Interfacing a TTL driver to a CMOS load with a pull-up resistor.

Resistor R has negligible effect on the low state output of the TTL driver. In its low state it will sink a current which will depend on the value of the pull-up resistor. As the TTL driver should not sink a current in excess of I_{OL} max, or 16 mA and V_{CC} may be as high as 5.25 V, the value of resistor R should be calculated as follows:

$$R_{\text{min}} = \frac{V_{CC}}{I_{OL}\ \text{max}} = \frac{5.25}{.016} = 328 \text{ ohms}$$

In the first place the choice of the maximum value of the pull-up resistor should be such as to allow a minimum sink current of 1.6 mA to flow through it. Therefore, R_{max} should not exceed

$$R_{\text{max}} = \frac{5.25}{1.6 \text{ mA}} = 3281 \text{ or } 3.3 \text{ K ohms}$$

The value of R_{max} is also a function of the input capacitance of the load. When the driver output is high the pull-up resistor may raise it enough to cut off the upper totem-pole transistor and charge the input capacitance of the load. Since this will raise the time constant, it will have the effect of slowing down the speed of operation. Therefore, the value of pull-up resistor should neither be too low nor too high. Normally any value between 1 K and 3.3 K ohm may be adequate.

A better interfacing solution depends on the use of a level shifter as shown in Fig. 7.3. In this arrangement, a CMOS level shifter, IC 40109, has been used. Its input side operates on a 5 V supply which is the same as for the TTL device, and the output side operates on a supply of 12 V which is the same as for the CMOS load. The pull-up resistor pulls up the high state output voltage of the TTL driver to about 5 V, which is compatible with the requirement of the CMOS Buffer. The output logic levels of the Buffer are compatible with the CMOS load.

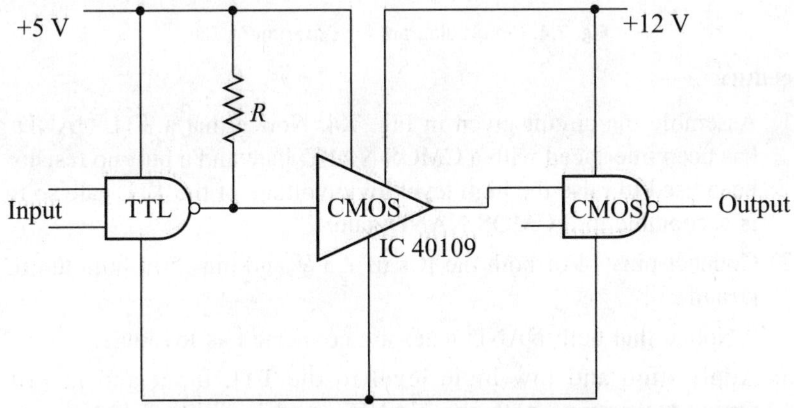

Fig. 7.3. TTL to CMOS interfacing using a level shifter.

Perhaps the most reliable solution to the interfacing problem is a family of logic devices 54/74 HCTXX, manufactured by Texas Instruments, as they are fully compatible with both TTL and CMOS devices.

EXPERIMENT 7.1 : INTERFACING TTL DRIVER TO CMOS LOAD WITH A PULL-UP RESISTOR

Objective

To assess the effectiveness of this interfacing technique.

Materials Required

Logic trainer

Volt-ohm-milliammeter

TTL IC 7400 : Quad, 2-input NAND gates

CMOS IC 4011 : Quad, 2-input NAND gates

Resistor 470 ohm

Resistor 1.2 K ohm

The circuit diagram for this experiment is given in Fig. 7.4.

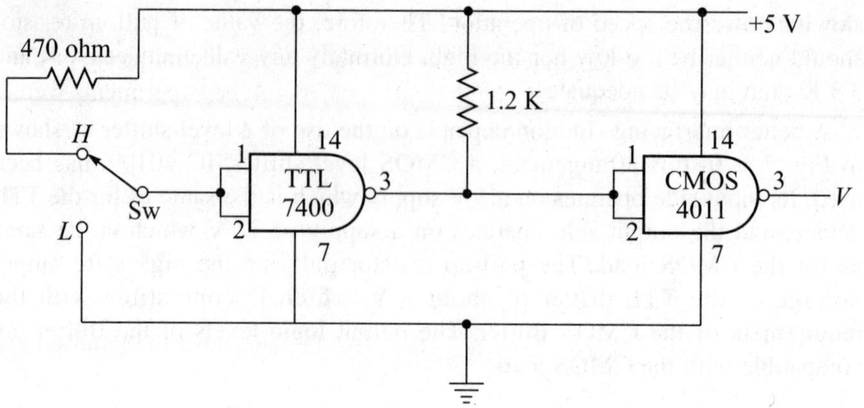

Fig. 7.4. Circuit diagram for Experiment 7.1.

Procedure

1. Assemble the circuit given in Fig. 7.4. Notice that a TTL NAND gate has been interfaced with a CMOS NAND gate and a pull-up resistor has been used to raise the high level logic voltage of the TTL gate so that it is acceptable to a CMOS NAND gate.
2. Connect pins 14 of both the ICs to + 5 V and pins 7 of both the ICs to Ground.

   * Notice that both NAND gates are connected as Inverters.
3. Apply high and low logic level to the TTL input and record the output voltage of both the NAND gates in Table 7.1. Your observations should tally with Table 7.1 (Appendix 3).

From your observations of the output voltage of the TTL NAND gate you will notice that the high and low logic levels of the TTL output are compatible with the required input logic levels of the CMOS NAND gate.

Table 7.1

TTL NAND gate		*CMOS NAND gate*
Input	*Output*	*Output*
High		
Low		

7.6 INTERFACING TTL DRIVER TO EXTERNAL LOADS

Almost all logic as well as linear systems are designed to perform some useful function such as lamps, relays, electric motors etc. In most cases, TTL and CMOS drivers cannot be directly connected to an external load. Some interfacing circuitry is almost always required. Here we will consider some simple interfacing applications.

7.6.1 Interfacing TTL Drivers to LEDs

Experiments have been designed to illustrate interfacing techniques. As the experiments are simple, detailed instructions to carry out the experiments were not considered necessary.

EXPERIMENT 7.2

Objective

To interface TTL driver to LED on high and low driver output.

Materials Required

Logic trainer
Volt-ohm-milliammeter
TTL IC 7400 : Quad, 2-input NAND gates
Resistor 330 ohm
Resistor 470 ohm
Resistor 560 ohm
LED (Red)

Procedure

1. Assemble the circuits given in Figs. 7.5 and 7.6.

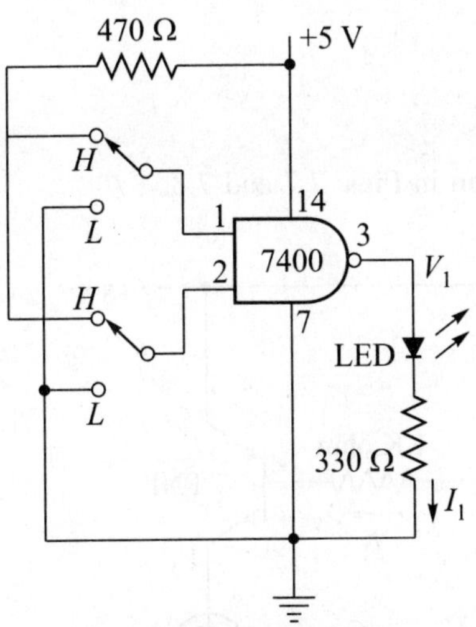

Fig. 7.5. For Experiment 7.2, LED lights up on high output.

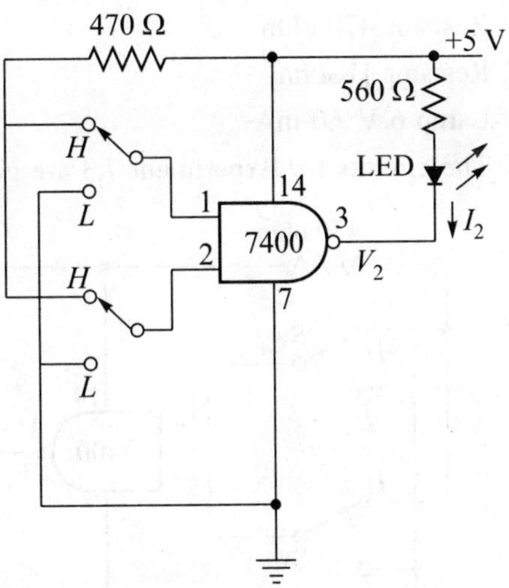

Fig. 7.6. LED lights up on low output.

2. Measure the voltages and currents marked in the diagrams.
3. Record your observations of voltages and currents in Table 7.2.

Table 7.2

Fig. 7.5	*Fig. 7.6*
V_1	V_2
I_1	I_2

4. Compare your observations with experimental Table 7.2 for Experiment 7.2 given in Appendix 3. When you refer to Table 7.2 in Appendix 3 you will notice that the high level and low level output voltages as well as the output currents are within the prescribed range.

EXPERIMENT 7.3 : TTL LAMP DRIVERS

Objective

To interface a lamp to a TTL driver.

Materials Required

Logic trainer

Volt-ohm-milliammeter

IC 7400 : Quad, 2-Input NAND gates

Transistor NPN : CL 100

Transistor PNP : CK 100

Resistor 470 ohm

Resistor 1K ohm

Lamp 6 V, 60 mA

The circuits for Experiment 7.3 are given in Figs. 7.7 and 7.8.

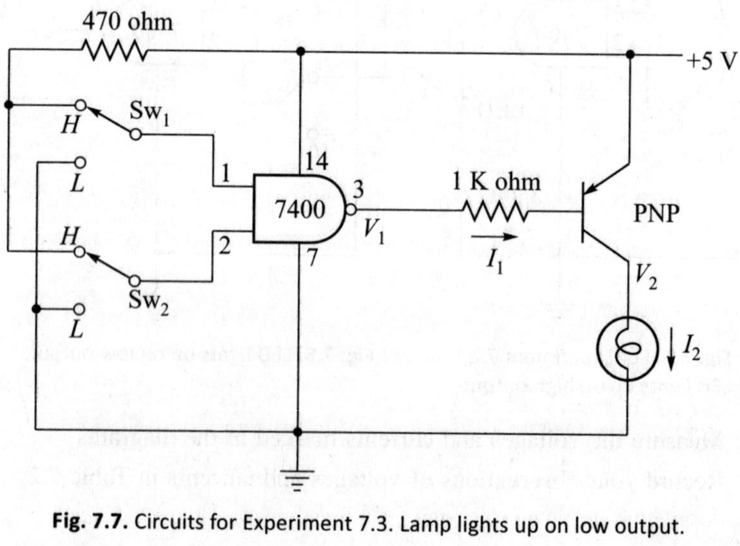

Fig. 7.7. Circuits for Experiment 7.3. Lamp lights up on low output.

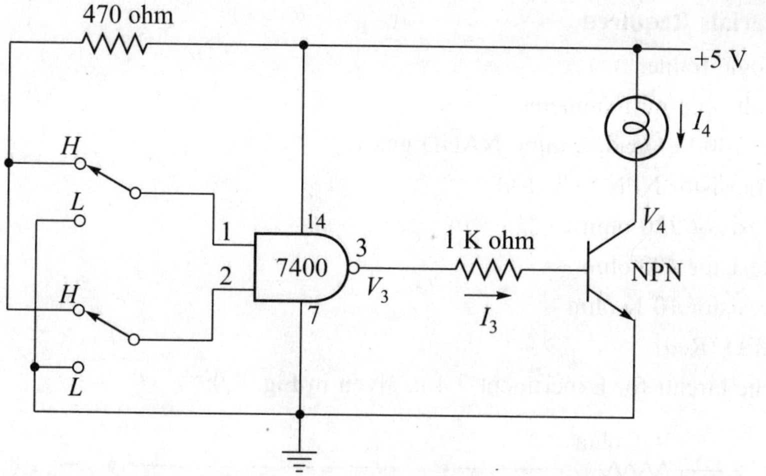

Fig. 7.8. Circuit for Experiment 7.3. Lamp lights up on high output.

Procedure

1. First assemble the circuit given in Fig. 7.7 and set up switches Sw_1 and Sw_2 for low output from the NAND gate.
2. Measure the currents and voltages marked in Fig. 7.7, record your result in Table 7.3 and compare it with Table 7.3 in Appendix 3.
3. Now assemble the circuit given in Fig. 7.8, and set up switches Sw_1 and Sw_2 for high output from the NAND gate. As before measure the voltages and currents, record your observations in Table 7.3 and compare your results with Table 7.3 in Appendix 3.

Table 7.3

Fig. 7.7	*Fig. 7.8*
V_1	V_3
V_2	V_4
I_1	I_3
I_2	I_4

EXPERIMENT 7.4 : TTL DRIVER—LOAD INTERFACE (LIMITED DRIVE CURRENT)

Objective

To consider an interfacing scheme for limited drive current.

Materials Required

Logic trainer

Volt-ohm-milliammeter

IC 7400 : Quad, 2-input NAND gates

Transistor NPN : CL 100

Resistor 220 ohm

Resistor 470 ohm

Resistor 10 K ohm

LED (Red)

The circuit for Experiment 7.4 is given in Fig. 7.9.

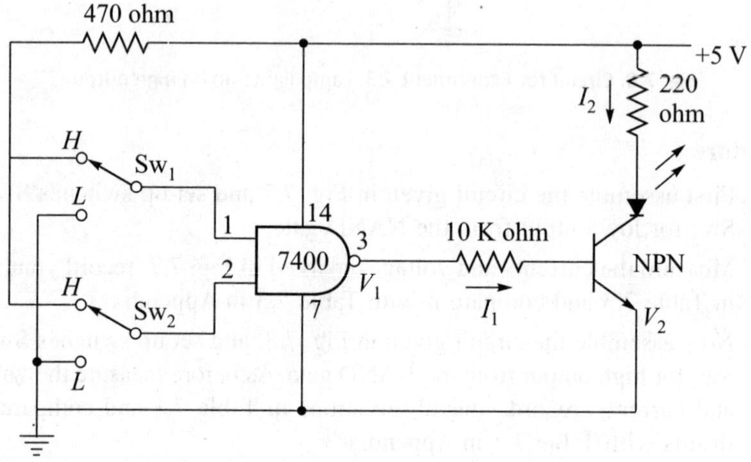

Fig. 7.9. Circuit diagram for Experiment 7.4.

Procedure

1. Set Sw_1 and Sw_2 so that the NAND gate output is high. The output current of the NAND gate will be limited by the 10 K ohm resistor. The LED will light up.
2. Measure the voltages and currents marked in the diagram and record your observations in Table 7.4.
3. Compare your results with Table 7.4 in Appendix 3.

Table 7.4

V_1	V_2
I_1	I_2

EXPERIMENT 7.5 : CMOS TO LAMP INTERFACE

Objective

To implement a circuit to interface a CMOS driver to a low power lamp.

Materials Required

Logic trainer

Volt-ohm-milliammeter

IC CMOS 4011 : Quad, 2-input NAND gates

Transistor NPN : CL 100

Resistor 470 ohm

Resistor 560 ohm

Lamp 6V, 60 mA

The circuit diagram for Experiment 7.5 is given in Fig. 7.10.

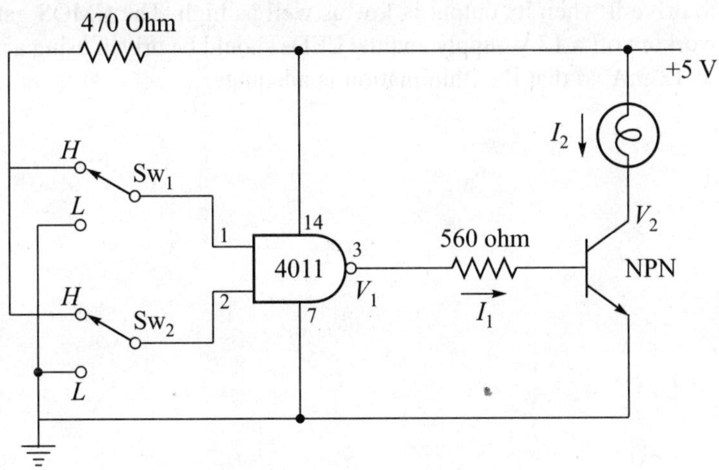

Fig. 7.10. Circuit diagram for Experiment 7.5.

Procedure

1. Adjust switches Sw_1 and Sw_2 for high output from the NAND gate. When you switch on the power supply the lamp will light up.
2. Measure the voltages and currents marked in the diagram, enter your result in Table 7.5 and compare it with Table 7.5 in Appendix 3.
3. Also notice that the voltages and currents are within the prescribed range.

Table 7.5

V_1	V_2
I_1	I_2

If a CMOS driver has to be interfaced with an inductive load requiring higher voltage and current, the following design changes can be made:

* Instead of a single transistor use a Darlington pair.
* Use a higher voltage power supply for the Darlington pair.
* Use a silicon diode across the inductive load.

PROBLEMS

7.1. In Experiment 7.3 indicate the positions of switches Sw_1 and Sw_2 so that the NAND gates have the required low and high outputs.

7.2. Draw a circuit to interface a CMOS gate to a lamp rated at 6 V, 60 mA so that it lights up when the CMOS gate output is low.

7.3. Since CMOS gates cannot properly drive LEDs directly they have to be driven using a transistor. Draw circuits which will enable a CMOS gate to drive it when its output is low as well as high. The CMOS gates are working off a 12 V supply and the LEDs should be driven using a current of 12 mA so that the illumination is adequate.

CHAPTER

8 Bistable Multivibrators (Flip-Flops)

8.1 INTRODUCTION

Flip-flops, also known as latches, are in fact multivibrators. The term bistable defines the most important characteristic of flip-flops and points to the fact that they are two-state devices which can be switched from one state to the other. In other words, a simple flip-flop can store one bit of data, binary 1 or 0. This emphasizes the fact that the basic function of a flip-flop is memory. A number of flip-flops put together constitute a register, which can store a multi-bit word. Therefore, flip-flops find considerable application in data storage, counting and timing operations.

8.2 RS NAND LATCH

A latch will continue to remain in any one of the two states in which it has been put until it is triggered into the other states. Latches can be built with NAND as well as NOR gates. We will consider a NAND latch for which a circuit diagram is given in Fig. 8.1 (*a*).

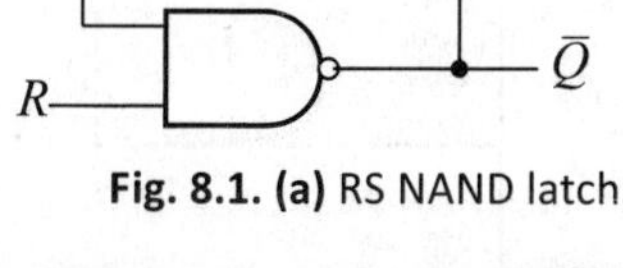

Fig. 8.1. (a) RS NAND latch.

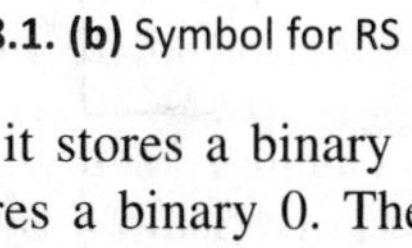

Fig. 8.1. (b) Symbol for RS latch.

The *S* input sets the latch and, when it is set, it stores a binary 1. The *R* input resets the latch and in the reset state it stores a binary 0. The outputs are *Q* and $\overline{Q}$ (complement). When the latch is set, *Q* is binary 1 and $\overline{Q}$ is binary 0. When it is reset, *Q* is binary 0 and $\overline{Q}$ is binary 1. This has been summed up in Table 8.1.

Table 8.1

Latch	Outputs	
	Q	$\overline{Q}$
Set	1	0
Reset	0	1

The output states of a NAND latch for all possible input conditions are given in Table 8.2.

Table 8.2
Truth Table for RS NAND Latch

Inputs		Outputs		Comment
S	R	Q	$\overline{Q}$	
0	0	1	1	Invalid
0	1	1	0	Set
1	0	0	1	Reset
1	1	X	$\overline{X}$	No change. Same as previous state; Set or Reset.

The timing diagram for a NAND latch is given in Fig. 8.2. The operation of a NAND latch can be summarized as follows:

(*a*) The output is invalid when both inputs are binary 0.

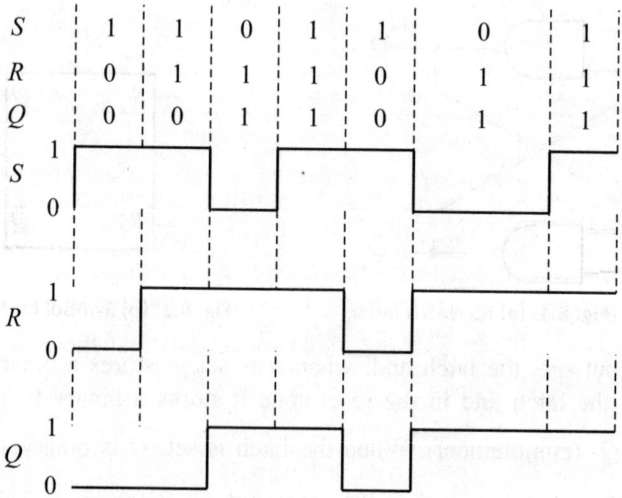

Fig. 8.2. Timing diagram for a NAND latch.

(*b*) To set a NAND latch you apply a binary 0 at the *S* input.

(*c*) To reset a NAND latch you apply a binary 0 at the *R* input.

(*d*) If both inputs to a NAND latch are binary 1 there is no change in the output of the latch.

EXPERIMENT 8.1 : RS NAND LATCH

Objective

To study the operation of a RS NAND latch.

Materials Required

Logic trainer

TTL IC 7400 : Quad, 2-input NAND gate

The circuit diagram for Experiment 8.1 is given in Fig. 8.3.

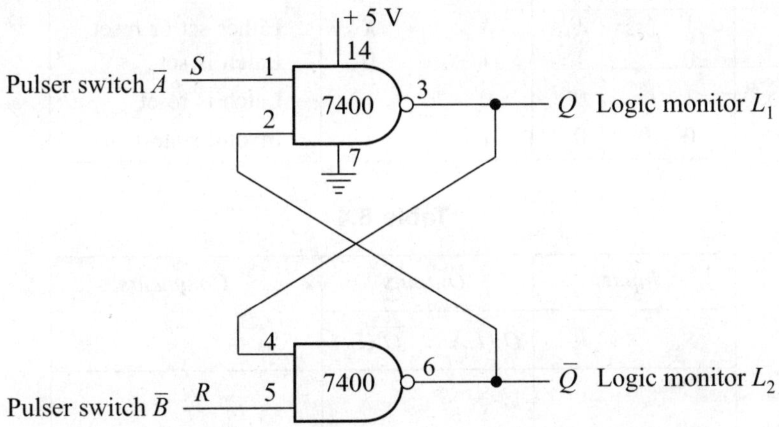

Fig. 8.3. Circuit diagram for Experiment 8.1.

Pulser switches are used in this experiment to enter momentary pulses at the Set and Reset inputs. Logic trainers normally have two pulser switches connected to two outlets one of which provides a normally high output and the other a normally low output. If output is taken from the normally high socket, depressing the pulser switch will cause a 1 → 0 transition and releasing it will cause a 0 → 1 transition. The outlets may be marked *A* (low) and *B* (low) and $\overline{A}$ (high) and $\overline{B}$ (high). For the purpose of this experiment the *S* and *R* inputs are connected to $\overline{A}$ (high) and $\overline{B}$ (high) sockets so that these two inputs are normally held high.

Procedure

1. Assemble the circuit given in Fig. 8.3.
2. Connect pin 14 of the IC to + 5 V and pin 7 to ground. On applying power to the latch you will notice that the latch will either be set or reset. This state is represented by *X* and $\overline{X}$ in the first row of Table 8.3.

3. Enter the combinations of S and R inputs by momentarily depressing switches $\overline{A}$ and $\overline{B}$ as required and note the states of the logic monitors L_1 and L_2. Enter your observations in Table 8.4. When you depress both the switches at the same time both S and R inputs go low and outputs Q and $\overline{Q}$ go high as shown in the last row of Table 8.3. This represents an invalid state. Your observations should tally with Table 8.3.

Table 8.3

Truth Table for NAND Latch

Inputs		*Outputs*		*Comments*
S	R	Q (L_1)	$\overline{Q}$ (L_2)	
1	1	X	$\overline{X}$	Either set or reset
0	1	1	0	Latch is set
1	0	0	1	Latch is reset
0	0	1	1	Invalid state

Table 8.4

Inputs		*Outputs*		*Comments*
S	R	Q (L_1)	$\overline{Q}$ (L_2)	

Observations on NAND Latch Operation

The following conclusions can be drawn from observations in Table 8.3:

1. The latch can be set (Q binary 1) by applying a momentary 0 at the S input.
2. The latch can be reset (Q binary 0) by applying a momentary 0 at the R input.
3. If the latch is already set it will remain set if a 0 is applied at the S input.
4. If the latch is reset it will remain reset if a 0 is applied at the R input.
5. The state of the latch can be changed only by applying a momentary 0 to the opposite input.
6. These observations lead to the obvious conclusion that the latch remembers the input at which the last momentary 0 was applied.

8.3 D LATCH

In an RS latch if both S and R inputs go low at the same time, the output is invalid. This situation can be avoided by adding an Inverter ahead of the S input as shown in Fig. 8.4. This modified latch is known as the D latch.

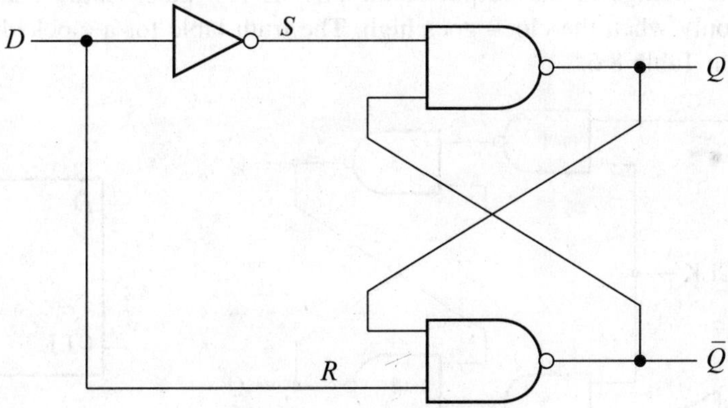

Fig. 8.4. D Latch.

The input at D drives the R input and its complement drives the S input. When a high input is applied at D the S input goes low which sets the latch. When the D input is low, the R input is also low, which resets the latch. Since the S and R inputs will always be in opposite states, the invalid state will never occur. The operation of the D latch has been summarized in Table 8.5.

Table 8.5
Truth Table for D Latch

Input	*Outputs*	
D	Q	$\bar{Q}$
0	0	1
1	1	0

8.4 CLOCKED D LATCH

In the latches which we have discussed so far the data input influences the output state the moment it is applied. It is at times necessary that the output should respond to the data input at a later time. This will ensure that only the required input data influences the output at a specific point in time, so that any other data at the input at other points in time does not influence the output. A modified D latch which fulfills this criterion, and known as the clocked D latch, is shown in Fig. 8.5 (*a*).

You will notice from the diagram that it closely resembles a D-latch with the difference that provision has been made to connect a clock to gates 1 and 2, the outputs of which are connected to an RS latch comprising gates 3 and 4. Virtually, therefore, the RS latch is activated by a D-latch with a clock. When the clock is low, the outputs of gates 1 and 2 go high and so there will be no change in the output of the NAND RS latch. Gates 1 and 2 are enabled only when the clock goes high. The truth table for a clocked D-latch is given in Table 8.6.

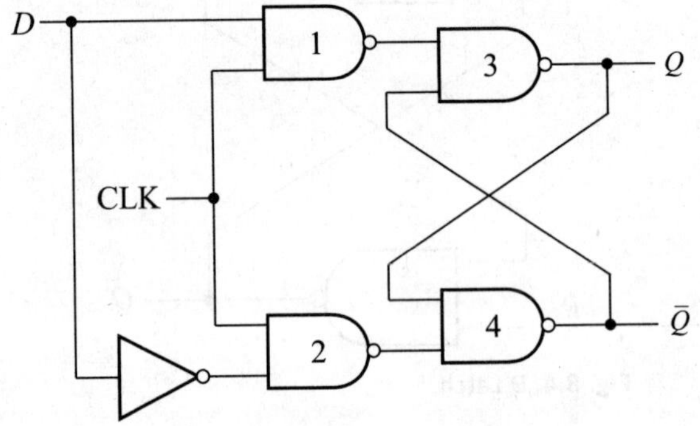

Fig. 8.5. (a) Clocked D Latch.

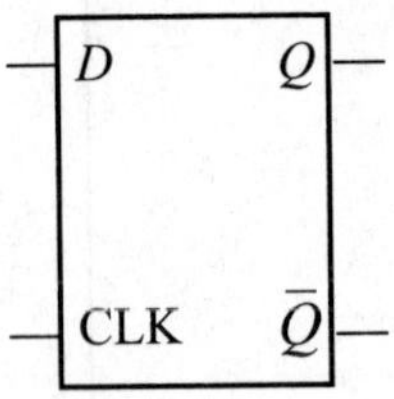

Fig. 8.5. (b) Symbol for Clocked D Latch.

Table 8.6

Truth Table for Clocked D Latch

Clock	*Input* *D*	*Output* *Q*
⊔	X	No change
⊓	0	0
⊓	1	1

EXPERIMENT 8.2 : CLOCKED D LATCH

Objective

To study the performance of a clocked D latch.

Materials Required

Logic trainer

TTL IC 7475 : Quad, clocked D latch

Resistor 470 ohm

The circuit diagram for the experiment is given in Fig. 8.6.

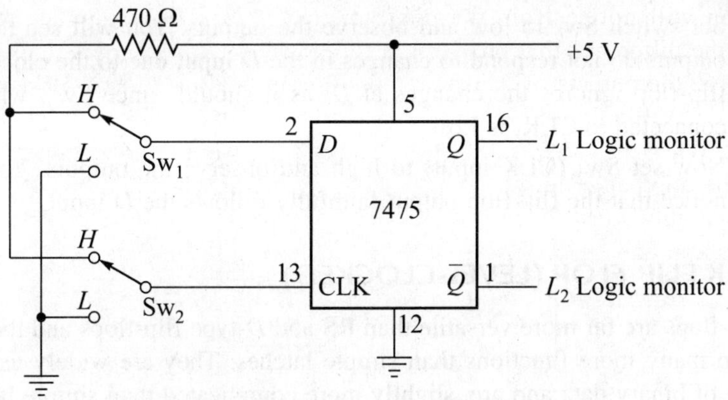

Fig. 8.6. Circuit diagram for Experiment 8.2.

Procedure

1. Assemble the circuit given in Fig. 8.6. Connect pin 5 of the IC to + 5 V and pin 12 to ground.
2. Using logic switches Sw_1 and Sw_2 apply logic voltage levels to inputs D and clock as indicated in Table 8.6 and record your observations for each set of inputs in Table 8.7. Your observations should tally with Table 8.6.

Table 8.7

Clock	*Input*	*Output*
CLK	*D*	*Q*

3. For a visual display of what happens with different sets of inputs replace switch Sw_1 by a 1 Hz clock input and connect a logic monitor, L_3, to monitor the 1 Hz clock signal. These changes have been shown in Fig. 8.7.

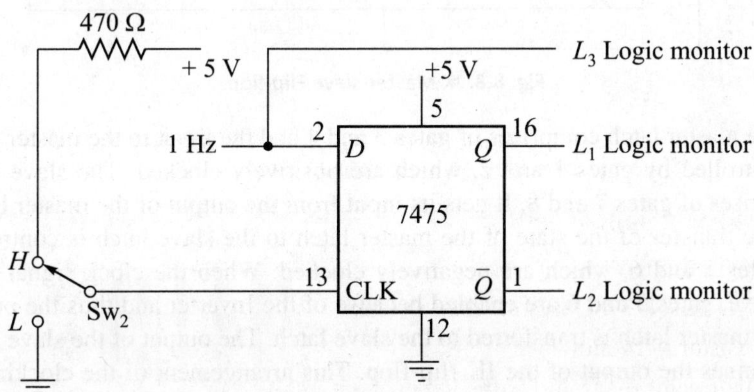

Fig. 8.7

4. Set switch Sw_2 to low and observe the outputs. You will see that the outputs do not respond to changes in the *D* input due to the clock. The flip-flop ignores the changes at *D*, as it should, since Sw_2, which is connected to CLK, is Low.

   Now set Sw_2 (CLK input) to high and observe the outputs. You will notice that the flip-flop output faithfully follows the *D* input.

8.5 JK FLIP-FLOP (LEVEL-CLOCKED)

JK flip-flops are far more versatile than RS and *D*-type flip-flops and they can perform many more functions than simple latches. They are widely used for storage of binary data and are, slightly more complicated than simple latches.

As you will see from the circuit diagram of a JK flip-flop, given in Fig. 8.8, it is a combination of two clocked latches, the first one is called the 'master latch', which feeds the second one called the 'slave latch'. The JK flip-flop is, therefore, better known as JK master-slave flip-flop. You will notice that the master and slave latches bear a very close resemblance to clocked *D* latches and also operate very much in the same way.

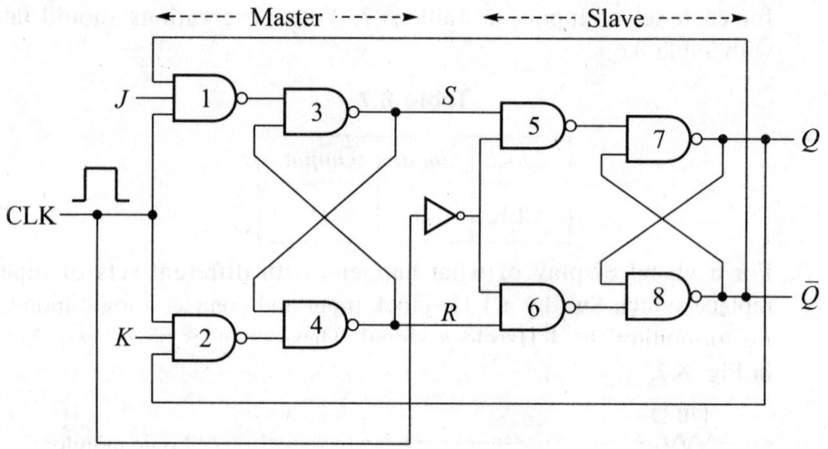

Fig. 8.8. JK Master-slave Flip-flop.

The master latch comprises of gates 3 and 4 and the input to the master latch is controlled by gates 1 and 2, which are positively clocked. The slave latch comprises of gates 7 and 8. It gets its input from the output of the master latch; but the transfer of the state of the master latch to the slave latch is controlled by gates 5 and 6, which are negatively clocked. When the clock signal goes negative, gates 5 and 6 are enabled because of the Inverter and thus the output of the master latch is transferred to the slave latch. The output of the slave latch determines the output of the JK flip-flop. This arrangement of the clocking of the two latches leads to the following conclusions:

1. When the clock goes high, gates 1 and 2 will be enabled and the master latch will become active. At the same time the Inverter will inhibit gates 5 and 6 as a result of which the slave latch will be disabled.
2. When the clock goes low, gates 1 and 2 will be inhibited and the master latch will be inactive. At the same time the Inverter will enable gates 5 and 6 as a result of which the slave latch is enabled. In short, when the master latch is disabled the slave latch is enabled.

The truth table for a JK flip-flop is given in Table 8.8.

Table 8.8

Truth Table for JK Flip-Flop

Inputs		*Outputs*		
		Before clock pulse		*After clock pulse*
J	*K*	*CLK*	*Q* (*t*)	*Q* (*t* + 1)
1	0	⎍	*X*	1
0	1	⎍	*X*	0
1	1	⎍	*X*	*X* toggle
0	0	⎍	*X*	No change

The four modes of operation of the JK flip-flop have been shown in Table 8.8. The table shows the state of the *Q* output prior to the arrival of the clock pulse, *t*, and at the end of the clock pulse (*t* + 1). *X* represents both the set and the reset states.

The operation of the flip-flop can be summed up as follows:

* To reset the flip-flop apply the following inputs:

  $J = 0$

  $K = 1$

  and thereafter apply a clock pulse which will reset the flip-flop on the trailing edge.

* To set the flip-flop apply the following inputs:

  $J = 1$

  $K = 0$

  and then apply a clock pulse which will set the flip-flop on its trailing edge.

* If you want the flip-flop to toggle, keep both *J* and *K* high and apply clock signals. The flip-flop will complement itself each time the clock switches from high to low. The flip-flop is said to toggle. The output waveform is given in Fig. 8.12.

* If you have to inhibit the flip-flop keep both *J* and *K* inputs low.

Preset and Clear functions are also incorporated in JK flip-flops. The truth table for this JK master-slave flip-flop is given in Table 8.9.

Table 8.9

Truth Table for JK Master-Slave Flip-Flop having Preset and Clear Functions

Inputs					*Output*
Asynchronous		*Synchronous*			
PR	*CLR*	*CLK*	*J*	*K*	*Q*
0	0	*X*	*X*	*X*	Invalid
0	1	*X*	*X*	*X*	1
1	0	*X*	*X*	*X*	0
1	1	*X*	0	0	No change
1	1	⎍	0	1	0
1	1	⎍	1	0	1
1	1	⎍	1	1	Toggle

While the Preset and Clear inputs are active the synchronous inputs have no effect on the output. They are active only when the Preset and Clear inputs are held high, as Preset and Clear inputs are active low. The JK flip-flop shown in Fig. 8.8 is level-clocked and, therefore, the master latch is active when the clock goes high. This makes it necessary to keep the *J* and *K* inputs constant as long as the clock is high. Because of this complication edge-triggering of JK flip-flops has become common. The symbols for positive and negative edge-triggered JK flip-flops are given in Figs. 8.9 and 8.10.

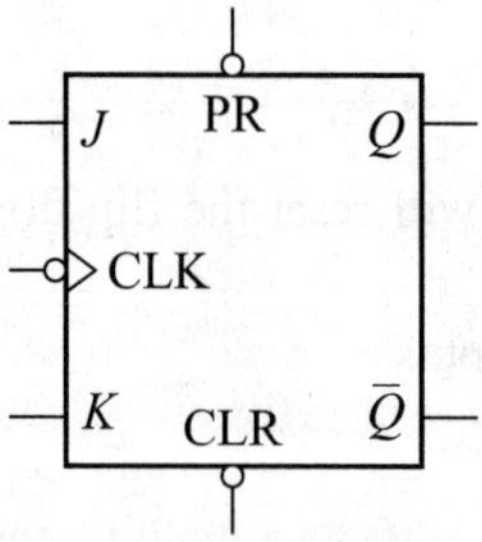

Fig. 8.9. Negative edge-triggered JK flip-flop.

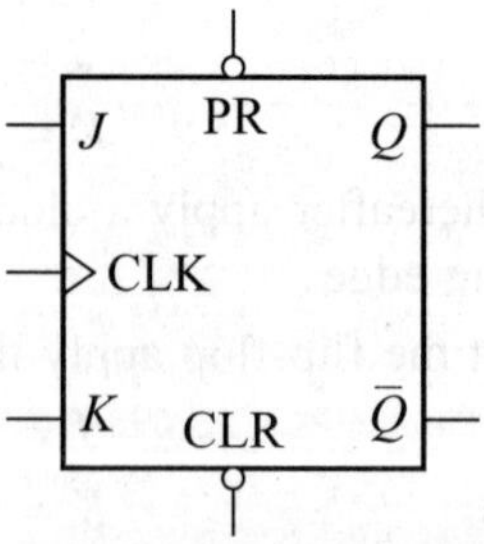

Fig. 8.10. Positive edge-triggered JK flip-flop.

EXPERIMENT 8.3 : JK FLIP-FLOP

Objective

To demonstrate the operation of JK flip-flop, IC 7476.

Materials Required

Logic trainer

TTL IC 7476

Frequency counter

Resistor 470 ohm

JK flip-flop IC 7476 contains two independent JK flip-flops which use level-clocking. It is important to remember that for normal operation the Preset and Clear inputs should be held high (tied positive).

The circuit diagram for this experiment is given in Fig. 8.11.

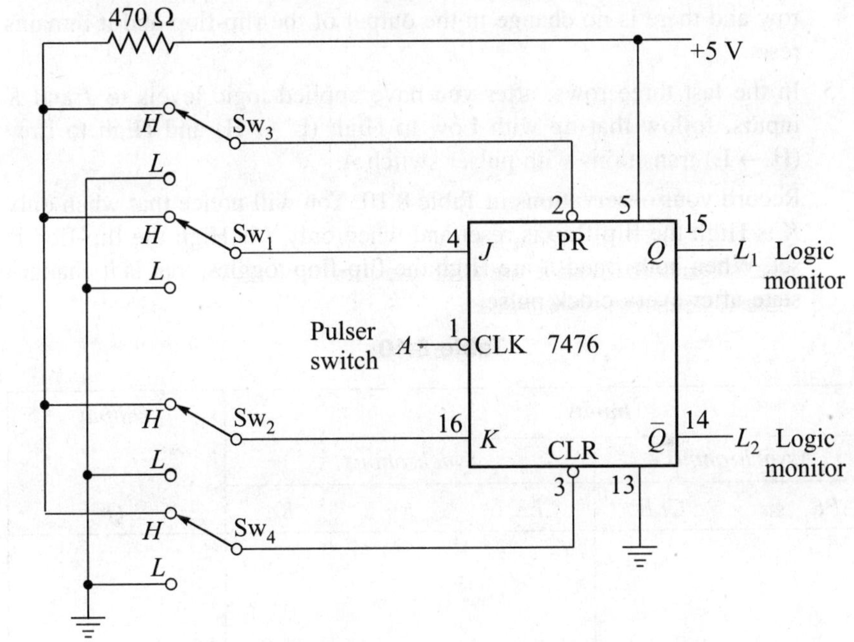

Fig. 8.11. Circuit diagram for Experiment 8.3.

Procedure

1. Only one of the JK flip-flops in the IC has been used for this experiment. Connect the flip-flop as shown in Fig. 8.11. The *J, K,* PR and CLR inputs are connected to logic switches Sw_1 to Sw_4. The clock input is connected to pulser switch socket *A,* which is normally low. Outputs Q and $\overline{Q}$ are connected to logic monitors. After you have made these connections, connect pin 5 of the IC to + 5 V and pin 13 to ground.

Asynchronous Function

2. First we will consider the asynchronous operation of the flip-flop. As *J* and *K* inputs have no effect on the asynchronous operation, the

position of switches Sw_1 and Sw_2 does not matter. With switches Sw_3- and Sw_4 apply logic levels to PR and CLR inputs as indicated in the first three rows of Table 8.9 and record your observations of the output in Table 8.10. Your observations should tally with Table 8.9. Notice that when both PR and CLR are low, the output is invalid. When only PR is low, the flip-flop is set and when only CLR is low, it is reset.

3. Notice that the flip-flop is already reset as shown in the third row of Table 8.9.
4. Now for the remaining four rows (synchronous operation) the PR and CLR inputs are held high and logic levels are applied to *J* and *K* inputs as in the last four rows of the table. No clock pulse is applied in the fourth row and there is no change in the output of the flip-flop and it remains reset.
5. In the last three rows, after you have applied logic levels to *J* and *K* inputs, follow that up with Low to High (L → H) and High to Low (H → L) transitions with pulser switch *A*.

   Record your observations in Table 8.10. You will notice that when only *K* is High the flip-flop is reset and when only *J* is High the flip-flop is set. When both *J* and *K* are High the flip-flop toggles, that is it changes state after every clock pulse.

Table 8.10

Inputs					*Output*
Asynchronous		*Synchronous*			
PR	*CLR*	*CLK*	*J*	*K*	*Q*

Toggle Function

6. When both *J* and *K* inputs are held High, the flip-flop toggles. The output waveform for this function is shown in Fig. 8.12. You will notice a very important relationship between the clock and output frequencies. The output frequency is half the clock frequency. Thus, when the flip-flop toggles it divides the clock frequency by 2. If the clock frequency is 100, the output frequency will be 50.

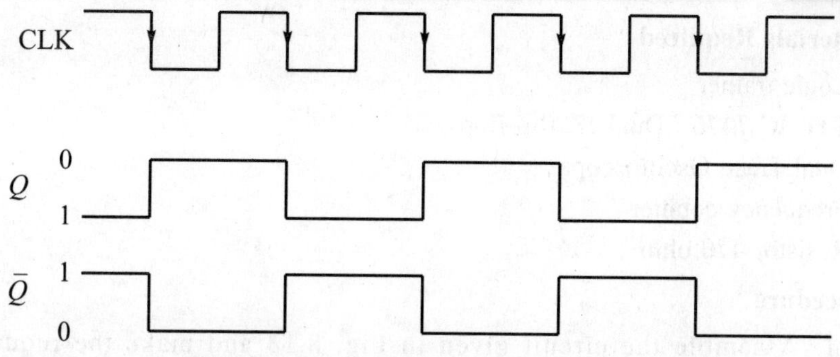

Fig. 8.12. Timing diagram for Toggle flip-flop.

Inhibit Function

7. When both J and K inputs are Low as in the fourth row of Table 8.9 there will be no change in the output state of the flip-flop even if the clock changes from High to Low or Low to High.

EXPERIMENT 8.4 : JK FLIP-FLOP

Objective

To verify the frequency-division function of the JK flip-flop.

The circuit diagram for the experiment is given in Fig. 8.13.

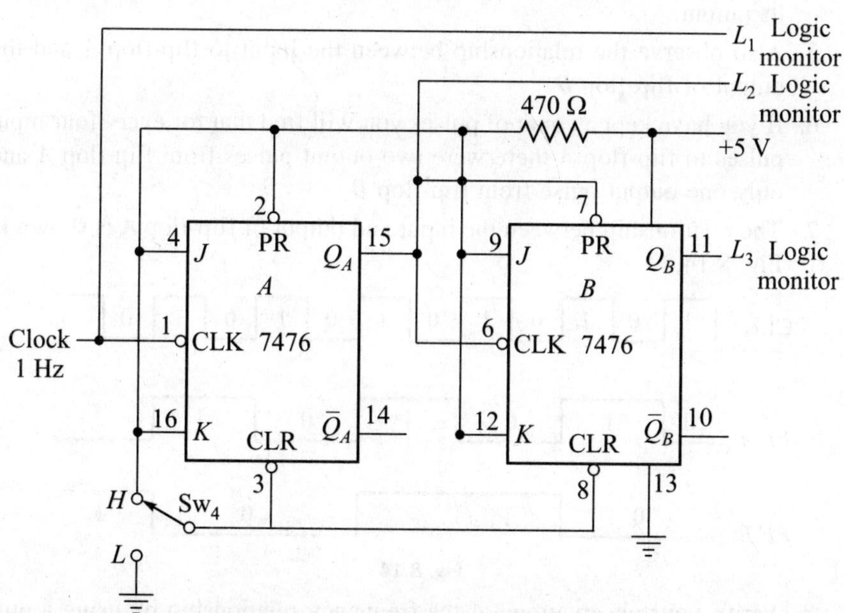

Fig. 8.13. Circuit diagram for Experiment 8.4.

Materials Required

Logic trainer
TTL IC 7476 : Dual JK flip-flop
Dual Trace Oscilloscope
Frequency counter
Resistor 470 ohm

Procedure

1. Assemble the circuit given in Fig. 8.13 and make the required connections. Notice that both the flip-flops in the IC have been used in this experiment. Both the *J* and *K* inputs of both the flip-flops will be at High logic level. Connect the 1 Hz clock signal to pin 1 of the IC. Connect pin 5 of the IC to + 5 V and pin 13 to ground. Observe that both the flip-flops are wired in the toggle mode.
2. Reset both flip-flops *A* and *B* by connecting switch Sw_4 to ground and then returning it to High to enable the flip-flops to function.
3. Now switch on the 1 Hz clock and observe the relationship between logic monitor L_1, (which is connected to clock) and L_2 by counting the number of input and output pulses.
4. Similarly keep a count of the relationship of the input to flip-flop *B* and its output.
5. Also observe the relationship between the input to flip-flop *A* and the output of flip-flop *B*.
6. If you have kept a count of pulses you will find that for every four input pulses to flip-flop *A* there were two output pulses from flip-flop *A* and only one output pulse from flip-flop *B*.
7. The relationship between the input and output of flip-flop *A* is shown in Fig. 8.14,

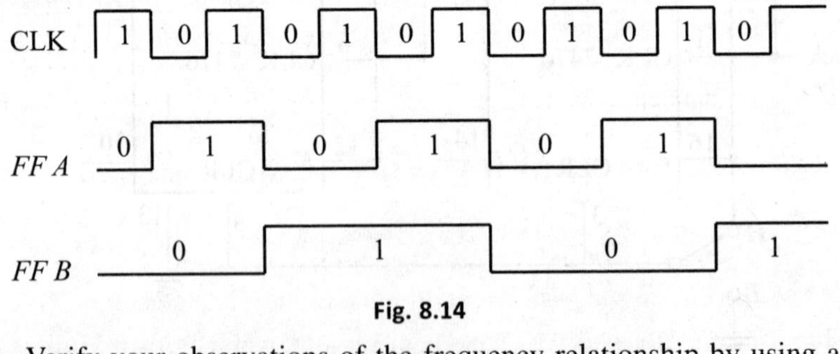

Fig. 8.14

8. Verify your observations of the frequency relationship by using a dual trace oscilloscope and connecting it to L_1 and L_2 and then L_2 to L_3 after changing the clock frequency to 1 kHz.
9. For further confirmation of the input and output frequencies use a frequency counter when the clock frequency is 1 kHz.

8.6 T FLIP-FLOP

When the J and K inputs of a JK flip-flop are tied together and are held high the flip-flop will complement itself each time the clock switches from high to low and, as stated earlier, the output frequency will be half of the clock frequency. This configuration of a JK flip-flop functions as a T flip-flop. It has only one input called the T input. Figure 8.15 (*a*) shows a JK flip-flop wired as a T flip-flop and Fig. 8.15 (*b*) shows the T flip-flop symbol.

JK flip-flops are widely used in sequential logic circuits in the toggle mode and for this reason a special symbol has been devised, Fig. 8.15 (*b*). RS and D flip-flops can also be converted to T flip-flops by wiring them as in Figs. 8.16 and 8.17.

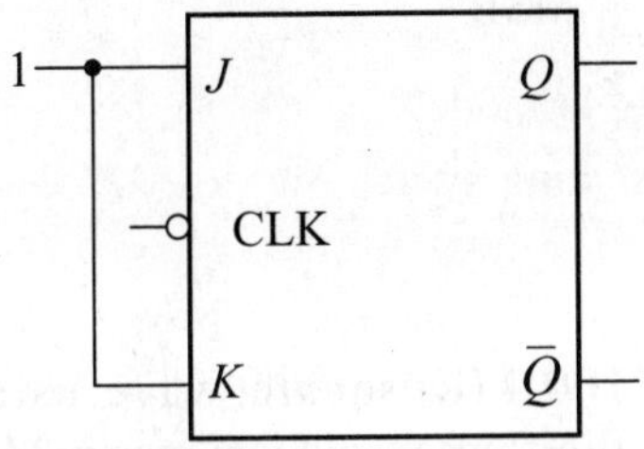

Fig. 8.15. (a) JK flip-flop converted to T flip-flop.

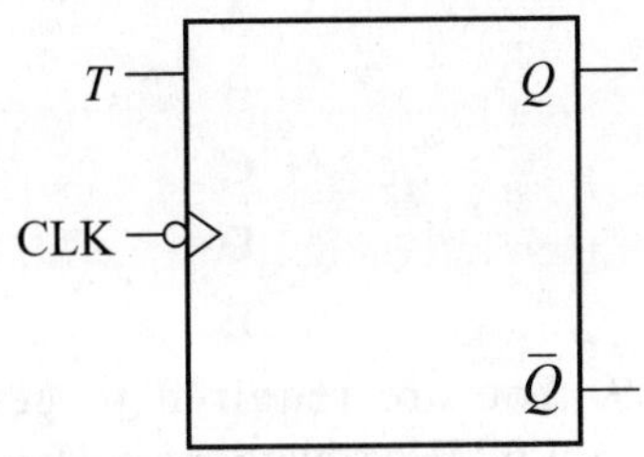

Fig. 8. 15. (b) Symbol for T flip-flop.

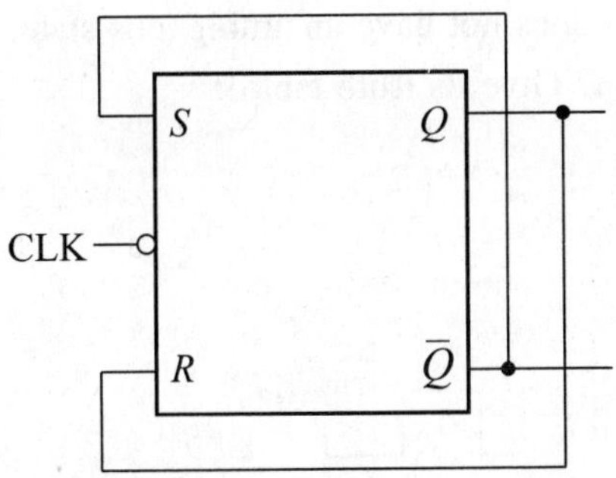

Fig. 8.16. RS flip-flop converted to T flip-flop.

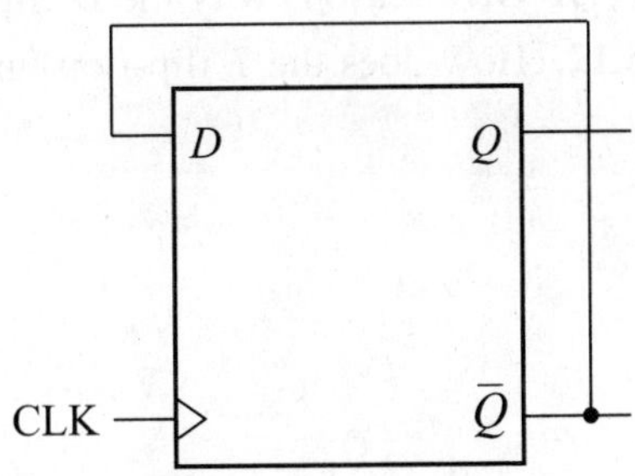

Fig. 8.17. D flip-flop converted to T flip-flop.

PROBLEMS

8.1. Draw the circuit diagram for a latch using only NOR gates.

8.2. The output of the NAND latch is invalid when both its inputs are binary 0. How does a NOR latch behave when both its inputs are binary 0?

8.3. What is the output of a NOR latch when a binary 1 is applied at the R input? How does a NAND latch function under similar conditions?

8.4. What happens to a NOR latch when both inputs are binary 1 ?

8.5. If both inputs to a NAND latch are binary 1 what is the state of the output?

8.6. In a JK flip-flop, *S*, *C* and *J* inputs are High and *K* input is Low. What will be the state of the output after

(*a*) one clock pulse?

(*b*) four clock pulses?

8.7. In a JK flip-flop, *J*, *K*, *S* and *C* inputs are High. If the clock input is 6 kHz square wave, what is the output?

8.8. The output of a storage register using four *D* flip-flops are as follows; what binary number is it storing?

A	Set	MSB
B	Set	
C	Set	
D	Reset	
E	Set	

8.9. You are required to generate a 100 kHz square wave using a 1.6 MHz square wave. How many JK flip-flops would you require? Draw a circuit to implement this and sketch the waveforms.

8.10. How many JK flip-flops will be required to store decimal number 163?

8.11. Give reasons why the *D* flip-flop does not have an ambiguous state.

8.12. How does the *T* flip-flop function? Give its truth table.

CHAPTER 9

Astable and Monostable Multivibrators

9.1 INTRODUCTION

Astable multivibrators keep oscillating between the Set and Reset states at a fixed frequency and thus generate a train of clock pulses which are required for synchronization and timing functions. These pulses are referred to as clock pulses which are essential for the operation of sequential logic circuits such as counters and shift registers.

Clock oscillators are used to generate a train of clock pulses. These pulses are rectangular or square in shape; but they are also required to meet some other essential requirements which are mentioned below:

* **Clock cycle time:** This is the period of one clock waveform during which all logic elements in a system must complete their operations. This should, therefore, remain constant.
* **Clock output logic levels:** Clock output logic levels should maintain the required high and low logic levels over a period of time.
* **Clock pulse rise and fall time:** The rise and fall time of pulses should be zero.
* **Pulses should be square in shape:** This implies that the duty cycle of pulses should be 50 per cent.

Choice of Logic Elements in Fabricating Clocks

It is desirable to use the same family of logic elements in building clock oscillators as those used in the system itself, so that they can be operated from the same power supply, which will ensure that the clock output is compatible with the system requirement.

Unlike the astable multivibrator, which has no stable state, the monostable multivibrator has a stable state which corresponds to the reset state of a flip-flop, when the Q output is logic 0 and the $\overline{Q}$ output is logic 1. When the monostable is triggered by an input signal, the output changes to the opposite state for a predetermined period which depends on the time constant of the circuit.

The astable, monostable and bistable multivibrators have been considered in Appendix 5 which deals exclusively with the application of 555 Timer for these devices and it is suggested that the student should go through this appendix before proceeding further with this chapter.

9.2 TTL CLOCK OSCILLATOR

A simple TTL clock oscillator is discussed here mainly with a view to make the student familiar with the basic design of clock oscillators using TTL gates. If you refer to Fig. 9.1, you will notice that the oscillator consists of two Inverters 1 and 2, with the output of one connected to the input of the other. Inverter 3 is used as a buffer which isolates the load from the frequency-determining components.

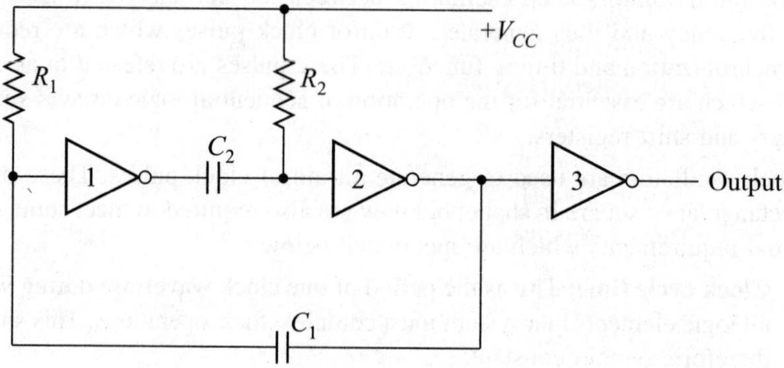

Fig. 9.1. TTL Clock Oscillator.

Assuming that C_2 begins to charge through R_2 when power is switched on, the input voltage of Inverter 2 will begin to rise, and the output will stay high till the input voltage to the Inverter reaches a high logic level, at which point the Inverter output will go low and will stay low for a duration depending on the time constant C_2R_2.

At this point, capacitor C_1 will begin to charge through R_1 when the input to Inverter 1 will begin to rise and its output will stay high till the input to the Inverter attains a high logic level, when the Inverter output will go low and will stay low for a period depending on the time constant C_1R_1.

Capacitors C_1 and C_2 will charge alternately and clock pulses will be generated. The frequency of oscillations can be determined from the following equations:

$$f = \frac{1}{t_1 + t_2} = \frac{1}{0.7\,R_1C_1 + 0.7\,R_2C_2}$$

If $\quad R_1 = R_2 = R$ and $C_1 = C_2 = C$

$$f = \frac{1}{1.4\,RC} = \frac{0.7}{RC}$$

where R is in megohm
C is in microfarad
and f is in Hz.

9.3 CRYSTAL-CONTROLLED CLOCK OSCILLATOR

The frequency of the oscillator we have just considered can be further improved by using a quartz crystal. We will consider this in Experiment 9.1 which follows:

EXPERIMENT 9.1 : CRYSTAL-CONTROLLED OSCILLATOR

Objective

To study the performance of a quartz-controlled TTL clock oscillator.

Materials Required

Logic trainer
Volt-ohm-milliammeter
TTL IC 7404 : Hex Inverter
I MHz Quartz crystal
Resistors 6.8 K (2)
Capacitors 100 μF (2)
Frequency counter
Oscilloscope.

The circuit diagram for the clock oscillator is given in Fig. 9.2.

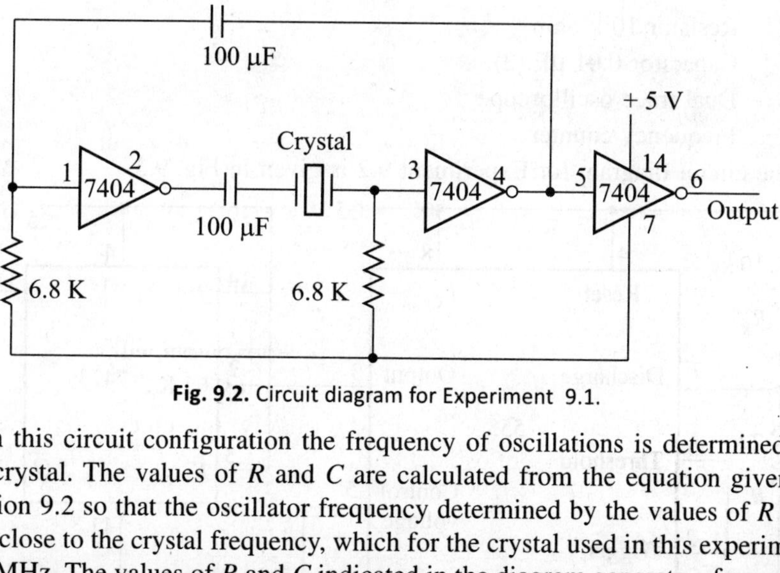

Fig. 9.2. Circuit diagram for Experiment 9.1.

In this circuit configuration the frequency of oscillations is determined by the crystal. The values of R and C are calculated from the equation given in Section 9.2 so that the oscillator frequency determined by the values of R and C is close to the crystal frequency, which for the crystal used in this experiment is 1 MHz. The values of R and C indicated in the diagram generate a frequency which will be close to the frequency of the quartz crystal.

Procedure

1. Assemble the circuit and connect pin 14 of the IC to + 5 V and pin 7 to ground.

2. Measure the output voltage at pin 6 of the IC using an R.F. Probe. It should be about 4 V.
3. Measure the output frequency with a frequency counter.

An important requirement is that the clock pulses should be square in shape which implies that the duty cycle of the pulses should be 50 per cent. We will see in Experiment 9.2 how this can be achieved.

A perfectly symmetrical square wave is realized in this experiment by connecting 555 Timer output to a JK Master-Slave flip-flop which toggles at every cycle. This will give a 2-phase clock output; but as the waveform in Fig. 9.4 shows, the output frequency will be half of the input frequency. Since the duty cycle will be 50 per cent, the output will be a perfectly symmetrical square wave.

EXPERIMENT 9.2 : SQUARE WAVE GENERATOR

Objective

To study how a square wave can be generated by connecting a binary divider to 555 Timer output.

Materials Required

Logic trainer
TTL IC 7473 : Dual JK level-triggered flip-flop
Timer IC 555
Resistors 2 k ohm
Resistor 10 k ohm
Capacitor 0.01 μF (2)
Dual trace oscilloscope
Frequency counter.

The circuit diagram for Experiment 9.2 is given in Fig. 9.3.

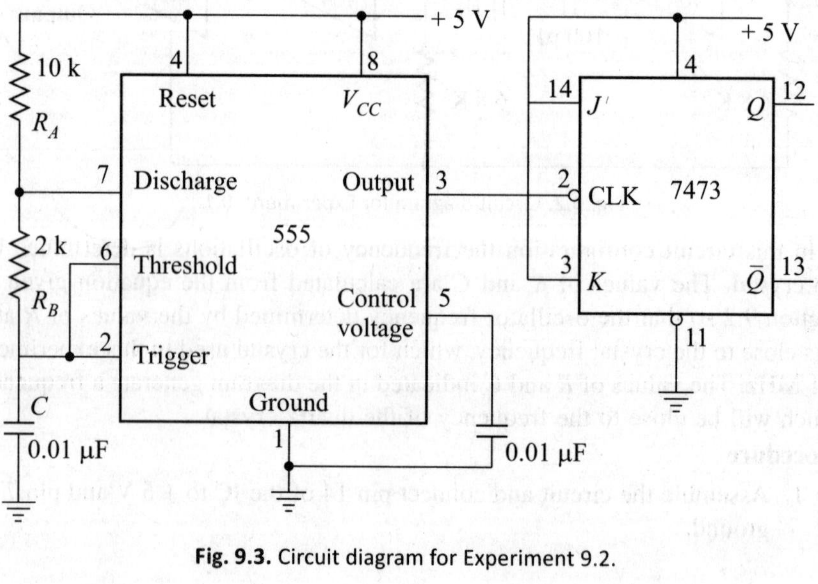

Fig. 9.3. Circuit diagram for Experiment 9.2.

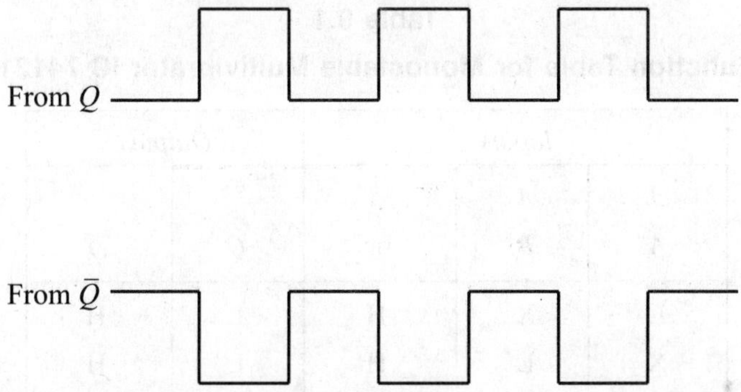

Fig. 9.4. Shows the output waveform of JK flip-flop in Fig. 9.3.

Procedure

1. Assemble the circuit given in Fig. 9.3. Notice that only one of the two JK flip-flops in IC 7473 has been used. For the values of R_A and R_B used in this experiment the output of Timer 555 will be a frequency of about 1 kHz and the duty cycle will be about 70 per cent.
2. Connect the inputs of a dual trace oscilloscope to the Q and $\overline{Q}$ outputs of the flip-flop. You will notice that the output pulses are square in shape, indicating that the duty cycle is 50 per cent.
3. Measure the output frequencies of the Timer and the JK flip-flop. The output frequency of the JK flip-flop will be about 500 Hz, that is half the output frequency of the Timer.

9.4 NON-RETRIGGERABLE MONOSTABLE : IC 74121

This monostable has three inputs, 3 and 4 which are active low and 5 which is active high and triggers the device on a positive edge low to high ($0 \rightarrow 1$) transition. Arrows in the function table (Table 9.1) indicate that the device is edge-triggered. It can be triggered in many ways. If inputs 3 or 4 or both are low, a low to high ($0 \rightarrow 1$) transition on input 5 will trigger it. When input 5 is held high, inputs 3 and 4 can also trigger it on a high to low transition ($1 \rightarrow 0$). Inputs 3 and 4 require rise times in excess of 1 V / micro-second, whereas input 5 responds to rise rates as low as 1 V / Sec.

IC 74121 has a 2 k ohm internal timing resistor and, when it is used, pin 9 is connected to pin 14. The output pulses realized on the application of these trigger pulses are shown in Table 9.1. When internal resistor is used, the external resistor R (Fig. 9.5) is not used.

Table 9.1

Function Table for Monostable Multivibrator IC 74121

Inputs			*Outputs*	
3 *A*	*4* *B*	*5* Sw_1	Q	$\overline{Q}$
L	X	H	L	H
X	L	H	L	H
X	X	L	L	H
H	H	X	L	H
H	↓	H	⊓	⊔
↓	H	H	⊓	⊔
↓	↓	H	⊓	⊔
L	X	↑	⊓	⊔
X	L	↑	⊓	⊔

In the above table (Table 9.1)

H = High

L = Low

X = Don't care

↓ = High to Low transition

↑ = Low to High transition.

Since this monostable is non-retriggerable, it implies that, after it has been triggered into the quasistable state, it must switch back to the stable state before it can be triggered again. The pulse width can be calculated from the following expression:

Pulse width = 0.694 *RC* seconds

where *R* is in ohm

and *C* is in farads

Depending on the values of *R* and *C* this IC is capable of producing pulses of widths from 40 ns to 40 seconds.

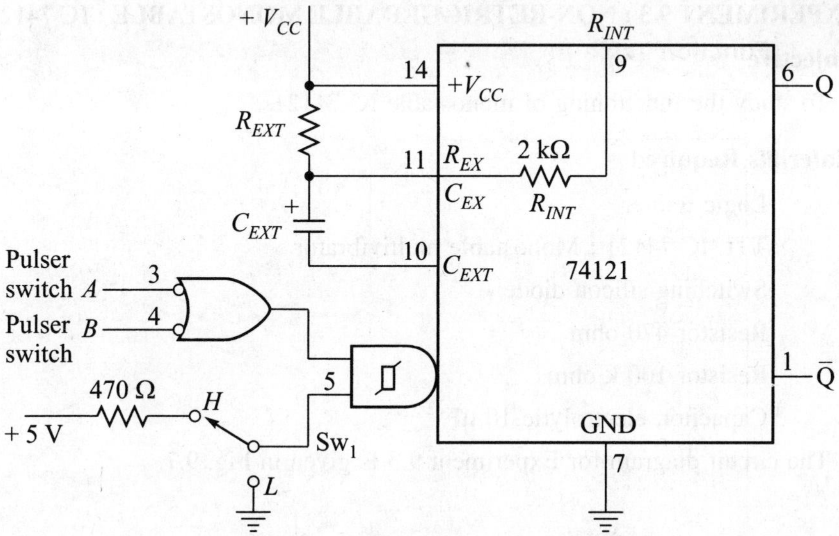

Fig. 9.5. Monstable multivibrator IC 74121.

This monostable is often used for generating single clock pulses in digital trainers and for general purpose work. For such applications the circuit given in Fig. 9.6 will be quite satisfactory. For the values shown in Fig. 9.6 the time duration of the output pulse will be about 7 milliseconds.

The internal resistor R_{int} of 2 k ohm built into the IC can be made effective by connecting pin 9 to pin 14, V_{CC}. If only the internal resistor is used without any external capacitor, a minimum pulse width of about 35 ns may be obtained. If an external resistor and capacitor are used, R_{ext} may have a range of 1.4 k ohm to 40 k ohm and C_{ext} may have a value of upto 1000 μF.

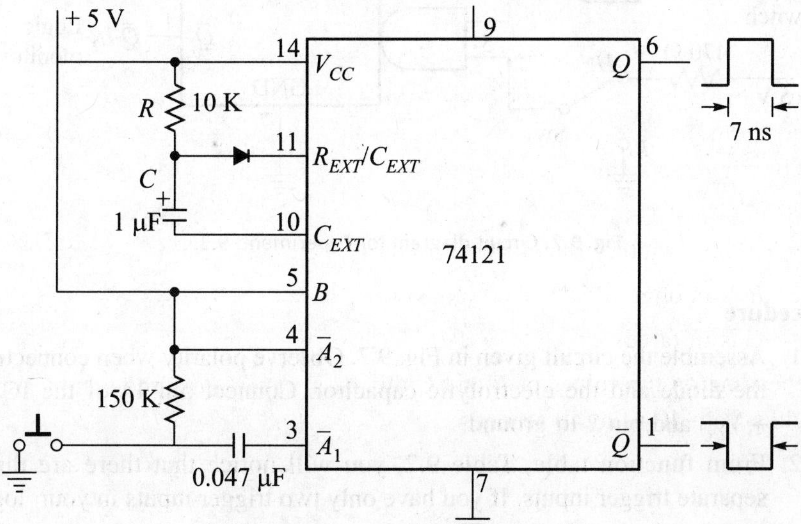

Fig. 9.6. Single clock pulse generator.

EXPERIMENT 9.3 : NON-RETRIGGERABLE MONOSTABLE : IC 74121

Objective

To study the functioning of monostable IC 74121.

Materials Required

Logic trainer
TTL IC 74121 : Monostable multivibrator
Switching silicon diode
Resistor 470 ohm
Resistor 100 k ohm
Capacitor, electrolytic 10 μF.

The circuit diagram for Experiment 9.3 is given in Fig. 9.7.

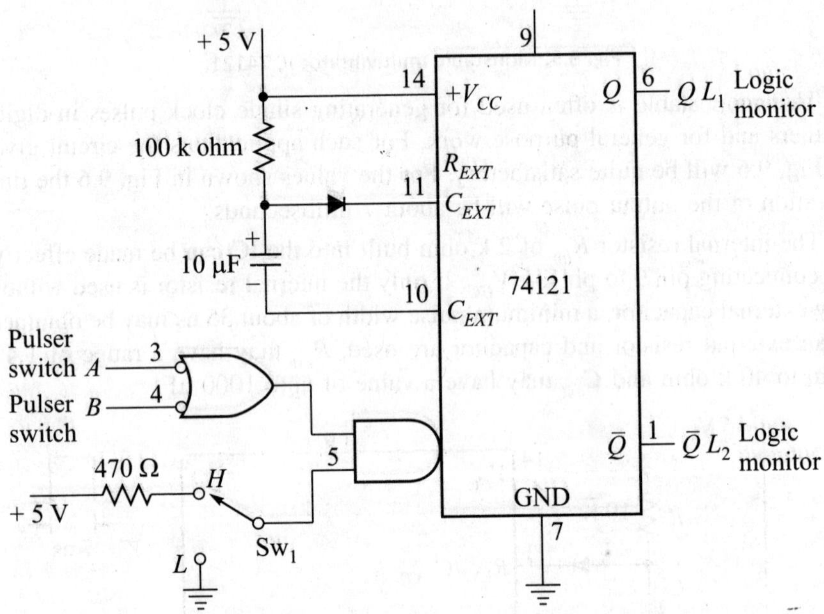

Fig. 9.7. Circuit diagram for Experiment 9.3.

Procedure

1. Assemble the circuit given in Fig. 9.7. Observe polarity when connecting the diode and the electrolytic capacitor. Connect pin 14 of the IC to $+ V_{CC}$ and pin 7 to ground.
2. From function table, Table 9.2, you will notice that there are three separate trigger inputs. If you have only two trigger inputs in your logic

trainer you can use a logic switch for one of the trigger inputs as has been shown in Fig. 9.7.

3. You will find from the function table that *H* refers to high logic voltage level and *L* refers to low logic voltage level and that arrows refer to trigger inputs. Therefore, if you initially connect pins 3 and 4 to pulser switch socket, which is normally low, you can apply a high to low transition from that position as well as a low voltage level; but for applying a high logic voltage level you will have to connect pins 3 and 4 to $\overline{A}$, which is normally high.
4. Using pulser switches *A* and *B* and logic switch Sw_1 apply logic voltage levels and trigger pulses and record your observations for each set of inputs in Table 9.2. Your observations should tally with Table 9.1.
5. Calculate the pulse width from the data marked in Fig. 9.7. It should tally with the result given in Appendix 3 for this experiment.
6. Notice that a diode has been incorporated to prevent a reverse voltage being applied to the capacitor.

Table 9.2

Function Table for Monostable Multivibrator IC 74121

Inputs			*Outputs*	
A	*B*	Sw_1	Q	$\overline{Q}$
L	X	H		
X	L	H		
X	X	L		
H	H	X		
H	↓	H		
↓	H	H		
↓	↓	H		
L	X	↑		
X	L	↑		

H = High voltage level

L = Low voltage level

X = Don't care

↑ = Trigger : Low to High transition

↓ = Trigger : High to Low transition.

9.5 RETRIGGERABLE MONOSTABLE : IC 74123

Monostable IC 74121 requires a finite time to recover from the previous trigger pulse because of the time the capacitor in the circuit takes to recharge. It is this recovery time which limits the duty cycle to about 90 per cent.

Monostable IC 74123 is retriggerable because the recovery time is practically zero, which makes a 100 per cent duty cycle possible. This means that if the resistor and capacitor, which are used externally, have the required values, which can produce a pulse of longer duration than the time interval between two trigger pulses, the output at Q can be a constant binary 1. The way it happens is like this. When a trailing edge occurs at the input of the monostable, an output pulse is produced and if the trailing edge of the next trigger pulse occurs before the output pulse can complete its duration, which depends on the external resistor and capacitor, the first timing interval is terminated and the second timing interval is initiated. Therefore there is no break in the output pulse.

Figure 9.8 shows that the monostable is triggered by the first trigger pulse. An output pulse is generated and when the second trigger pulse occurs, it retriggers the monostable on its trailing edge. The output pulse has not completed its duration even when the trailing edge of the third trigger pulse occurs. The monostable is, therefore, retriggered. As there is no pulse after the third trigger pulse, the output pulse terminates after it has completed its duration. Notice that the output pulse completes its normal duration, t_2, following the trailing edge of trigger pulse 3. Also notice that t_2 is larger than t_1.

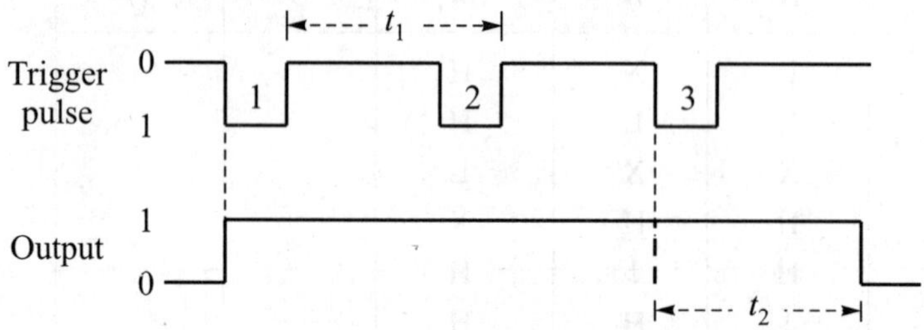

Fig. 9.8

This special feature of monostable IC 74123 enables it to detect a missing pulse. The monostable will remain in binary 1 output state as the timing period of the output pulse is larger than the interval between input pulses. If there is a missing input pulse the monostable will time out and the output will go low.

Operating Procedures for IC 74123

As this IC does not have an internal resistor, an external resistor as well as a capacitor are required to be connected. A functional diagram for IC 74123 is given in Fig. 9.9 and the function table is given in Table 9.3. Pin 16 is to be connected to + V_{CC} (+ 5 V) and pin 8 to ground.

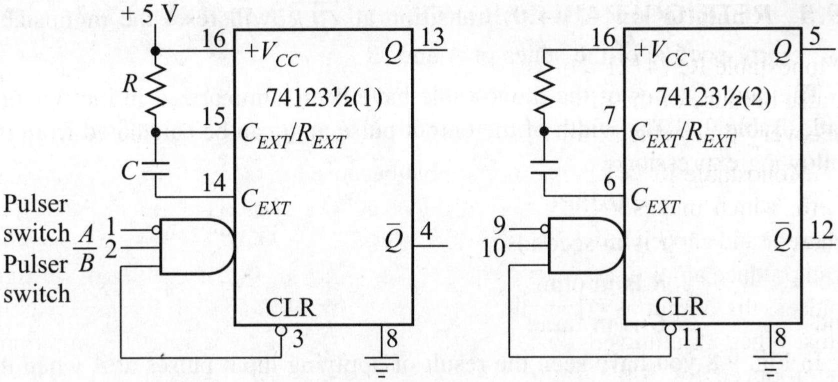

Fig. 9.9. Logic diagram for IC 74123.

A function table for IC 74123 is given in Table 9.3.

Table 9.3
Function Table for IC 74123

Inputs			*Outputs*	
A	$\overline{B}$	$\overline{CLR}$	*Q*	$\overline{Q}$
↓	H	H	⊓	⊔
L	↑	H	⊓	⊔
L	H	↑	⊓	⊔
X	X	L	⊔	⊓
X	X	↑	⊔	⊓

H = High voltage level
L = Low voltage level
X = Don't care
↑ = Trigger : Low to High transition
↓ = Trigger : High to Low transition.

The monostable can be triggered by applying input pulses at A, $\overline{B}$ or $\overline{CLR}$.

1. A high to low ($1 \rightarrow 0$) transition at input A will trigger the circuit when $\overline{B}$ and $\overline{CLR}$ are held high. The output at Q will become binary 1.
2. A low to high ($0 \rightarrow 1$) input at $\overline{B}$ will also trigger the circuit when A is held low and $\overline{CLR}$ is held high.
3. The monostable can also be triggered by a low to high ($0 \rightarrow 1$) transition at $\overline{CLR}$ when A is held low and $\overline{B}$ is held high.

4. A high to low ($1 \rightarrow 0$) transition at $\overline{CLR}$ will reset the monostable irrespective of the states of A and $\overline{B}$.

The output states of the monostable have been summarized in the function table, Table 9.3. The width of the output pulse at Q can be calculated from the following expression:

$$t = 0.33RC$$

where t is in seconds

R is in ohm

and C is in farad

In Fig. 9.8 you have seen the result of applying input pulses at A when the duration of the trigger pulse was shorter than the duration of the output pulse.

Figure 9.9 shows the output pulses when inputs are applied at A and the time interval between trigger pulses is larger than the duration of the output pulse. Inputs $\overline{B}$ and $\overline{CLR}$ are required to be held high.

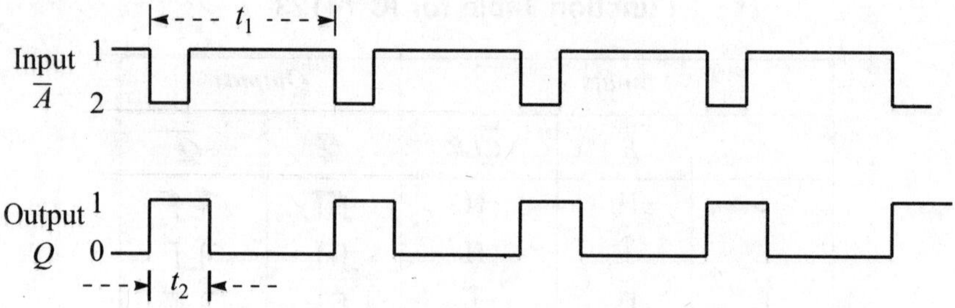

Fig. 9.10. Output waveform when the duration of the pulse is shorter than the interval between trigger pulses.

EXPERIMENT 9.4 : RETRIGGERABLE MONOSTABLE : IC 74123

Objective

* Study the operating characteristics of IC 74123.
* Study their operation when they are interconnected.
* Consider their application in sequencing and timing functions.
* Consider their application in producing delayed pulses.

Materials Required

Logic trainer
TTL IC 74123 : Retriggerable monostable
Switching silicon diodes (2)
Resistor 470 ohm
Resistor 18 k ohm
Resistor 100 k ohm
Capacitor, electrolytic 470 μF (2).

The circuit for Experiment 9.4 is given in Fig. 9.11.

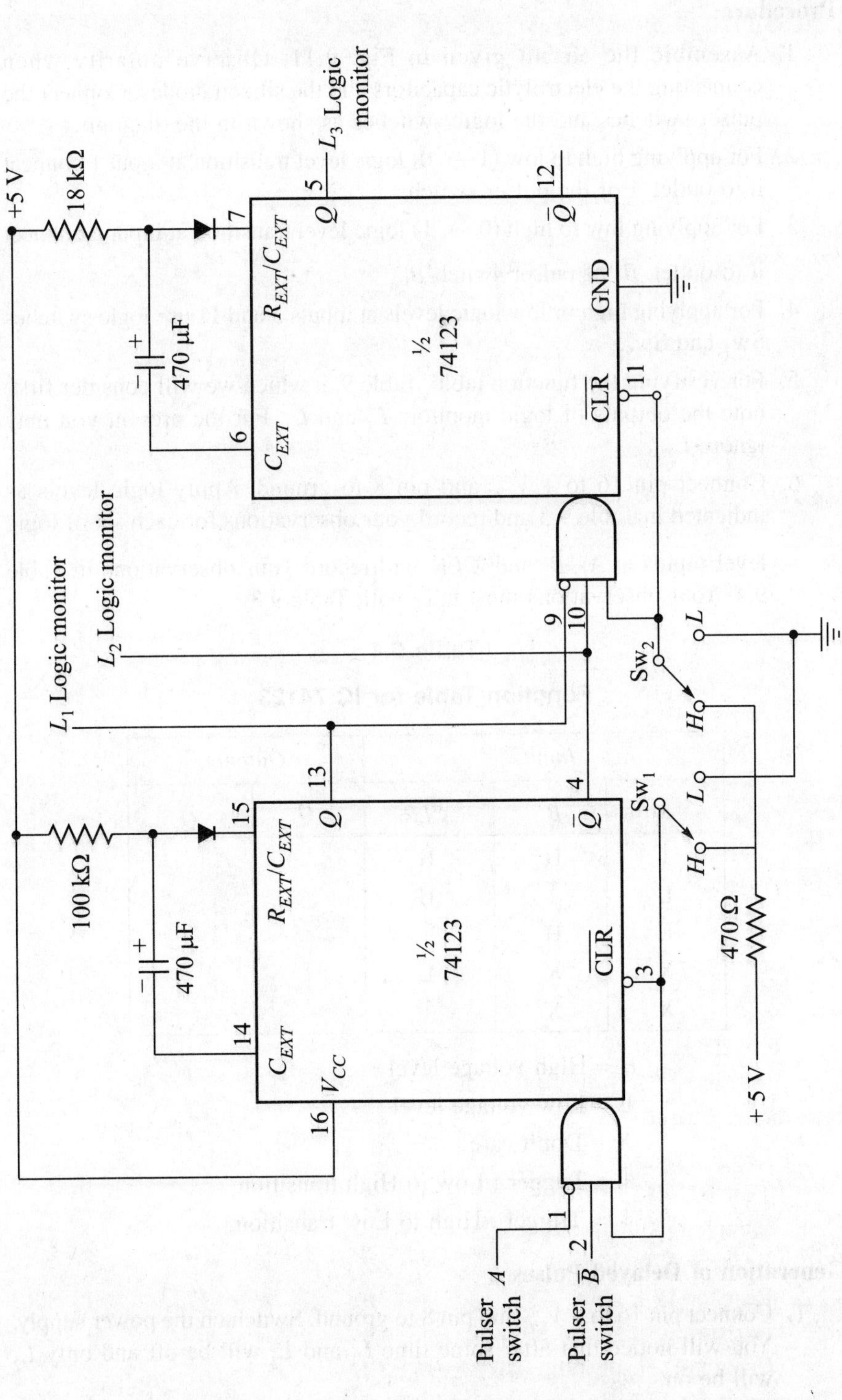

Fig. 9.11. Circuit diagram for Experiment 9.4.

Procedure

1. Assemble the circuit given in Fig. 9.11. Observe polarity when connecting the electrolytic capacitors and the silicon diodes. Connect the pulser switches and the logic switches as shown in the diagram.
2. For applying high to low (1 → 0) logic level transition at input 1 connect it to outlet *A* of the pulser switch.
3. For applying low to high (0 → 1) logic level transition at input 2 connect it to outlet $\overline{B}$ of pulser switch *B*.
4. For applying high or low logic levels at inputs 3 and 11 use logic switches Sw_1 and Sw_2.
5. For verifying the function table, Table 9.3, which we will consider first, note the outputs of logic monitors L_1 and L_2. For the present you may ignore L_3.
6. Connect pin 16 to + V_{CC} and pin 8 to ground. Apply logic levels as indicated in Table 9.3 and record your observations for each set of logic level inputs at *A*, $\overline{B}$ and $\overline{CLR}$ and record your observations in Table 9.4. Your observations must tally with Table 9.3.

Table 9.4

Function Table for IC 74123

Inputs			*Outputs*	
A	$\overline{B}$	$\overline{CLR}$	*Q*	$\overline{Q}$
↓	H	H		
L	↑	H		
L	H	↑		
X	X	L		
X	X	↑		

H = High voltage level
L = Low voltage level
X = Don't care
↑ = Trigger : Low to High transition
↓ = Trigger : High to Low transition.

Generation of Delayed Pulses

1. Connect pin 16 to + V_{CC} and pin 8 to ground. Switch on the power supply. You will notice that after some time L_1 and L_3 will be off and only L_2 will be on.

2. Apply a high to low transition with pulser switch A. You will notice that L_1 is switched on, while L_2 and L_3 will be off. Use your watch to record the time for which L_1 remains on. It will be on for about 15 seconds.
3. Calculate the pulse duration time for the values of R and C used for monostable 1. You will find that it is about 15 seconds.
4. As soon as L_1 goes off a high to low transition will be applied at input 9 of monostable 2 by the low-going output of monostable 1.
5. Note the time for which L_3 remains on. It should be just about 2.3 seconds.
6. Calculate the pulse duration time for monostable 2 from the values given in Fig. 9.11. It should be about 2.3 seconds.
7. If, during your observations, you wish to cut short the duration of the output pulse of monostable 1, apply a logic 0 to input 3 with logic switch Sw_1. This will terminate the duration of the pulse.
8. If you wish to cut short the duration of both the pulses from monostables 1 and 2, apply low logic levels to both inputs 3 and 11.

The above observations show how a delay function can be implemented. The first monostable produces a delay of about 15 seconds, when the second monostable is switched on for about 2 seconds.

Sequencing and Timing

If a number of devices are to be activated in sequence, more monostables can be used to form a chain.

PROBLEMS

9.1. Determine values for the resistors and capacitors used in the TTL clock oscillator shown in Fig. 9.1 so that it oscillates at a frequency of 100 kHz.

9.2. Calculate the frequency of oscillations and duty cycle (high output time) for a 555 Timer astable oscillator from the following data:

$$R_A = 50 \text{ k ohm}$$
$$R_B = 5 \text{ k ohm}$$
$$C = 0.05\ \mu\text{F}$$

9.3. Calculate the values of R_B and C for a 555 Timer astable oscillator from the following data:

$f = 1$ MHz

Duty cycle = 30% (low output time)

$R_A = 5000$ ohm

9.4. Calculate the pulse width for a monostable oscillator using IC 74121 when R_{ext} = 10 k ohm and C_{ext} = 1 μF.

9.5. A monostable oscillator using IC 74121 is required to produce a pulse width of 20 ms. Calculate the value of the capacitor if the internal resistor is used.

9.6. Design a missing pulse detector for pulses which occur 72 times a minute.

9.7. Calculate the pulse width for an astable oscillator using IC 74123 from the following data:

$$R_{ext} = 15 \text{ k ohm}$$
$$C_{ext} = 5000 \ \mu\text{F}$$

9.8. In Fig. 9.11, pin 10 of the IC has been connected to pin 4. Explain why this has been done?

9.9. What changes should be made in the circuit of Fig. 9.11 so that it functions as an astable oscillator?

9.10. Draw pulse waveforms for the circuit given in Fig. 9.11.

9.11. Design a circuit using IC 74123 so that it produces the pulse waveforms given in Fig. P-9.11.

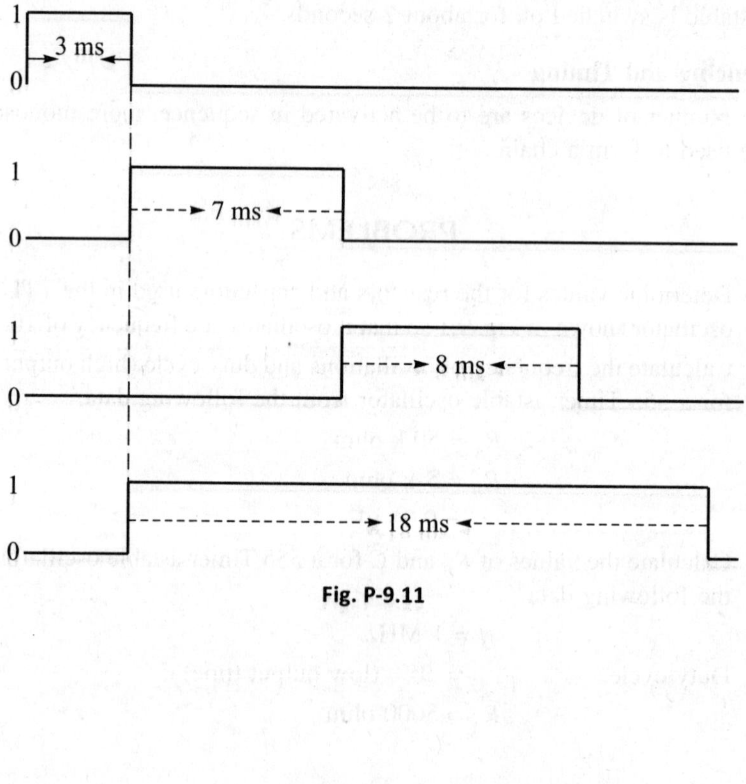

Fig. P-9.11

CHAPTER 10

Arithmetic Logic Circuits

10.1 INTRODUCTION

For carrying out arithmetic operations we require logic circuits which can analyze and implement arithmetic commands such as addition, subtraction etc. In this chapter, therefore, we will consider logic building blocks which can carry out simple arithmetic operations. However, before we do that it will be useful if we recapitulate the process by which we add and subtract binary numbers. This will help us in designing arithmetic building blocks.

10.2 ADDITION

We will consider the addition of two binary numbers to understand how the rules of addition, as we already know them, are applied in practice. It is shown in Fig. 10.1.

				LSB ↓	
1	0	0	0	←	Carries
x	1	1	0	0	Augend A
x	1	0	1	0	Addend B
	0	1	1	0	Sum

Fig. 10.1

In Fig. 10.2, we have shown the order in which the addition has been carried out and the sum and carry generated in the process.

Column	*A*	*B*	*Sum*	*Carry*
1	0	0	0	0
2	0	1	1	0
3	1	0	1	0
4	1	1	0	1

Fig. 10.2

You must have noticed that there are no carries from the first three columns. There is a carry-out from the fourth column which is the carry-in for the fifth column to the left. You will also observe that when there is no carry, as in the first four columns, only two digits are required to be added; but when there is a carry-in, as in the fifth column, three digits are required to be added. For the present, we will consider a circuit for the addition of two digits and later turn our attention to a circuit which will add two digits and the carry-in.

Based on our discussion above the rules for binary addition are given in Table 10.1.

Table 10.1

Augend A	*Addend* B	*Sum* Σ	*Minterm designation*	*Carry* C_{in}	*Minterm designation*
0	0	0		0	
0	1	1	$\overline{A} \cdot B$	0	
1	0	1	$A \cdot \overline{B}$	0	
1	1	0		1	$A \cdot B$

The minterm designation of the sum which is $\overline{A} \cdot B \oplus A \cdot \overline{B}$ represents the output of a 2-input XOR gate, and the carry is represented by the output of an AND gate. The sum and carry equations can, therefore, be written as follows:

$$\text{Sum} = A \oplus B$$

$$\text{Carry} = A \cdot B$$

If the same two inputs A and B are applied to XOR and AND gates the output of the XOR gate will represent the sum and the AND gate will represent the carry. Based on the above consideration the circuit given in Fig. 10.3 will perform the required function of adding two digits. This circuit is known as the *half-adder*.

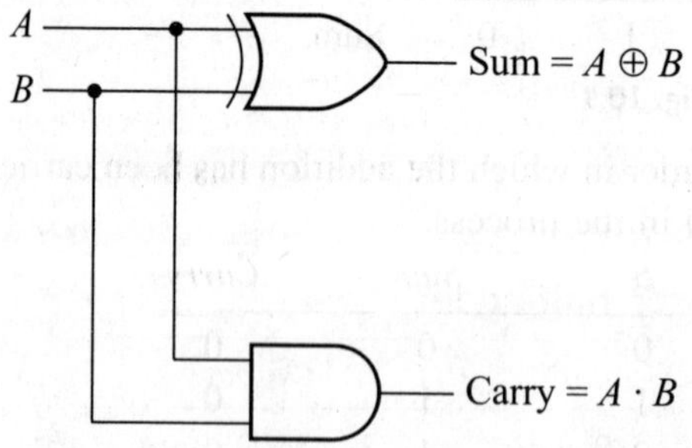

Fig. 10.3. Half-adder.

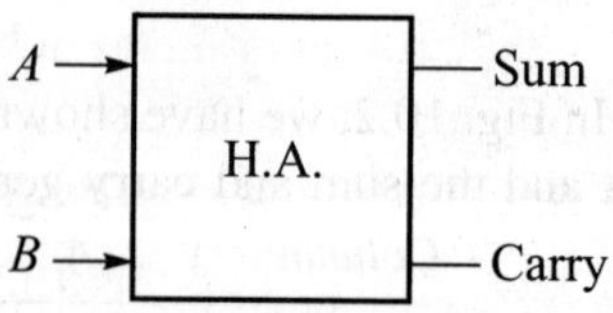

Fig. 10.4. Symbol for Half-adder.

EXPERIMENT 10.1 : HALF-ADDER

Objective

To assemble and study the operation of a half-adder.

Materials Required

Logic trainer
IC 7408 : Quad, 2-input AND gate
IC 7486 : Quad, Exclusive-OR (XOR) gate
Resistor 470 ohm.

The circuit diagram for Experiment 10.1 is given in Fig. 10.5.

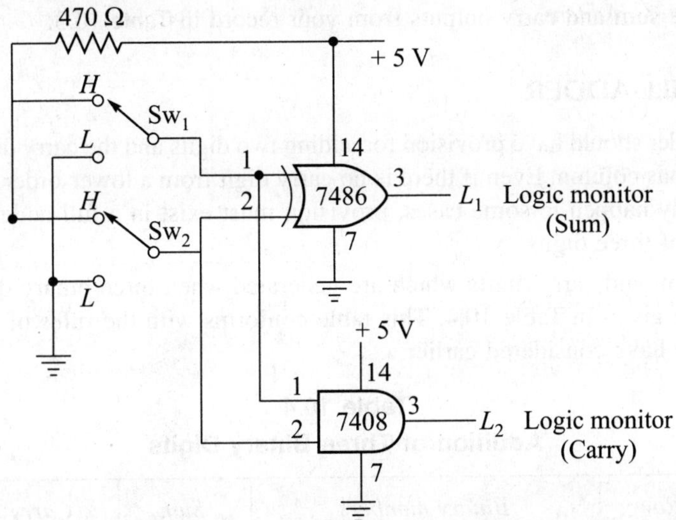

Fig. 10.5. Circuit diagram for Experiment 10.1.

Procedure

1. Assemble the circuit given in Fig. 10.5. Connect pin 14 of both the ICs to + 5 V and pin 7 to ground.
2. With logic switches Sw_1 and Sw_2 apply logic voltage levels to A and B for each set of inputs as given in Table 10.2 and record your observations in Table 10.3.

Table 10.2

Truth Table for Half-adder

Inputs		*Output*	*Output*
A	*B*	L_1 *(XOR) gate* *Sum*	L_2 *(AND) gate* *Carry*
0	0	0	0
0	1	1	0
1	0	1	0
1	1	0	1

Table 10.3

Inputs		*Outputs*	
A	*B*	*Sum*	*Carry*

3. Your observations must tally with Table 10.2. Write the equations for the sum and carry outputs from your record in Table 10.3.

10.3 FULL-ADDER

A full-adder should have provision for adding two digits and the carry digit from the previous column. Even if there is no carry digit from a lower-order column, which may happen in some cases, provision must exist in a full-adder for the addition of three digits.

The sum and carry digits which are generated when three binary digits are added are given in Table 10.4. This table conforms with the rules of addition which we have considered earlier.

Table 10.4
Addition of Three Binary Digits

Row	*Binary numbers*			*Sum*	*Carry*
	A	*B*	*C*		C_0
1	0	0	0	0	0
2	0	0	1	1	0
3	0	1	0	1	0
4	0	1	1	0	1
5	1	0	0	1	0
6	1	0	1	0	1
7	1	1	0	0	1
8	1	1	1	1	1

Now if you refer to Table 6.4 in Appendix 3, which is the truth table for a 3-input XOR gate, and take the sum of digits representing the input logic levels, you will find that the output column of this table truly represents the sum of the digits; but it does not represent the carry. Therefore, the sum of three digits, except the carry, can be represented by the output of a 3-input XOR gate as shown in Table 10.4.

Both, the sum and carry outputs can be realized by the full-adder circuit shown in Fig. 10.6 (*a*). The carry output is obtained by feeding inputs *A* and *B*, *B* and *C* and *A* and *C* to three 2-input AND gates and connecting their output to a 3-input OR gate. The output of the OR gate gives the required carry output. The output of the AND gates will be 1 when both inputs are 1. Table 10.5 summarizes these input states.

Table 10.5

Row from Table 10.4	*Inputs*			*AND gate*		*OR gate*
	A	*B*	*C*	*Input*	*Output*	*Output*
4	0	1	1	*B*, *C*	1	1
6	1	0	1	*A*, *C*	1	1
7	1	1	0	*A*, *B*	1	1
8	1	1	1	*A*, *B*	1	
				A, *C*	1	1
				B, *C*	1	

You will observe from Table 10.5 that the OR gate output is the same as the required carry output. The circuit of Fig. 10.6 (*a*) will, therefore, perform the function of a full-adder.

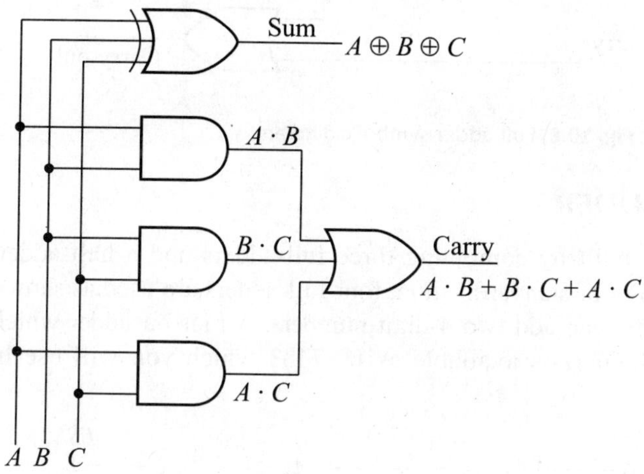

Fig. 10.6. (a) Full-adder.

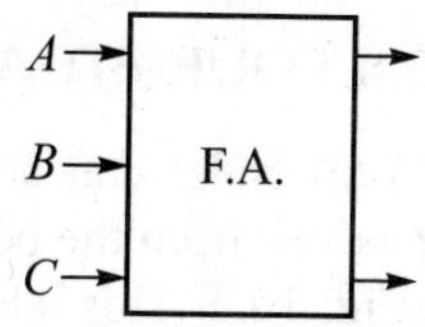

Fig. 10.6. (b) Symbol for Full-adder.

A full-adder circuit can also be designed in other ways. For instance, you can combine two half-adders to make a full-adder as in Fig. 10.7. In symbolic form, a binary full-adder using two half-adders and an OR gate is shown in Fig. 10.8.

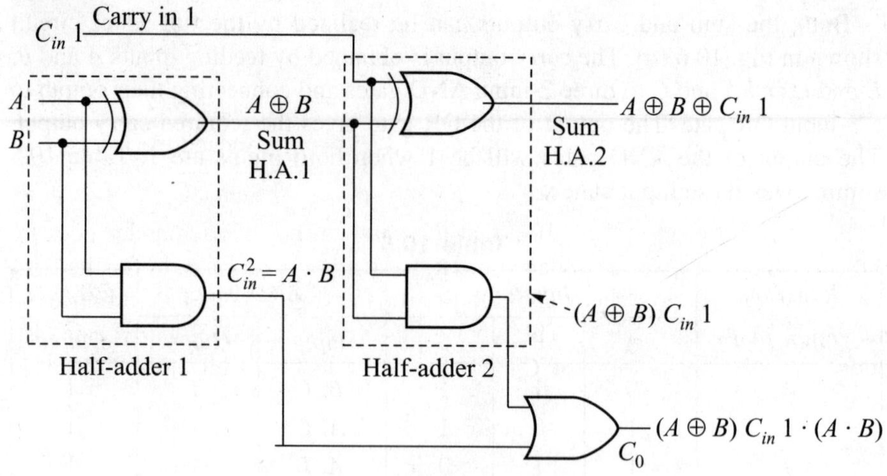

Fig. 10.7. Full-adder logic diagram.

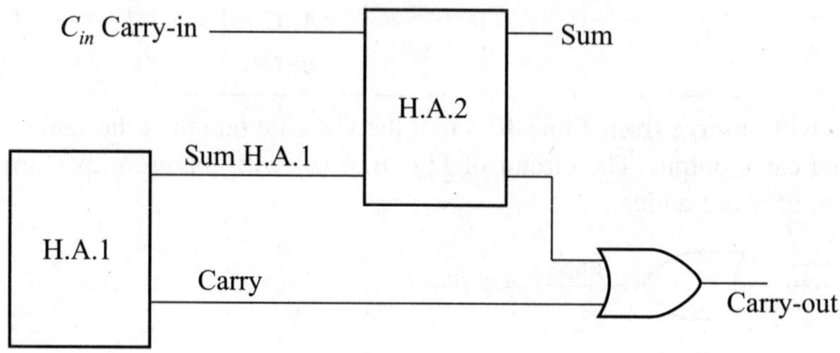

Fig. 10.8. Full-adder symbolic diagram.

10.4 FOUR-BIT ADDER

A 4-bit adder can be built by combining three full-adders and a half-adder. However, from the point of standardization, four full-adders are used as shown in Fig. 10.9. This adder can add two 4-digit numbers. A four-bit adder which consists of four full-adders is available as IC 7483 which you will use in Experiment 10.2.

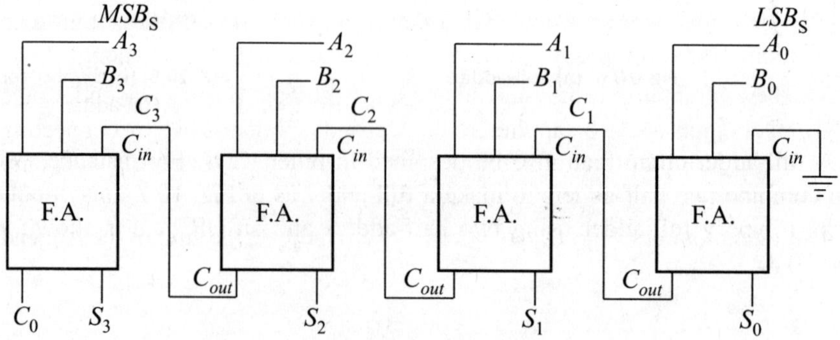

Fig. 10.9. Four-bit adder using four full-adders.

The addition sequence for this adder is as follows:

C_4	C_3	C_2	C_1	C_0	Carry
	A_3	A_2	A_1	A_0	Augend
	B_3	B_2	B_1	B_0	Addend
	S_3	S_2	S_1	S_0	Sum

You will notice from Fig. 10.9 that the carry output of each adder goes to the carry input of the next adder to the left. As there is no carry-in for the first adder, its input is grounded. The addition of the digits A_0 and B_0 produces a sum of S_0 and a carry of C_1. The other digits will be added similarly. The last adder will produce a carry of C_4 which will be lost in 4-bit arithmetic. In a similar way, 8-bit and 16-bit adders can be built.

EXPERIMENT 10.2 : FOUR-BIT ADDER

Objective

To carry out addition of 4-digit numbers using IC 7483.

Materials Required

Logic trainer

TTL IC 7483 Four-bit full-adder

NPN Transistor BC 108A

Volt-ohm-milliammeter

LED (Red)

Resistor 150 ohm

Resistor 470 ohm

Resistor 1 k ohm.

The circuit diagram for Experiment 10.2 is given in Fig. 10.10.

Procedure

1. Assemble the circuit given in Fig. 10.10. As there is no carry from a lower-order column, connect pin 13 of the IC to ground. Connect pin 5 of the IC to + 5 V and pin 12 to ground.
2. The transistor is used as a logic monitor for the carry output as four logic monitors, normally provided in logic trainers, will not be enough.
3. The 4-bit input *A* is the Augend and the 4-bit input *B* is the Addend. Logic level inputs are to be applied to the Augend or Addend bits by connecting the appropriate bits either to ground or to + V_{CC} through a 470 ohm resistor.
4. Logic monitors L_1 to L_4 have been used to indicate the sum of the Augend and Addend.

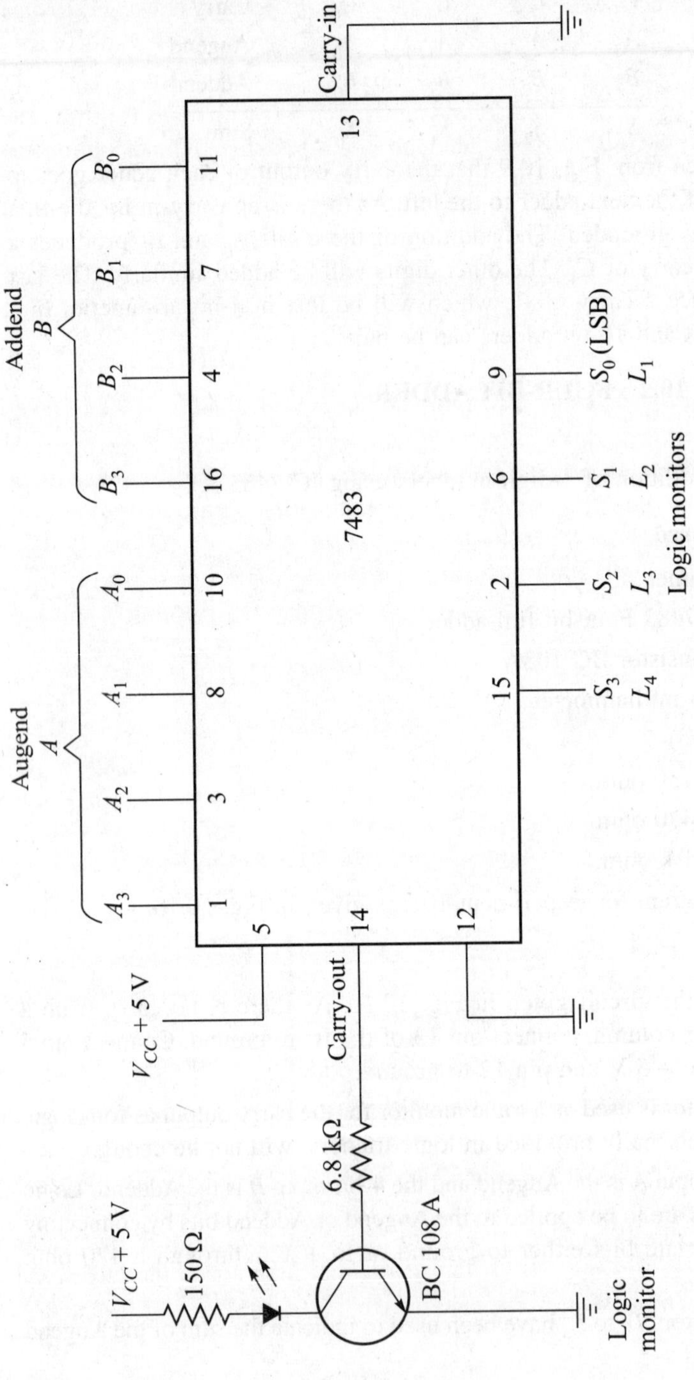

Fig. 10.10. Four-bit adder.

5. Since our experiment will involve only 4-bit arithmetic, the carry-out is of no significance and will be lost in 4-bit arithmetic. However, a logic monitor (a transistor) has been connected to pin 14 to show the carry-out from the 4-bit adder.
6. You may now proceed to add unsigned binary numbers as follows. The Augend and Addend to be added are given in Table 10.6. A column has been provided for the sum and another for the carry-out. Enter the Augend and Addend by applying appropriate logic inputs as suggested in para 3. Observe the logic monitors and enter the sum and carry outputs in Table 10.6. Your observations should tally with Table 10.6 in Appendix 3.
7. Remember that a carry-out of 1 will indicate that the capability of the adder has been exceeded and an error has crept in. You may verify this by adding two binary numbers the total of which exceeds 1111, as shown in the last addition which indicates a carry of 1 and a sum of 1 which is obviously incorrect. In each case of addition, verify the result of the experiment by manual addition.

Table 10.6

S. No.	*Augend*				*Addend*				*Sum*				*Carry*
	A_3	A_2	A_1	A_0	B_3	B_2	B_1	B_0	S_3 L_4	S_2 L_3	S_1 L_2	S_0 L_1	L_5
1	0	1	0	1	1	0	0	0					
2	1	0	0	1	0	1	0	0					
3	1	0	0	0	0	1	1	0					
4	1	1	0	0	0	0	1	0					
5	0	0	0	1	1	1	0	0					
6	1	0	1	1	0	1	1	0					

10.5 BINARY SUBTRACTION

It will be helpful if you review the rules for binary subtraction dealt with in Chapter 1. We will consider these rules again and their application to logic circuits.

As there are only two binary digits, 0 and 1, there are only four combinations of these digits. We will consider the result of subtraction in these four cases which are considered in Fig. 10.11. Pay particular attention to the DIFF. and BORROW digits and how they are generated.

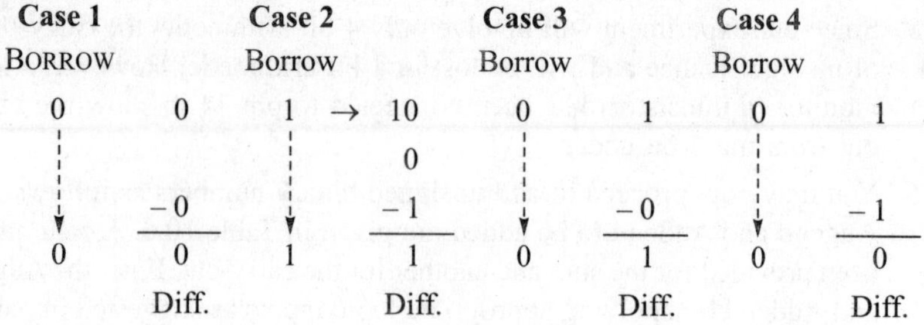

Fig. 10.11

The result of subtractions has been marked 'DIFFERENCE' (DIFF.). You will notice from the above illustrations that there is a BORROW only in Case 2. This is because 1 cannot be subtracted from 0 and that is why a 1 has been borrowed from the previous column and it assumes the value of 10 (or decimal 2) in that column. The result of subtracting 1 from 10 is therefore 1, but there is also a BORROW of 1. There are no borrows in the other subtractions.

10.5.1 Half-subtracter

We have dealt with four cases of subtractions of single binary digits. These lay down the ground rules for binary subtraction and have been summarized in Table 10.7.

Table 10.7

Minuend	*Subtrahend*	*Difference*	*Borrow*
A	*B*		
0	0	0	0
0	1	1	1
1	0	1	0
1	1	0	0

You will notice from Table 10.7 that

* the Diff. is 1, when A is 0 and B is 1 or when
* A is 1 and B is 0.
* in the remaining cases the Diff. is 1.

The equation for the Diff. is, therefore, as follows:

$$\text{Diff.} = A \cdot \overline{B} + \overline{A} \cdot B = A \oplus B$$

So far as the Borrow is concerned, you must have noticed that Borrow is 1 only when A is 0 and B is 1. The equation for Borrow is, therefore, as follows:

$$\text{Borrow} = \overline{A} \cdot B$$

The equation for Diff. can be implemented with an XOR gate and the Borrow can be implemented by an AND gate and an Inverter as in Fig. 10.12 (*a*). The symbol for a half-subtracter is given in Fig. 10.12 (*b*).

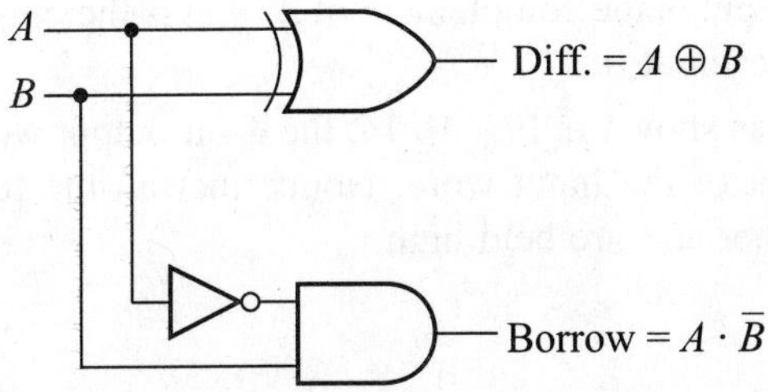

Fig. 10.12. (a) Half-subtracter.

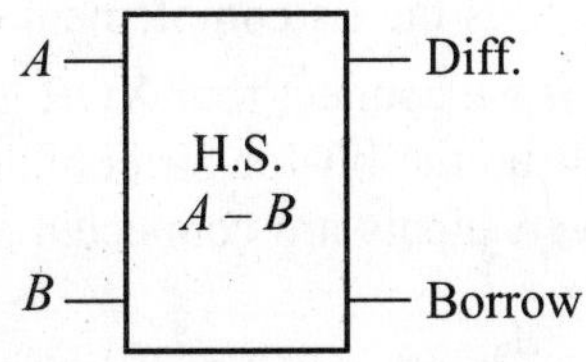

Fig. 10.12. (b) Logic symbol.

A full-subtracter can be made with two half-subtracters and an OR gate on the same lines as a full-adder. It has to take into account minuend A, subtrahend B and generate the Diff. and Borrow outputs. Therefore, we need a logic circuit with three inputs and two outputs.

10.5.2 2's Complement Adder-Subtracter

For carrying out subtraction we have to convert the subtrahend to its 2's complement form. We can do this by first inverting the subtrahend and then adding a 1 to it. This is done with a controlled-inverter which will convert the subtrahend to its l's complement form. An XOR gate is used as a controlled-inverter, with one of its inputs, labelled 'Invert', held low or high as required. The output of this XOR gate (Fig. 10.13) is given in Table 10.8 with invert held 'high' and 'low'.

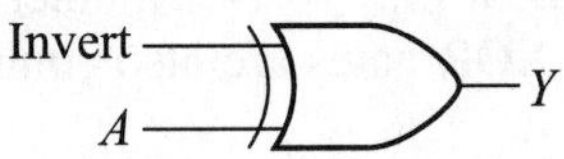

Fig. 10.13. Controlled-inverter.

Table 10.8
Controlled-Inverter

Inputs		*Output*	*Comment*
A	*Invert*	*Y*	
0	0	0	Output same as A
1	0	1	
0	1	1	Output is the complement of A
1	1	0	

The following conclusions can be drawn from Table 10.8:

* When Invert is low, the output is the same as *A*. In other words, Invert has no effect on *A*.
* When Invert is high, the output is the complement of *A*, that is the output is the l's complement of the input *A*.

If we connect four XOR gates as shown in Fig. 10.14, the 4-bit output word will be the bit-for-bit complement of the input word. Notice that all the four Invert inputs are connected together and are held high.

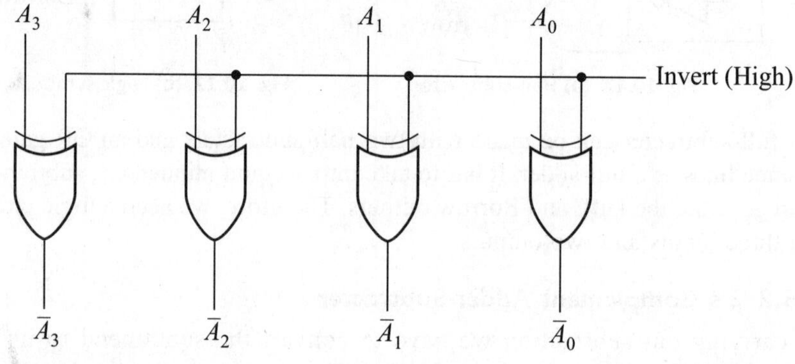

Fig. 10.14

A 4-bit adder-subtracter which uses the 2's complement method of subtraction is given in Fig. 10.15. The four full-adders shown in the diagram are available as IC 7483 and four XOR gates are available as IC 7486. The important point to note is that the *B* inputs of the full-adders are connected to a controlled-inverter in the same way as in Fig. 10.14. Another point worth noting is that the Invert inputs of all the XOR gates are also connected to the carry input of the first full-adder.

Addition

When addition is carried out the Invert is kept low so that bits B_3, B_2, B_1, B_0 pass on to the full-adders without any change and take part in the addition process in the normal way. Note that when the Invert is low, the carry-in for the first full-adder is 0, as the carry input for the first adder is connected to the Invert input, which is held low when addition is carried out. The four bits are, therefore, added up as given in Section 10.4.

A 4-bit adder-subtracter using ICs 7483 and 7486 is shown in Fig. 10.16.

When the 4-bit adder-subtracter is used for additions, it can add unsigned binary numbers like the circuit given in Fig. 10.10 and the output can accommodate unsigned binary numbers from 0 to 15 only. If two such 4-bit adders are used, it will acquire 8-bit capability and it can represent unsigned binary numbers from 0 to 255.

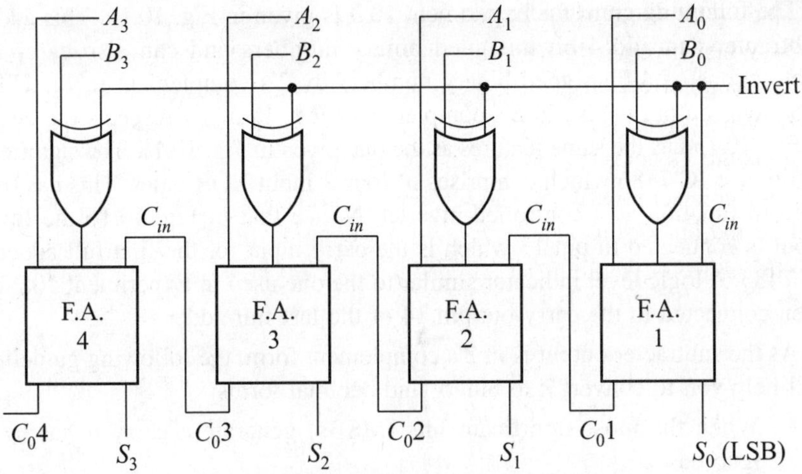

Fig. 10.15. Four-bit Adder-subtracter.

Subtraction

During subtraction the Invert is held high, which complements all the B inputs to the full-adders. Therefore, the B inputs to the full-adders are in l's complement form. As the Invert is also connected to the carry input of the first full-adder, and it is already held high, this results in 1 being added to the l's complement of the B input number. Thus, the resultant is the 2's complement of the B input.

Subtraction is effected by adding the 2's complement of the B input, so derived, to the A input as follows:

Minuend	A_3	A_2	A_1	A_0	
Subtrahend	$\hat{B}_3$	$\hat{B}_2$	$\hat{B}_1$	$\hat{B}_0$	2's complement of B
Difference	S_3	S_2	S_1	S_0	

EXPERIMENT 10.3 : FOUR-BIT ADDER-SUBTRACTER

Objective

To add and subtract binary numbers on a 4-bit adder-subtracter.

Materials Required

Logic trainer
TTL IC 7483 : Four-bit full-adder
TTL IC 7486 : Quad, Exclusive-OR gates
NPN Transistor BC 108A
Resistor 150 ohm
Resistor 470 ohm
Resistor 1 k ohm
LED (Red)

The circuit diagram for Experiment 10.3 is given in Fig. 10.16. This adder-subtracter can add 4-bit unsigned binary numbers and can also carry out subtractions of 4-bit signed binary numbers by 2's complement method. The subtracter output is also in 2's complement form. In many respects the circuit in Fig. 10.15 has the same features as the one given in Fig. 10.16. The significant addition is IC 7486 which comprises of four 2-input XOR gates. This has been used to function as a controlled-inverter. Notice that in Fig. 10.16 the Invert input is connected to pin 13 which is the carry input for the first full-adder in IC 7483. A logic level indicator similar to the one used in Experiment 10.2 has been connected to the carry-out pin 14 of the last full-adder.

As the subtracter output is in 2's complement form, the following guidelines will help you to convert it to binary and decimal forms:

* When the most-significant bits (MSBs) generate a carry it has to be ignored.
* As a result of subtraction, if the MSB in the result is a 1, the answer is in the negative and it is in 2's complement form. It has to be converted to binary form for a meaningful result.
* The answer will be positive if the MSB in the result is a 0.
* If the result of subtraction exceeds the limit of 2's complement representation the result will be wrong. The range will be exceeded when the MSB of A and $(-B)$ are of the same sign and the MSB of the result has the opposite sign because the answer will be beyond the range.

Procedure

1. Assemble the circuit given in Fig. 10.16. Connect pin 14 of IC 7486 and pin 5 of IC 7483 to + 5 V and pin 7 of IC 7486 and pin 12 of IC 7483 to ground. Also connect the free end of the 150 ohm resistor to + 5 V.
2. Logic level voltages are to be applied to the A and B inputs by connecting the appropriate bits either to ground or to $+V_{CC}$ through a 470 ohm resistor.
3. Logic monitors L_1 to L_4 have been used to indicate the output of the system (addition or subtraction). Logic monitor L_5 is used to monitor the overflow.

Addition

4. For addition of unsigned binary numbers follow the procedure given in Experiment 10.2. Keep Invert in the low position for carrying out additions.

Subtraction

5. For subtractions, keep Invert in the high position. Inputs A_3, A_2, A_1, A_0 and B_3, B_2, B_1, B_0 which represent the minuend and subtrahend

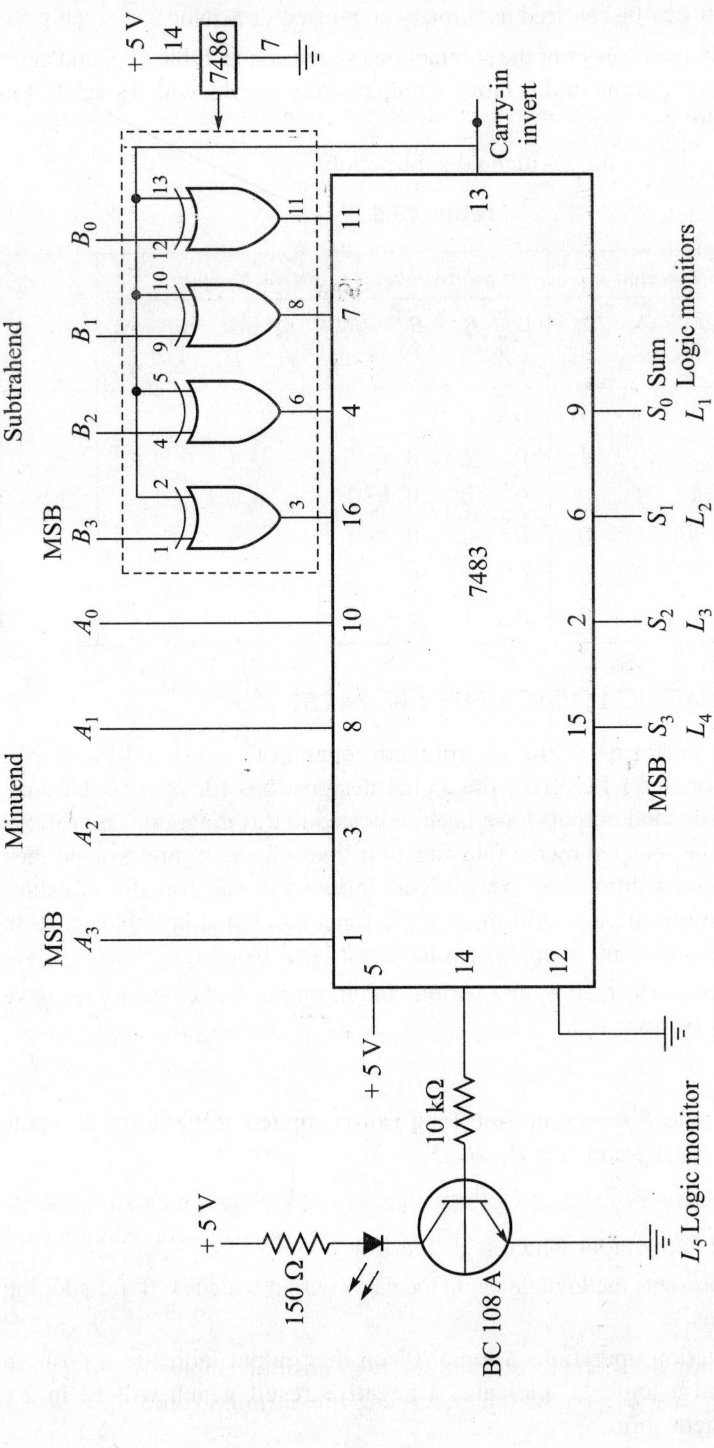

Fig. 10.16. Four-bit Adder-subtracter.

respectively should be in sign-magnitude form. The required binary numbers can be realized as already suggested at S. No. 2.

6. You may now carry out the subtractions suggested in Table 10.9 and enter your observations in this table. Compare your results with Table 10.9 in Appendix 3.
7. Check your results by manual subtraction.

Table 10.9

S. No.	*Minuend*				*Subtrahend*				*Sum*				*Carry*
	A_3	A_2	A_1	A_0	B_3	B_2	B_1	B_0	S_3 L_4	S_2 L_3	S_1 L_2	S_0 L_1	L_5
1	0	1	1	1	0	0	1	1					
2	0	1	0	1	0	1	0	0					
3	0	1	1	0	0	0	0	1					
4	0	0	1	0	0	0	0	1					
5	0	1	0	0	0	1	0	1					
6	0	1	0	0	0	1	1	0					

10.6 ARITHMETIC LOGIC UNIT : IC 74181

This ALU can perform 16 binary arithmetic operations on two 4-bit words. Figure 10.17 for this ALU gives the signal designations for active-high data. The several inputs and outputs have been indicated in this diagram. The various operations can be selected by the four function-lines (S_0, S_1, S_2 and S_3) and they include addition, subtraction, decrement, increment and transfer. Besides arithmetic operation it can perform 16 logic functions listed in Table 10.10 in the logic mode plus some more which have not been listed.

The functions performed by the various input, output and control lines have been discussed below:

Inputs

* Inputs *A* and *B* represent 4-bit long words applied at the *A* and *B* inputs (A_0, A_1, A_2, A_3) and (B_0, B_1, B_2, B_3).

Outputs

C_n is the carry input which is active-low.

* C_{n+4} represents the logic level on the carry output which is '0' for addition operations.

  In subtraction operations a logic '0' on this output indicates a positive result and a logic '1' indicates a negative result which will be in 2's complement form.

* F represents the 4-bit output word on F_0, F_1, F_2, F_3 outputs.
* $A = B$ represents the equality output. A logic '1' on this line shows that $A = B$. (Refer to Comparator function for more details).

$\overline{G}$ represents the Carry Generate output

and $\overline{P}$ represents the Carry Propagate output.

The Generate and Propagate outputs are needed when several 74181 ALUs are used alongwith IC 74182, look-ahead carry-generator to increase speed of operation.

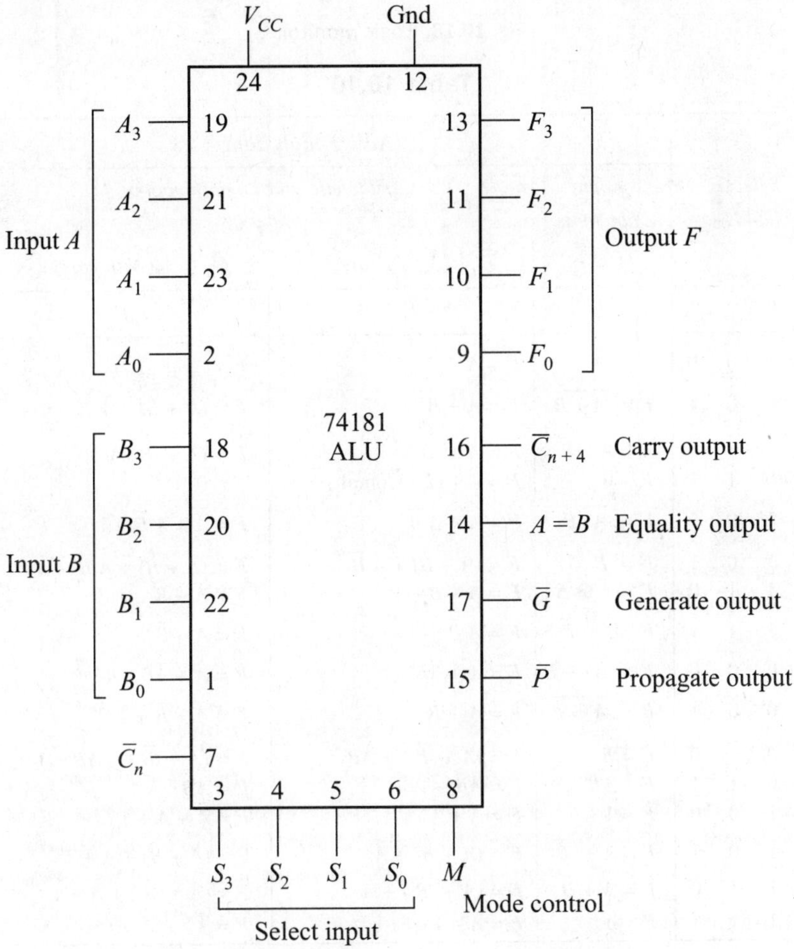

Fig. 10.17. Pin connections for Arithmetic Logic Unit IC 74181 and signal designations for active-high data.

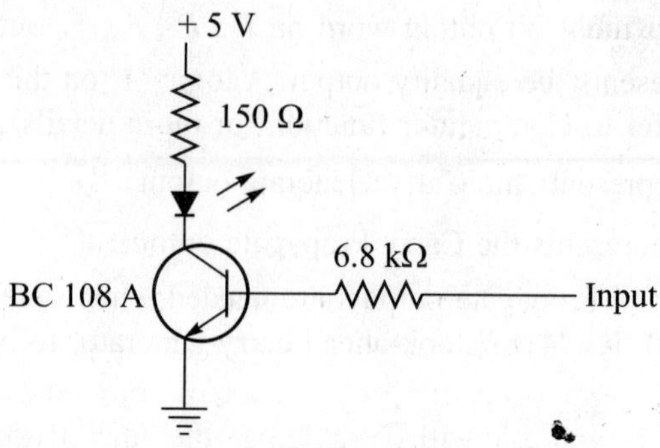

Fig. 10.18. Logic monitor.

Table 10.10

Select inputs				Active-high data		
				Logic functions $M = 1$	*Arithmetic and Logic functions* $M = 0$ $\overline{C_n}$ = 1 (*No carry*)	$\overline{C_n}$ = 1 (*With carry*)
S_3	S_2	S_1	S_0	F	F	F
0	0	0	0	$F = \overline{A}$	$F = A$	$F = A + 1$
0	0	0	1	$F = \overline{A + B}$	$F = A + B$	$F = (A + B) + 1$
0	0	1	0	$F = A \cdot \overline{B}$	$F = A + \overline{B}$	$F = (A + \overline{B}) + 1$
0	0	1	1	$F = 0$	$F = -1$ (2's Compl.)	$F = 0$
0	1	0	0	$F = \overline{A \cdot B}$	$F = A + A\overline{B}$	$F = A + A\overline{B} + 1$
0	1	0	1	$F = \overline{B}$	$F = (A + B) + A\overline{B}$	$F = (A + B) + A\overline{B} + 1$
0	1	1	0	$F = A \oplus B$	$F = A - B - 1$	$F = A - B$
0	1	1	1	$F = A \cdot \overline{B}$	$F = A\overline{B} - 1$	$F = A \cdot \overline{B}$
1	0	0	0	$F = \overline{A} + B$	$F = A + AB$	$F = A + AB + 1$
1	0	0	1	$F = \overline{A \oplus B}$	$F = A + B$	$F = A + B + 1$
1	0	1	0	$F = B$	$F = (A + \overline{B}) + AB$	$F = (A + \overline{B}) + AB + 1$
1	0	1	1	$F = AB$	$F = AB - 1$	$F = AB$
1	1	0	0	$F = 1$	$F = A + A$	$F = A + A + 1$
1	1	0	1	$F = A + \overline{B}$	$F = (A + B) + A$	$F = (A + B) + A + 1$
1	1	1	0	$F = A + B$	$F = (A + \overline{B}) + A$	$F = (A + \overline{B}) + A + 1$
1	1	1	1	$F = A$	$F = A - 1$	$F = A$

A : Represents 4-bit long word applied at the A inputs.

B : Represents 4-bit long word applied at the B inputs

F : Represents 4-bit long output word on F_3, F_2, F_1, and F_0 outputs

$+$: Represents logical OR operation.

$+$: Represents addition.

$\overline{C_n}$: Represents the logic level on the carry input. It is active-low.

Arithmetic Operations

* **Addition of 4-bit Words**

For addition of 4-bit words at A and B inputs the Mode and Select inputs should be as follows:

$$M = 0 \text{ and}$$
$$\text{Select inputs} = 1001$$

The F output will be $A + B$.

If there is a carry-in from the lower order of four bits, C_n will be low indicating a carry-in and a 1 will be added to the sum. If this addition also produces a carry, the C_{n+4} output will go low.

* **Addition of Larger Words at Low Speed**

When words larger than four bits are not required to be added at high speed, two such ALUs can be cascaded by connecting the C_{n+4} output of the lower order device to the C_n input of the higher order device.

* **Addition of Larger Words at High Speed**

For the addition of larger words at high speed the Propagate output $\overline{P}$ and the Generate carry output $\overline{G}$ can be connected to IC 74182 which provides for a look-ahead carry-generator and uses the $\overline{P}$ and $\overline{G}$ signals to anticipate and generate carry input signals.

* **Subtraction**

The inputs should be programmed as follows:

$$\text{Select input} = 0\ 1\ 1\ 0$$
$$\text{Mode input} = 0$$
$$C_n = 1$$

The ALU subtracts the word on the B inputs from the word on the A inputs. Subtraction is performed by generating the l's complement of the B word and adding the result to the word on the A inputs. This gives a result of $A - B - 1$. To get the desired result, $A - B$, 1 has to be added to the result by making the C_n input low. The C_{n+4} output gives the sign. If it is 0 the result of subtraction is positive and it is 1 for negative result. The magnitude is given by the F outputs if the result is positive and in 2's complement form if it is negative.

* **Decrementing Words**

If the A word is to be decremented by 1 ($A - 1$), the inputs should be as follows:

$$\text{Select inputs} = 1\ 1\ 1\ 1$$
$$\text{Mode control} = 0$$
$$C_n \text{ input} = 1$$

* **Incrementing Words**

If the Select inputs are programmed as follows the word on the output will be A

Select inputs = 0 0 0 0

Mode control = 0

$C_n = 1$

If C_n is held low the word on the output will be $A + 1$.

* **Magnitude Comparison**

The magnitude of two binary words can be compared by observing the $A = B$ output when the inputs are programmed as follows:

Select input = 0 1 1 0

Mode control = 0

$C_n = 0$

If $A = B : A = B$ output will be high.

If $A > B : C_{n+4}$ output will be high and $A = B$ output will be low.

If $B > A : C_{n+4}$ output will be low.

Logic Operations

This ALU can perform the OR, XOR, Comparator, AND, OR, NAND, NOR besides ten other logic operations. We will consider some of these to show how these functions are performed. For the AND function the inputs have to be programmed as follows:

Select input = 1 0 1 1

Mode control = 1

Let us assume that the words to be ANDed are as follows:

Word A	:	A_3	A_2	A_1	A_0	1	0	0	1
Word B	:	B_3	B_2	B_1	B_0	1	0	1	1
Output F	:	F_3	F_2	F_1	F_0	1	0	0	1

The output F will be as shown on the right. With the inputs selected the two words will be ANDed. All the logic operations are done on a bit-by-bit basis by this ALU. In this case the output will be 1 0 0 1. You will find this operation easy to comprehend if you consider that the two bits in each position are being processed by a 2-input AND gate.

Let us consider the XOR operation for which the Select input has to be 0 1 1 0 and $M = 1$.

Word A	:	A_3	A_2	A_1	A_0	1	1	0	0
Word B	:	B_3	B_2	B_1	B_0	0	1	1	0
Output F	:	F_3	F_2	F_1	F_0	1	0	1	0

As before the two bits in each position are XORed resulting in the output given on page 128.

EXPERIMENT 10.4 : ARITHMETIC LOGIC UNIT

Objective

To study the operation of arithmetic logic unit IC 74181.

Materials Required

Logic trainer
TTL IC 74181 : Arithmetic Unit
NPN Transistors : BC 108A (2)
LED (2)
Resistors 150 ohm (2)
Resistor 470 ohm
Resistors 6.8 k ohm (2).

Refer to Fig. 10.17 for all connections to be made to the inputs and outputs and follow the instructions given here for the purpose.

Procedure

1. Connections to the Select inputs to obtain high and low logic levels should be made to logic switches Sw_1 through Sw_4.
   High logic levels required for input words A and B should be made to + 5 V through a 470 ohm resistor. Low logic levels required for words A and B can be obtained by connecting them to ground.
2. Outputs F_0 through F_3 should be connected to logic monitors L_1 through L_4.
3. Logic monitors are also required for outputs C_{n+4} and $A = B$ and as there are normally only four logic monitors in logic trainers, they may be assembled as shown in Fig. 10.18.
4. High and low logic levels required for $\overline{C}_n$ (pin 7) and Mode control (pin 8) may also be obtained as suggested above for words A and B.
5. After you have made all the connections, go over them again to make sure that there are no mistakes. Now connect pin 24 to + V_{CC} and pin 12 to ground.
6. The experiments to be carried out have been listed in Tables 10.11, 10.12 and 10.13. Carry out these experiments and enter your observations in these tables.
7. Compare your observations with tables in Appendix 3 bearing identical numbers.

Experiment 10.4

S. No.	Select inputs	M	C_n	Word A	Word B	Output
	S_3 S_2 S_1 S_0			A_3 A_2 A_1 A_0	B_3 B_2 B_1 B_0	F_3 F_2 F_1 F_0

Additions **Table 10.11**

S. No.	Select inputs	M	C_n	Word A	Word B	Output
1				0 0 1 1	0 0 0 1	
2				0 0 0 1	0 0 1 0	
3				0 1 0 0	0 0 1 0	

Subtractions **Table 10.12**

S. No.	Select inputs	M	C_n	Word A	Word B	Output
1				1 0 0 1	0 0 1 0	
2				1 1 0 1	0 1 1 1	
3				0 1 1 0	0 0 1 1	

Magnitude Comparison **Table 10.13**

S. No.	Select inputs	M	C_n	Word A	Word B	C_{n+4} Output	$A = B$ Output	
1				1 0 1 0	0 1 0 0			
2				1 0 1 0	1 0 0 0			
3				1 0 0 1	1 0 0 1			

PROBLEMS

10.1. Draw a circuit diagram of a full-subtracter using XOR and NAND gates.

10.2. Draw a full-adder circuit using only NAND gates.

10.3. Design an 8-bit adder-subtracter using two ALUs IC 74181.

10.4. Implement a 1's complement circuit using IC 7486.

10.5. Describe a simple test to check IC 7483.

10.6. How will you use IC 74181 to multiply a binary number by 2?

CHAPTER 11

Shift Registers

11.1 INTRODUCTION

Shift registers are sequential circuits and, as they employ flip-flops, they possess memory. However, memory is not the only requirement. Shift registers are primarily required to store binary data momentarily until it is utilized. Data is sometimes required to be presented in a manner which may be different from the way it is temporarily stored in a shift register. Shift registers can present data to a device in a serial or parallel form, quite different from the manner in which it is fed to it. These devices can also shift data left or right.

Shift registers have also found application as circulating shift registers when the data can be kept circulating. In this chapter, we will consider the various modes of operation of shift registers listed below:

* Serial Input / Serial Output
* Serial Input / Parallel Output
* Parallel Input / Serial Output
* Parallel Input / Parallel Output.

11.2 UNIVERSAL SHIFT REGISTER : IC 7495A

This shift register has provision for serial and parallel outputs and can perform right shift and left shift functions. It has 4-bit capability and it can be loaded by serial as well as parallel data inputs. The logic diagram for this shift register is given in Fig. 11.1. You will notice from the diagram that the Mode control not only controls the input gating to the flip-flops but it also controls the two clocks.

Shift-right Function

When the Mode control is low, all gates 1 are enabled which enables the shift register to perform the shift-right function. When serial data is applied at pin 1, it passes through gates 1 and 2 to flip-flop *A*. The output of flip-flop *A* is connected in a similar way to the input of flip-flop *B, B* to *C* and *C* to *D*. The low level on Mode control disables gate 6 and enables gate 4. A clock pulse applied to clock 1 passes through gate 5 and controls all the flip-flops. When clock pulses are applied in this mode, data can be shifted right from flip-flop *A* to *B, B* to *C*, and so on.

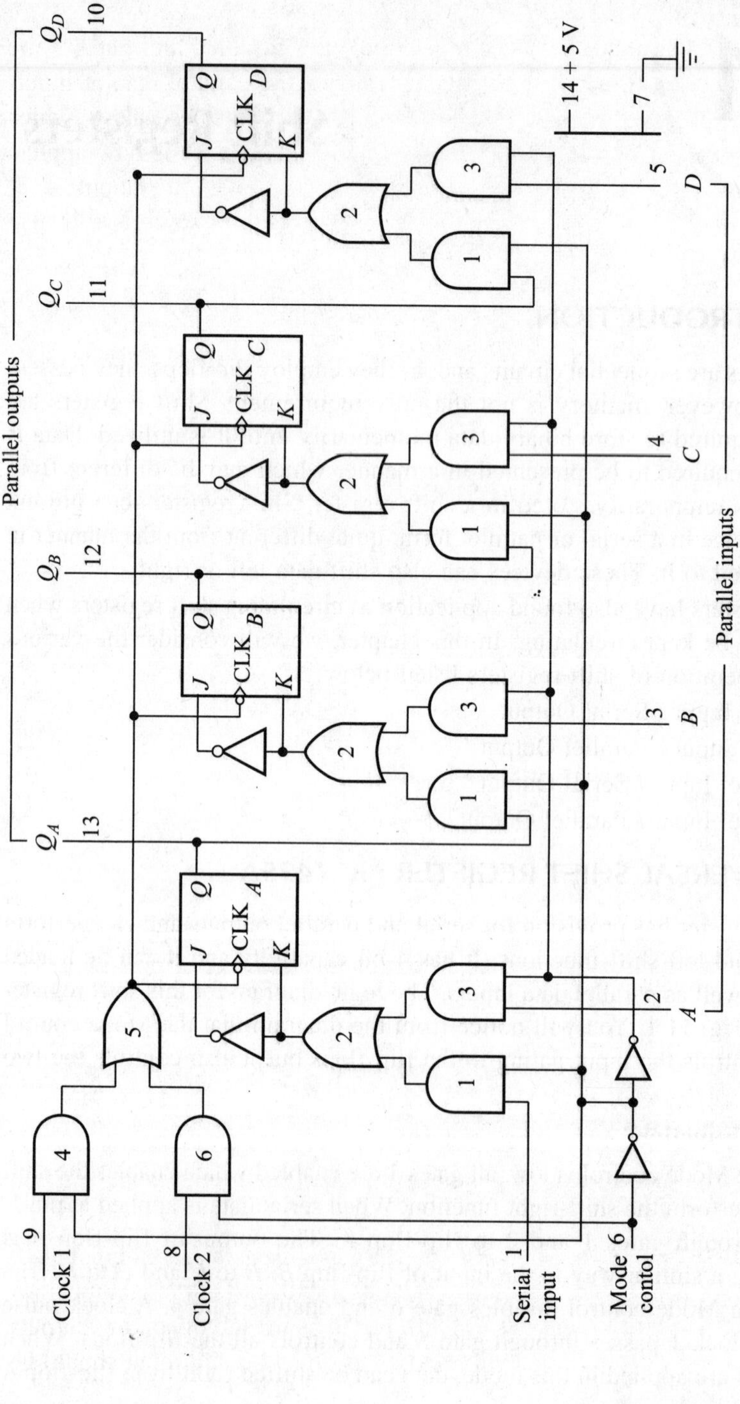

Fig. 11.1. Logic diagram of IC 7495A shift register.

Parallel-in/Parallel-out Function

When the Mode control is high, all gates 1 are disabled and gates 3 are enabled. In this mode the shift register recognizes the parallel data at inputs *A B, C* and *D*. High logic level on Mode control also enables gate 6, which now enables clock 2 to control the flip-flops. If a clock pulse is now applied to clock 2, flip-flops will be actuated and the external 4-bit word at inputs *A, B, C* and *D* will be loaded into the parallel output. This procedure will also enable you to clear the register or to load any 4-bit word into it.

Left-Shift Function

For the shift-left operation the following connections are required to be made as shown in Fig. 11.3. Q_D to *C*, Q_C to *B* and Q_B to *A*. Serial data is applied at input *D* (pin 5). The Mode control is held high and clock pulses are applied at clock 2 which shifts the data from right to left. If the data is to be shifted right all that you have to do is to feed serial data at serial input (pin 1) and hold the Mode control low. If clock pulses are now applied using clock 1 the data at the input will be shifted right.

While experimenting with this shift register it will be of help to you if you bear in mind its following characteristic features:

(1) **Mode Control Held Low**

(*a*) The register recognizes the serial input at pin 1 and ignores the parallel inputs at *A, B, C* and *D*.

(*b*) Clock 2 is disabled and clock 1 is enabled.

(*c*) Clock 1 pulses will shift data to the right.

(2) **Mode Control Held High**

(*a*) The register recognizes the parallel inputs at *A, B, C* and *D* and ignores the serial inputs at pin 1.

(*b*) Clock 1 is disabled and clock 2 is enabled.

(*c*) A clock 2 pulse will load the data at the parallel inputs into the register.

(3) **Left-Shift/Right-Shift Mode**

(*a*) With Mode control low the register recognizes the serial input at pin 1. Clock 1 is enabled and clock 1 pulses will shift serial input data into the register and move it right.

(*b*) With Mode control high the register recognizes the serial input at *D*. Clock 2 is enabled and pulses at clock 2 will shift the serial input data into the register and move it left. For the shift-left operation, outputs and inputs should be connected as in Fig. 11.3 and input should be applied at pin 5.

EXPERIMENT 11.1 : SHIFT REGISTER : IC 7495A

Objective

To demonstrate the following functions of shift register IC 7495A:

1. Clearing register
2. Serial input/Parallel output
3. Parallel input/Parallel output
4. Parallel input/Serial output.

Materials Required

Logic trainer
IC 7495A : 4-bit shift register
Resistor 470 ohm.

The circuit diagram for Experiment 11.1 is given in Fig. 11.2.

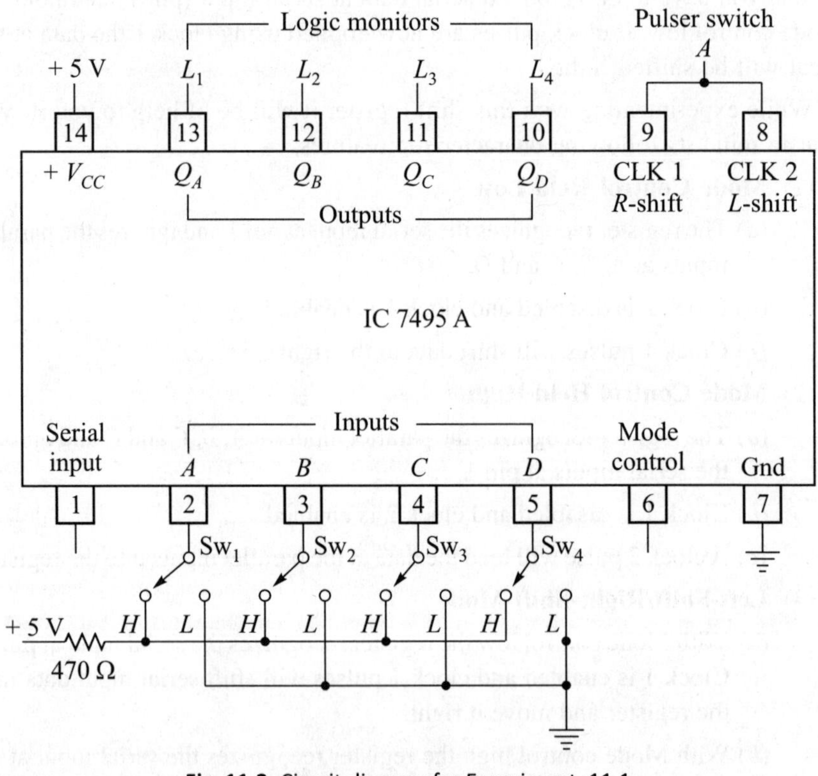

Fig. 11.2. Circuit diagram for Experiment 11.1.

Procedure

1. Mount the IC 7495A on the logic trainer and make the required connections as shown in Fig. 11.2. Connect pins 2, 3, 4 and 5 of the

IC to logic switches Sw_1, Sw_2, Sw_3 and Sw_4 in the same way as in Fig. 8.11 (Chapter 8) for applying high and low logic levels at these inputs.

2. The serial input (pin 1) and Mode control (pin 6) also require high and low inputs; but as there are normally only four logic switches in most logic trainers, low inputs can be applied by connecting these inputs to ground and high inputs can be applied by connecting them to + V_{CC} through a 470 ohm resistor.
3. Pins 8 and 9 are to be connected to Pulser switch *A*.
4. Logic monitors should be connected as indicated in the diagram. Connect pin 7 of the IC to ground and pin 14 to + 5 V.

Clearing Function

5. Proceed as follows to clear the register:
   * Set the Mode control switch to low (for right shift).
   * Set the serial input switch Sw_3 to low.
   * Set parallel inputs *A*, *B*, *C* and *D* to logic 0.
   * When you switch on the power supply the register may come up in any state. To clear the register apply clock pulses with pulser switch till the output is 0 0 0 0.

Serial Input/Parallel Output

After the register has been cleared, any 4-bit serial number can be loaded into the register.

6. Proceed as follows:
   * Set the Mode control switch Sw_4 to low.
   * Set the serial input switch Sw_3 to high.
   * Apply a clock pulse which will shift the serial input 1 into the register: Q_A will be 1.
   * Return serial input switch Sw_3 to low and apply three clock pulses. The register will show an output of 0 0 0 1.

This procedure converts a serial input into a parallel output. You can load any 4-bit number into the register in this way. To clear the register, set the serial input to low and apply four clock pulses.

Parallel Input/Parallel Output

If the shift Register is set up as follows, a parallel input will produce a parallel output. In other words, any parallel 4-bit word can be loaded into the register by the following procedure.

7. Proceed as follows:
   * Set the Mode control to high (this enables the register to recognize parallel inputs and ignore serial inputs).

* Apply the following inputs at A, B, C and D using switches Sw_1 to Sw_4.

A	B	C	D
1	0	1	1

* As you apply a clock pulse the word will be loaded into the register (it is not necessary to clear the register before loading a parallel word).

Parallel Input/Serial Output

Without disturbing the word that you have just loaded into the register, that is 1 0 1 1, you can shift it out serially as follows.

8. Proceed as follows:
   * Set the Mode control to low.
   * Set serial input, pin 1, to low.
   * As you apply clock pulses notice that the word will be shifted out serially from Q_D and after four clock pulses the register will be cleared since the serial input is low.

EXPERIMENT 11.2 : RIGHT-SHIFT/LEFT-SHIFT FUNCTION : IC 7495A

Objective

To demonstrate the following functions of shift register IC 7495A :

1. Right-shift/Left-shift
2. Arithmetic operation.

Materials Required

Logic trainer

IC 7495A : 4-bit shift register

Resistor 470 ohm.

The circuit diagram for Experiment 11.2 is given in Fig. 11.3.

Procedure

1. Mount the IC 7495A on the board and make the connections as shown in Fig. 11.3. Connect the logic monitors to pins marked L_1, L_2, L_3 and L_4 as before. Connect pulser switch A to pin 9 and pulser switch B to pin 8.
2. Connect the parallel outputs to parallel inputs as in the diagram, that is connect Q_D to Q_C, Q_C to Q_B and Q_B to Q_A.
3. Connect switches Sw_1 to Sw_4 as shown in the diagram.
4. Notice that the serial inputs for left-shift will be applied with switch Sw_4 (pin 5) and the serial inputs for right-shift will be entered with switch Sw_2 (pin 1).

5. Connect pin 14 of the IC to + 5 V and pin 7 to ground.

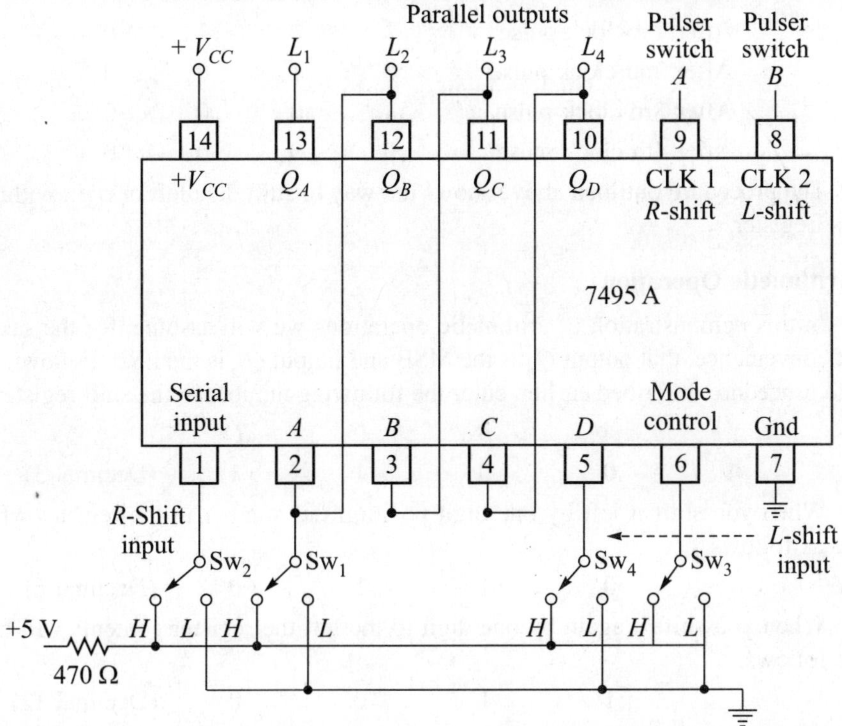

Fig. 11.3. Circuit diagram for Experiment 11.2.

Right-Shift Function

6. Set Mode control switch Sw_3 (pin 6) to low for right-shift. Set the serial input Sw_2 (pin 1) to low and apply clock pulses with pulser switch *A* to clear the register.
7. Enter a 1 at the serial input for right-shift with switch Sw_2 (pin 1) and apply one clock pulse with pulser switch *A*. The register will now read 1 0 0 0.
8. Enter a 0 at the serial input for right-shift with switch Sw_2 (pin 1) and apply three clock pulses. The register will now read 0 0 0 1.

Left-Shift Function

9. Set Mode control switch Sw_3 (pin 6) to high for left-shift and enter a 0 at the serial input for left-shift with switch Sw_4 (pin 5). Apply one clock pulse with pulser switch *B*. The register will now read 0 0 1 0.
10. Enter a 1 at the serial input with Sw_4 (pin 5) and apply one clock pulse with pulser switch *B*. The register will now read 0 1 0 1.

11. Enter a 0 at the serial input with Sw_4 (pin 5) and apply clock pulses with pulser switch *B*. The register will now read as follows:

After 1st clock pulse	1	0	1	0
After 2nd clock pulse	0	1	0	0
After 3rd clock pulse	1	0	0	0
After 4th clock pulse	0	0	0	0

The procedure outlined above shows the way to shift data left or right within the register.

Arithmetic Operation

In this demonstration of arithmetic operations we will assume, for the sake of convenience, that output Q_A is the MSB and output Q_D is the LSB. Following the procedure described earlier, enter the following number in the shift register:

A	*B*	*C*	*D*	
0	0	1	1	(Decimal 3)

When you shift it left by one digit position the number in the register will be as follows:

0 1 1 0 (Decimal 6)

When you shift it again by one digit to the left the register contents will be as follows:

1 1 0 0 (Decimal 12)

Shifting the binary number left by one digit is equivalent to multiplying it by 2 and similarly shifting it right by one digit will be equivalent to dividing the number by 2.

Now let us assume that there is a binary point after the first two digits. The number will now read

1 1 0 0 (Decimal 3)

After shifting it right by one digit position the number will be as follows:

0 1 1 0 0 (Decimal 1.5)

If you shift it right once more the new number will be

0 0 1 1 0 0 (Decimal 0.75)

11.3 RECIRCULATING SHIFT REGISTER

A diagram for this register is given in Fig. 11.4. If a 4-bit data is fed into this register and its output is fed back into the serial input, as shown in the diagram, and shift pulses are applied, the contents of the flip-flops, after each shift pulse, will be as shown in Table 11.1. After exactly four shift pulses the contents of the flip-flops will be the same as at the initial stage as the number of flip-flops is four. If there are '*n*' flip-flops, it will require '*n*' shift pulses for the register to return to the original state.

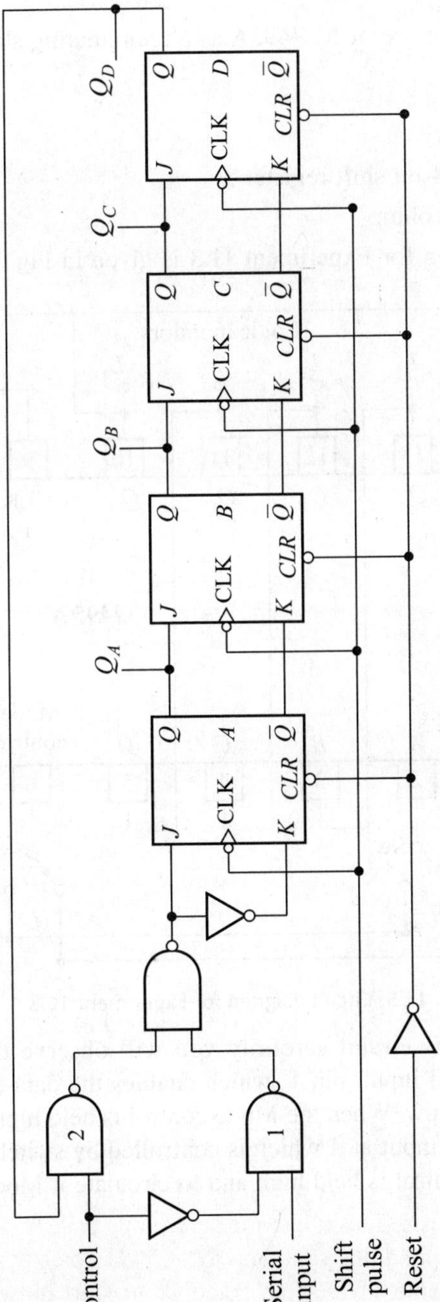

Fig. 11.4. Recirculating shift register.

EXPERIMENT 11.3 : RECIRCULATING SHIFT REGISTER : IC 7495A

Objective

To demonstrate the use of IC 7495A as a recirculating shift register.

Materials Required

Logic trainer

IC 7495A : 4-bit shift register

Resistor 470 ohm.

The circuit diagram for Experiment 11.3 is given in Fig. 11.5.

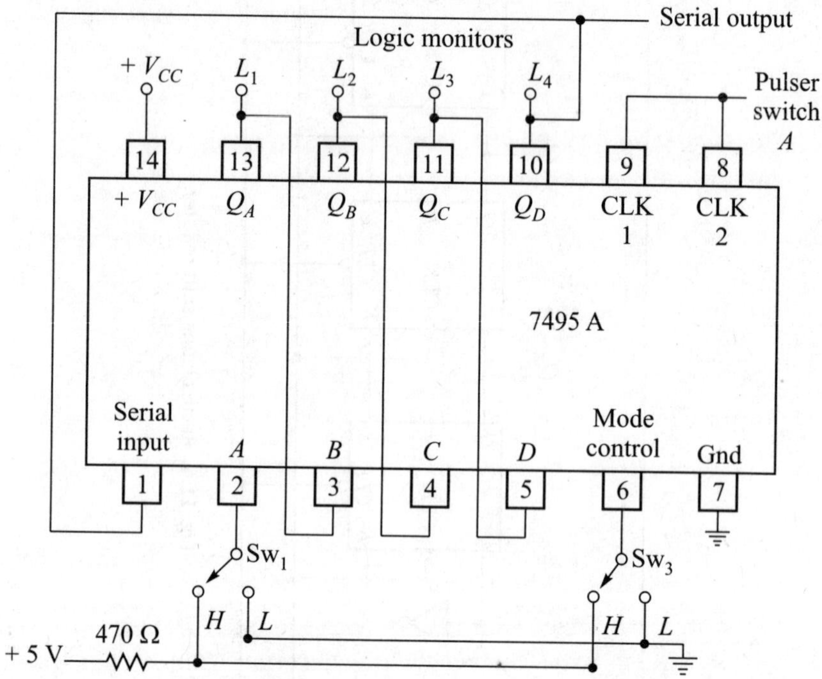

Fig. 11.5. Circuit diagram for Experiment 11.3.

If you examine this circuit carefully you will observe that output Q_D is connected to the serial input, pin 1, which enables the data to circulate if the Mode control is held low. When the Mode control is held high the register will recognize the parallel input at *A* which is controlled by switch Sw_1. Therefore, to feed data Mode control is held high and to circulate it Mode control is held low.

Procedure

1. Mount IC 7495A on the board and make the connections as shown in Fig. 11.5. Carefully observe how the outputs are connected to pins 2 to 5. Connect pin 14 of the IC to + 5 V and pin 7 to ground.

2. To clear the register keep Sw_1 low, Sw_3 high and thereafter apply four clock pulses.
3. Load 0 0 1 1 in the register. To do this keep Sw_1 and Sw_3 high and apply two clock pulses. The register will read 1 1 0 0 at this stage.
4. Now, keeping Sw_1 low and Sw_3 high apply two clock pulses. The register is now loaded with 0 0 1 1.
5. You can now recirculate the contents of the register by keeping Sw_3 low and applying clock pulses.

   You will notice the following changes, as given in Table 11.1, after each clock pulse. You will notice that the fourth clock pulse restores the initial state of the data in the register.
6. Connect a 1 Hz clock signal in place of the pulser switch *A* and observe how the data keeps recirculating in the register.

Table 11.1

Serial Input/Parallel Output Function

Clock pulses	*Outputs*			
	Q_A	Q_B	Q_C	Q_D
Initial state	0	0	1	1
After 1st clock pulse	1	0	0	1
After 2nd clock pulse	1	1	0	0
After 3rd clock pulse	0	1	1	0
After 4th clock pulse	0	0	1	1

CHAPTER 12

Counters

12.1 INTRODUCTION

Counters, like registers, are sequential circuits and, as they employ flip-flops, they possess memory. A binary counter can count the number of pulses applied at its input. When clock pulses are applied at the input of a counter, the flip-flops in the counter undergo a change of state in such a manner that the binary number stored in the flip-flops can represent the number of clock pulses applied at the counter input. However, there are some variations in the method of application of clock pulses to flip-flop inputs.

Briefly speaking there are two types of counters; asynchronous and synchronous types. In asynchronous counters all the flip-flops do not change state at the same time. In these counters each stage in the counter is triggered by the clock pulse which it receives from the output of the previous stage. These counters are also known as *ripple counters,* as the trigger pulse ripples through the flip-flops. Since, in this process, each flip-flop contributes a delay the total delay is additive.

A better speed of operation can be achieved by so connecting the flip-flops that they are all triggered by the same clock input. These counters are called *synchronous* or *parallel counters.* Sometimes counters are required not only to count up but also count down.

In this chapter, we will consider up and down counters as well as some special counters such as Ring and Johnson counters and also Modulus counters.

12.2 BINARY RIPPLE UP-COUNTER

We will now consider a 4-bit binary ripple up-counter which belongs to the class of asynchronous counters. This counter is capable of 2^4 or 16 states. A circuit diagram for this counter is given in Fig. 12.1. You will notice from the diagram that the normal output, Q, of each flip-flop is connected to the clock input of the next flip-flop. Also notice that the J and K inputs of all the flip-flops are held high to enable them to toggle (change state) at every transition of the input pulse from $1 \rightarrow 0$. All the flip-flops are connected in this manner.

After this counter has counted up to 1 1 1 1, the next trigger input will recycle it back to 0 0 0 0 and it will start counting again in the same sequence.

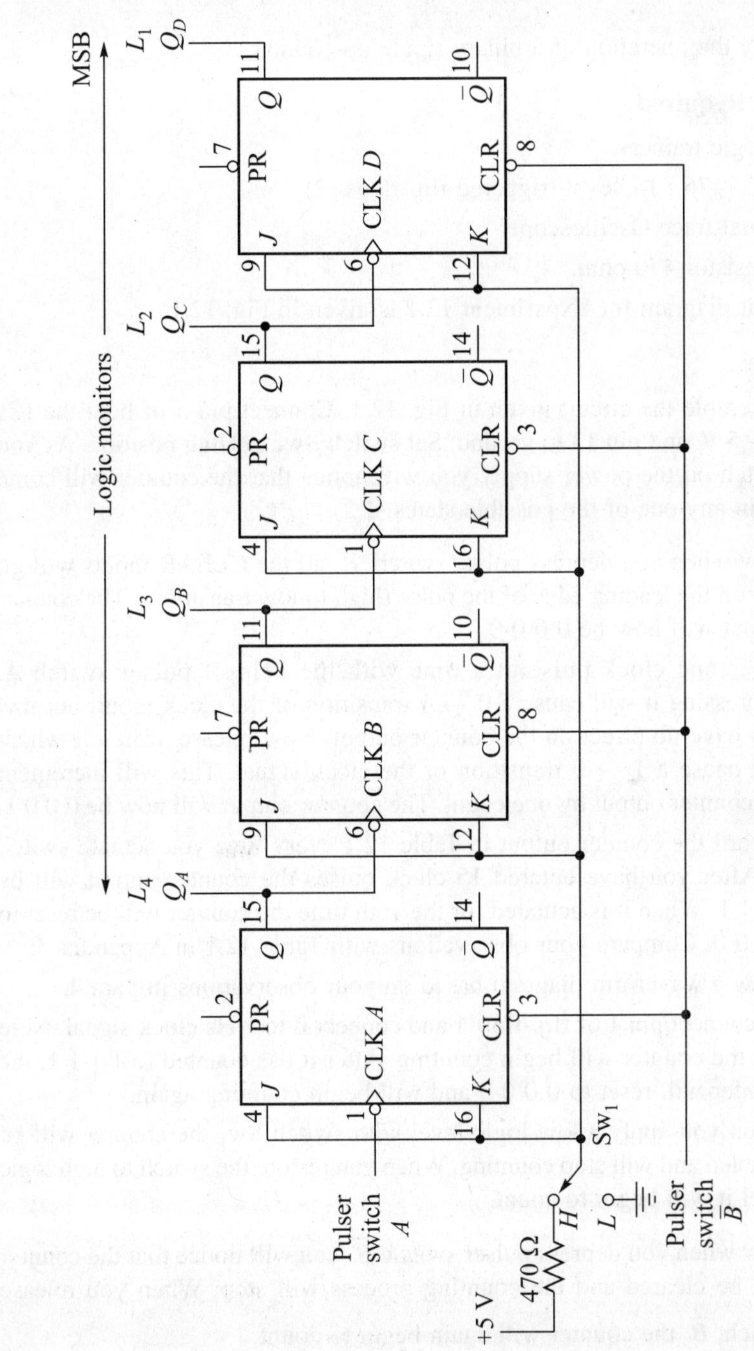

Fig. 12.1. Circuit diagram for Experiment 12.1.
(Binary up-counter)

EXPERIMENT 12.1 : BINARY UP-COUNTER

Objective

To study the operation of a binary ripple up-counter.

Materials Required

Logic trainer

IC 7476 : JK level-triggered flip-flops (2)

Dual-trace Oscilloscope

Resistor 470 ohm.

A circuit diagram for Experiment 12.1 is given in Fig. 12.1.

Procedure

1. Assemble the circuit given in Fig. 12.1. Connect pin 5 of both the ICs to + 5 V, and pin 13 to ground. Set switch Sw_1 to High position. As you switch on the power supply you will notice that the counter will come up in any one of the possible states.
2. Now when you depress pulser switch $\overline{B}$ all the CLEAR inputs will go low on the leading edge of the pulse (high to low transition). The counter output will now be 0 0 0 0.
3. Enter one clock pulse at a time with the help of pulser switch *A*. Depressing it will cause a 0 → 1 transition of the clock input; but this will have no effect on the counter output. Now release switch *A* which will cause a 1 → 0 transition of the clock signal. This will increment the counter output by one count. The counter output will now be 0 0 0 1.
4. Record the counter output in Table 12.1 every time you actuate switch *A*. After you have entered 15 clock pulses the counter output will be 1 1 1 1. When it is actuated for the 16th time the counter will be reset to 0 0 0 0. Compare your observations with Table 12.1 in Appendix 3.
5. Draw a waveform diagram based on your observations in para 4.
6. Disconnect pin 1 of flip-flop *A* and connect it to 1 Hz clock signal. Note that the counter will begin counting. After it has counted to 1 1 1 1, the counter will reset to 0 0 0 0 and will begin counting again.
7. When you apply a low logic level with switch Sw_1 the counter will be disabled and will stop counting. When you restore the switch to high logic level it will begin to count.
8. Now when you depress pulser switch $\overline{B}$ you will notice that the counter will be cleared and the counting process will stop. When you release switch $\overline{B}$ the counter will again begin to count.

9. Change the frequency of the clock signal to 1 kHz and connect one of the inputs of a dual-trace Oscilloscope to the clock signal and the other input to the counter output Q_A. You will notice that the frequency of the output at Q_A is half the clock frequency.
10. Repeat step 9 with outputs at Q_B, Q_C, Q_D connected in place of Q_A and comment on the relationship between the input and output frequencies.

Table 12.1

Input	*Outputs*			
Pulse count *n*	Q_D L_1	Q_C L_2	Q_B L_3	Q_A L_4
0 (Clear)				
1				
2				
3				
4				
5				
6				
7				
8				
9				
10				
11				
12				
13				
14				
15				
16 (0)				

12.3 BINARY RIPPLE DOWN-COUNTER

The binary up-counter increases the count by one each time a pulse occurs at its input, whereas the binary down-counter decreases the count by one each time a pulse occurs at its input. A diagram for a down-counter is given in Fig. 12.2, which is slightly different from Fig. 12.1 (binary up-counter). The only difference you will notice is that in the down-counter the complement output $\overline{Q}$, instead of the normal output, Q, is connected to the clock input of the next flip-flop. The counter output which is relevant even in the down-counter is the normal output of the flip-flops.

EXPERIMENT 12.2 : BINARY DOWN-COUNTER

Objective

To study the operation of a binary ripple down-counter.

Materials Required

Logic trainer

IC 7476 : JK level-triggered flip-flops (2)

Resistor 470 ohm

A circuit diagram for Experiment 12.2 is given in Fig. 12.2.

Procedure

1. Assemble the circuit given in Fig. 12.2. Connect pin 5 of both the ICs to + 5 V, and pin 13 to ground.
2. Reset all the flip-flops with pulser switch $\overline{B}$. The counter output will now be 0 0 0 0.
3. When you apply the first clock pulse to flip-flop *A* with pulser switch *A*, flip-flop *A* will set and its complement output will become 0. This will set flip-flop *B* and thus all the flip-flops will also set. The counter output will now be 1 1 1 1.
4. Now when you apply the second clock pulse to flip-flop *A*, it will reset and its complement output will become 1 which will not affect the other flip-flops. The counter output will now be 1 1 1 0.
5. After the 16th input pulse, the counter will be reset and its output will be 0 0 0 0.
6. You begin the experiment by resetting the counter with pulser switch $\overline{B}$. Enter one clock pulse at a time and record your observations in Table 12.2. Your observations should tally with Table 12.2 in Appendix 3.
7. Disconnect pin 1 of flip-flop *A* from pulser switch *A* and connect it to 1 Hz clock signal. The counter will begin counting. Notice that the counter output will decrease by one count at a time until the counter shows a count of 0 0 0 0 after the 16th input pulse. After the 17th clock signal the counter will display a count of 1 1 1 1. The counter will continue to count as before.

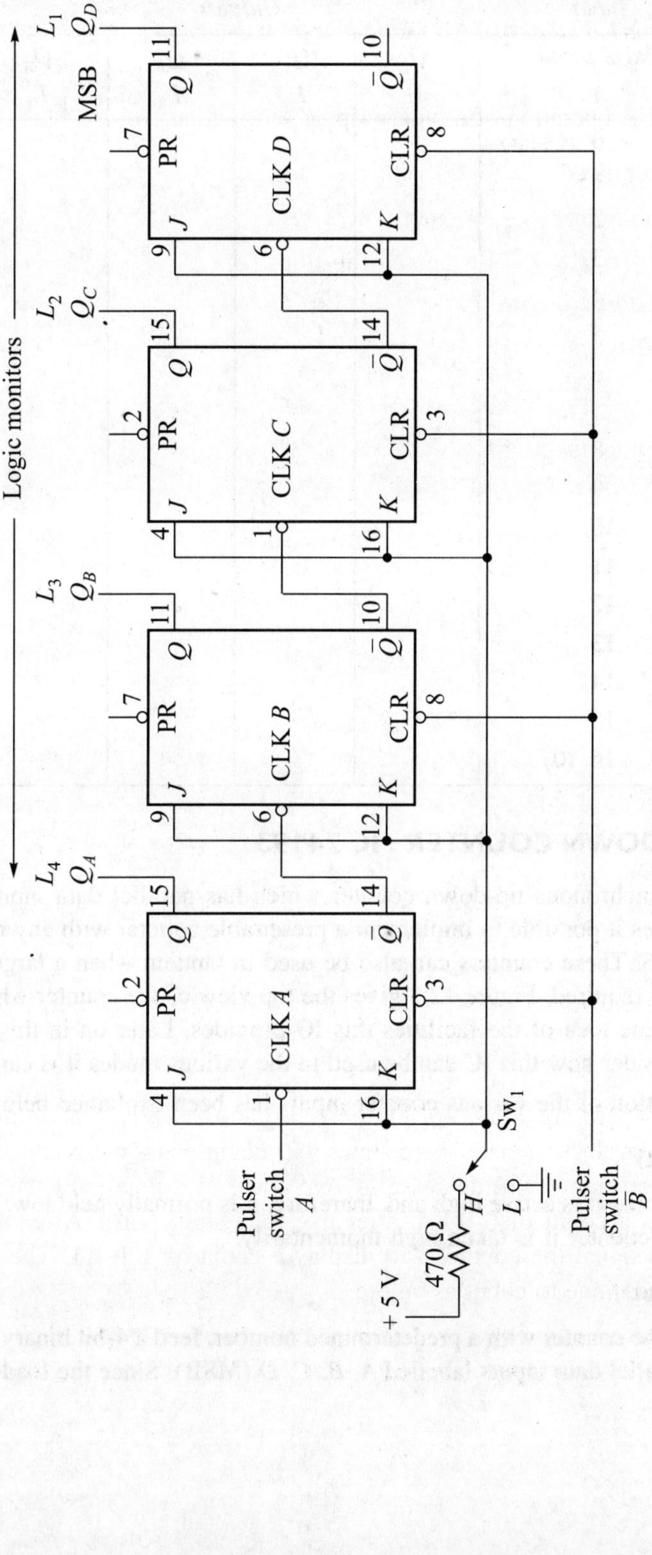

Fig. 12.2. Circuit diagram for Experiment 12.2 (Binary down-counter).

Table 12.2

Input	Outputs			
Pulse count *n*	Q_D L_1	Q_C L_2	Q_B L_3	Q_A L_4
0 (Clear)				
1				
2				
3				
4				
5				
6				
7				
8				
9				
10				
11				
12				
13				
14				
15				
16 (0)				

12.4 UP-DOWN COUNTER : IC 74193

This is a synchronous up-down counter which has parallel data inputs. This feature makes it possible to implement a presettable counter with any modulus from 2 to 16. These counters can also be used in tandem when a larger count capability is required. Figure 12.3 gives the top view of the counter which will give you some idea of the facilities this IC provides. Later on in this section we will consider how this IC can be used in the various modes it is capable of.

The function of the various counter inputs has been explained below:

Clear(Reset)

The clear input is active high and, therefore, it is normally held low. To clear or reset the counter it is taken high momentarily.

Load (Preset)

To load the counter with a predetermined number, feed a 4-bit binary number into the parallel data inputs labelled *A, B, C, D* (MSB). Since the load input is active low, take the LOAD input low momentarily to shift the parallel number into the counter. Both the CLEAR and LOAD inputs are asynchronous and they will therefore override all synchronous counting functions.

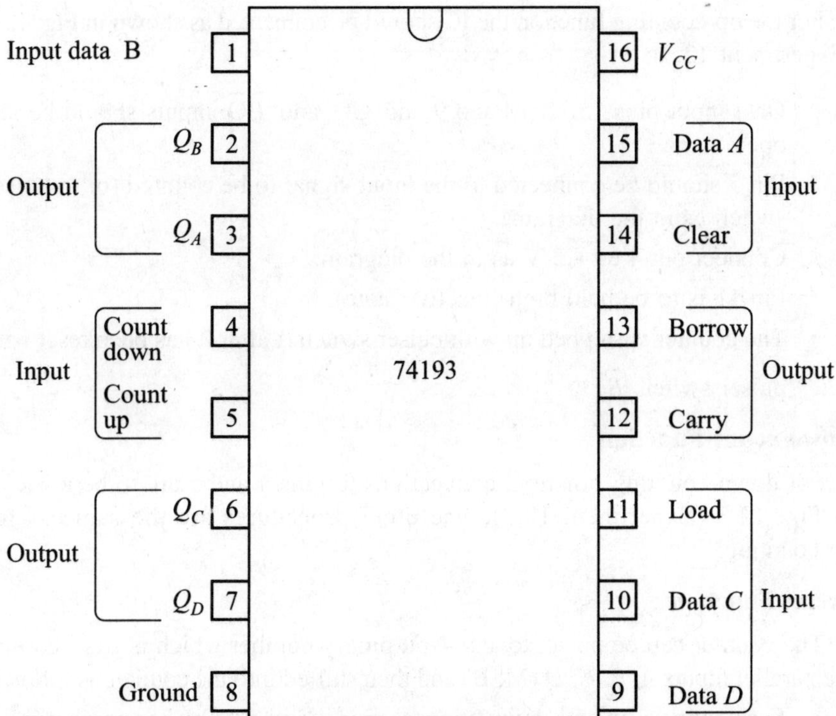

Fig. 12.3. Pin connections for IC 74193.
(Up-down counter)

Carry-Out ($\overline{CO}$) and Borrow-Out ($\overline{BO}$) Inputs

These inputs are used to drive the following IC 74193 when a larger count capability is required. While cascading these counters, the $\overline{CO}$, pin 12 and $\overline{BO}$, pin 13 outputs of a previous counter are connected to the UP, pin 5, and DOWN, pin 4, inputs respectively of the next counter in the chain.

In the up-counting mode the voltage at $\overline{CO}$, pin 12, remains high until the maximum count that is 1 1 1 1 is reached when its voltage drops very low. At the next input pulse when transition from 0 → 1 occurs, the counter output becomes 0 0 0 0 and the following counter is incremented from 0 0 0 0 to 0 0 0 1 and at the same time the voltage at pin 12 also goes up.

In the down-counting operation the voltage at $\overline{BO}$, pin 13, stays high until the minimum count 0 0 0 0 is reached when the voltage at pin 13 drops very low. At the next input pulse, when transition from 0 →1 occurs, the counter output becomes 1 1 1 1 and the following counter is decremented. At the same time the voltage at pin 13 goes up.

Up-count Function

For the up-counting function the IC should be connected as shown in Fig. 12.4 (Experiment 12.3):

* Data input pins 15, 1, 10 and 9 and $\overline{CO}$ and $\overline{BO}$ inputs should be left open.
* Pin 5 should be connected to the input signal to be counted (or to pulser switch as in the diagram).
* Connect pin 4 to + 5 V as in the diagram.
* Pin 11 is to be held high (inactive state).
* The counter is stepped up with pulser switch *A* after it has been reset with pulser switch $\overline{B}$.

Down-count Function

For down-counting function, connections to pins 4 and 5 are to be made as in Fig. 12.5 (Experiment 12.3). The other procedures are the same as for up-counting.

Preset Function

The counter can be preset to any 4-bit binary number which is first fed into the parallel inputs *A, B, C, D* (MSB) and then shifted into the counter as follows:

* For counting up and down connect pins 4 and 5 as mentioned before.
* Connect the CLEAR input, pin 14 to ground.
* Feed the desired binary number into the *A, B, C, D* inputs and take the $\overline{\text{LOAD}}$, pin 11, low temporarily to shift the number into the counter.
* The counter can now count up or down from that binary number. It is not necessary to reset the counter before presetting it.

The counter can also be preset by connecting the inputs of a 4-input NAND gate to the counter output $(Q_A\ Q_B\ Q_C\ Q_D)$ and the NAND gate output to $\overline{\text{LOAD}}$ input, pin 11. If you feed a number, say 1 0 1 0 and apply a clock signal to counter input, pin 5, the counter will begin to count and, when it reaches the maximum count, (decimal 15) that is 1 1 1 1, the NAND gate output will go low and number 1 0 1 0 will be shifted into the counter, which will again start counting from the preset number and the same sequence will be repeated. As state 1 1 1 1 is used to preset the counter it is no longer a stable state. The stable states are 10, 11, 12, 13 and 14 and the modulus (the number of discrete states) of the counter will be 5. The counter modulus in the up-counting mode for any preset number '*n*' is given by

$$\text{Modulus} = 16 - n - 1$$

In this case, $\text{Modulus} = 16 - 10 - 1 = 5.$

In the down-counting mode the counter will count down from the preset number. If the preset number is 1 0 1 0 (decimal 10), as before, the counter will count down as follows: 10, 9, 8, 7, 6, 5, 4, 3, 2, 1, 0. In this case, the counter modulus will be as follows:

$$\text{Modulus} = n + 1$$

In this case, $\text{Modulus} = 10 + 1 = 11$

When the counter is preset in this manner, as number 15 is not a stable state, it should not be preset to this number as the counter will get latched up in state 15.

EXPERIMENT 12.3 : UP-DOWN COUNTER : IC 74193

Objective

To study the following functions of IC 74193:

(1) Up-counter

(2) Down-counter.

Materials Required

Logic trainer
IC 74193 : Binary Up-down counter
Volt-ohm-milliammeter
Resistor 1 k ohm
Resistor 470 ohm.

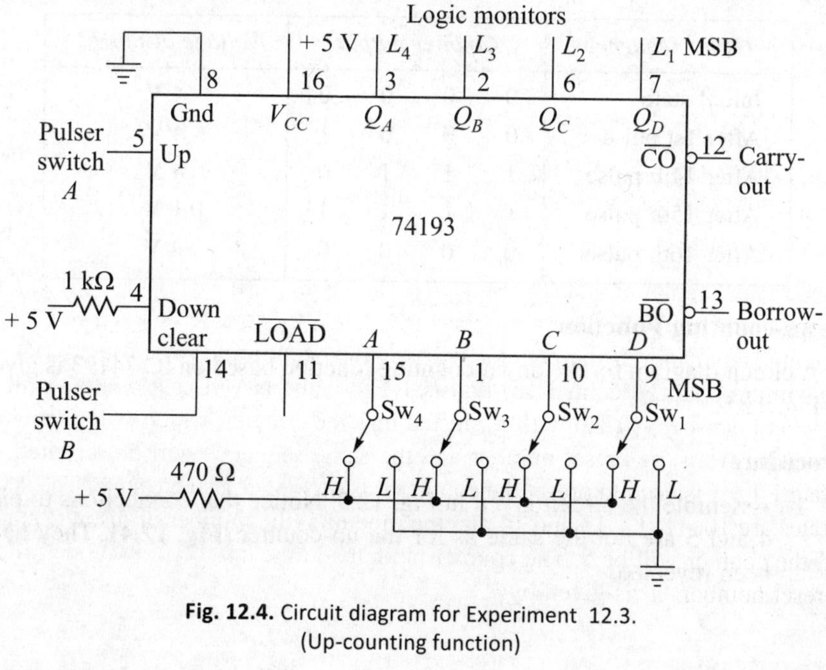

Fig. 12.4. Circuit diagram for Experiment 12.3. (Up-counting function)

Procedure

Up-counting Function

1. Assemble the circuit given in Fig. 12.4. Connect the logic switches and logic monitors as shown in the diagram. Connect pulser switch *A* to pin 5 and pulser switch *B* to pin 14. Connect the resistors as indicated. Connect the free ends of the two resistors to + 5 V.
2. Connect a voltmeter to pin 12.
3. Connect pin 16 of the IC to + 5 V and pin 8 to ground.
4. Switch on the power supply. The counter may come up in one of the possible states. Reset the counter to 0 0 0 0 with pulser switch *B*.
5. Record the voltage at pin 12 which should be about 4 V.
6. Apply input pulses to the counter with pulser switch *A*. Note that every time the pulser switch is depressed the counter will be stepped up by one count. Also record the voltage at pin 12 at every count.
7. Notice that when the counter shows a count of 1 1 1 1, that is the maximum count, the voltage at pin 12 will drop to 0.1 V.
8. At the next input pulse the counter output will drop to 0 0 0 0 and the voltage at pin 12 will go high indicating a transition of the carry-out signal from 0 →1, which can step up the next counter if there is one in position.
9. Table 12.3 will help you to follow the sequence of events.

Table 12.3

Count sequence	*Counter output*				*Voltage at pin 12*
Initial state	0	0	0	0	4 V
After 1st pulse	0	0	0	1	4 V
After 14th pulse	1	1	1	0	4 V
After 15th pulse	1	1	1	1	0.1 V
After 16th pulse	0	0	0	0	4 V

Down-counting Function

A circuit diagram for the down-counting function based on IC 74193 is given in Fig. 12.5.

Procedure

1. Assemble the circuit given in Fig. 12.5. Notice that connections to pins 4 and 5 are not the same as for the up-counter (Fig. 12.4). They have been reversed.

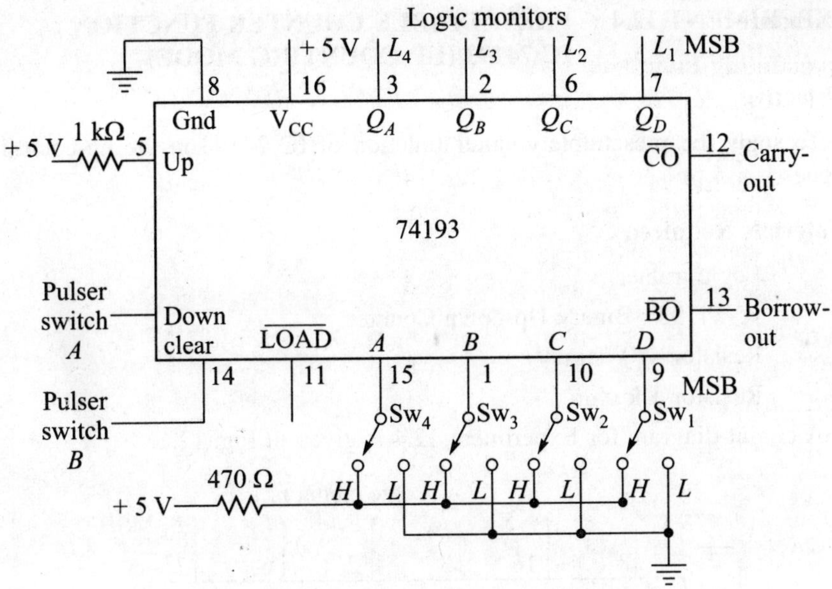

Fig. 12.5. Circuit diagram for Experiment 12.3.
(Down-counting function)

2. Connect a voltmeter between pin 13 and ground and also connect pin 8 of the IC to ground. Connect pin 16 of the IC and the free ends of 1 k ohm and 470 ohm resistors to + 5 V.
3. Switch on the power supply and reset the counter to 0 0 0 0 with pulser switch *B*. Apply a pulse input with pulser switch *A* at the DOWN input, pin 4, and notice that the counter output is now 1 1 1 1. Also note the voltage at pin 13, which should be high. Now apply pulses at the DOWN input, pin 4, with pulser switch *A*.
4. After 15 pulses the counter will register 0 0 0 0 and at the same time the voltage at pin 13 will drop to about 0.1 V. At the next pulse input the counter output will jump to 1 1 1 1 and, if another similar counter is connected in cascade, it will be decremented by one count at the same time during transition from 0 → 1 at pin 13. Table 12.4 gives the sequence of events.

Table 12.4

Count sequence	*Counter output*				*Voltage at pin 13*
Initial state	1	1	1	1	4 V
After 1st pulse	1	1	1	0	4 V
After 2nd pulse	1	1	0	1	4 V
After 15th pulse	0	0	0	0	0.1 V
After 16th pulse	1	1	1	1	4 V

EXPERIMENT 12.4 : PRESETTABLE COUNTER FUNCTION : IC 74193 (UP-COUNTING MODE)

Objective

To study the presettable counter function of IC 74193 in the up-counting mode.

Materials Required

Logic trainer
IC 74193 : Binary Up-down Counter
Resistor 470 ohm
Resistor 1 k ohm.

A circuit diagram for Experiment 12.4 is given in Fig. 12.6.

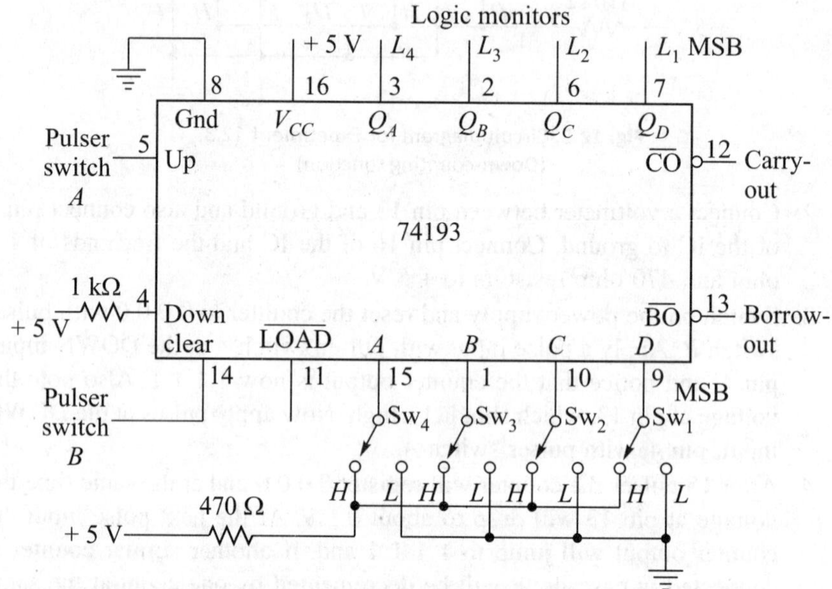

Fig. 12.6. Circuit diagram for Experiment 12.4. (Up-counting mode)

Procedure

1. Assemble the circuit given in Fig. 12.6. Make the ground connections shown in the diagram and connect pin 16 of the IC and the free ends of 470 ohm and 1 k ohm resistors to + 5 V. Connect the pulser switches *A* and *B* as in the diagram. Make the rest of the connections.
2. Set the logic switches to represent the following binary numbers. Each time depress the pulser switch *B* and note the counter output in Table 12.5. Notice that the number is shifted into the counter when pin 11 goes low.

Tabel 12.5

Input				Output			
Logic switches				*Logic monitors*			
Sw_1	Sw_2	Sw_3	Sw_4	L_1	L_2	L_3	L_4
1	0	0	1				
1	1	1	0				
1	1	1	1				
1	0	1	0				

You will notice that the counter output is the same as the input to the logic switches. The last binary number is 1 0 1 0 (decimal 10).

3. Now apply pulse inputs using pulser switch *A*. The counter output will be stepped up as follows after each pulse.

Table 12.6

Count sequence	*Binary number*				*Decimal number*
Initial count	1	0	1	0	10
After 1st pulse	1	0	1	1	11
After 2nd pulse	1	1	0	0	12
After 3rd pulse	1	1	0	1	13
After 4th pulse	1	1	1	0	14
After 5th pulse	1	1	1	1	15
After 6th pulse	0	0	0	0	0

Notice that counting begins from the preset number and the 6th pulse resets the counter and it starts counting again as before.

EXPERIMENT 12.5 : PRESETTABLE COUNTER FUNCTION : IC 74193 (DOWN-COUNTING MODE)

Objective

To study the presettable function of IC 74193 in the down-counting mode.

Materials Required

Logic trainer

IC 74193 : Binary Up-down Counter

Resistor 470 ohm

Resistor 1 k ohm

A circuit diagram for Experiment 12.5 is given in Fig. 12.7.

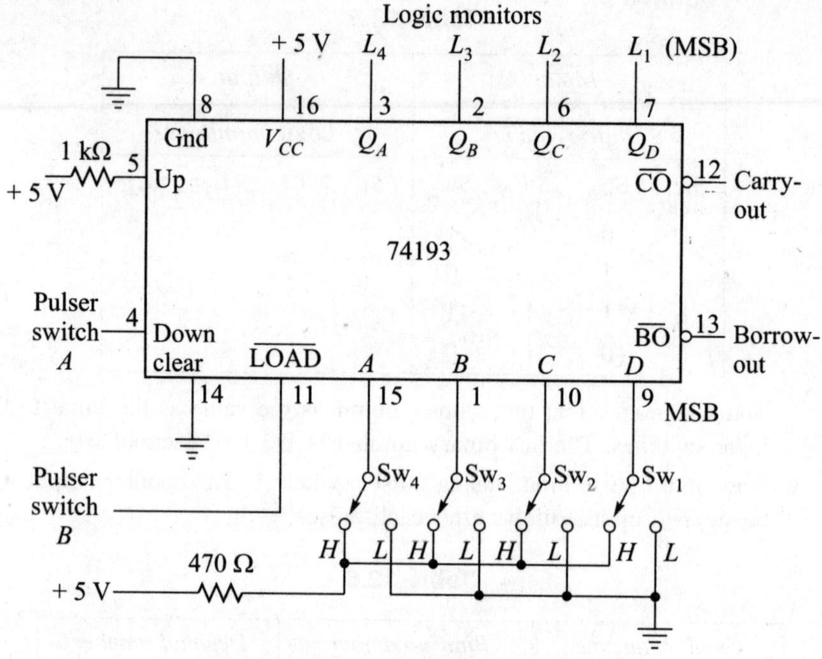

Fig. 12.7. Circuit diagram for Experiment 12.5. (Down-counting mode)

Procedure

1. Assemble the circuit given in Fig. 12.7. Connect pin 16 of the IC and the free ends of 470 ohm and 1 k ohm resistors to + 5 V. Make the ground connections as in the diagram.
2. Switch on the power supply and follow the procedure for down-counting mentioned earlier.
3. Preset the counter to 1 0 1 0 and begin the count down. You will notice that the 10th input pulse will reset the counter to 0 0 0 0 and the 11th pulse will result in a count of 1 1 1 1. Thereafter the counter will start counting down again from this number.

12.5 MODULO *N* COUNTER USING IC 74193

In Section 12.4, you have seen how IC 74193 can be preset to count UP or DOWN. In other words, you have learnt to use it as a modulo counter. There is a simpler way of implementing a modulo counter with this IC. We will consider this in Experiment 12.6.

EXPERIMENT 12.6 : MODULO COUNTER USING IC 74193

Objective

To demonstrate the implementation of a modulo counter with IC 74193.

Materials Required

Logic trainer
IC 74193 : Binary Up-down Counter
Resistor 470 ohm
Resistor 1 k ohm.

A circuit diagram for Experiment 12.6 is given in Fig. 12.8.

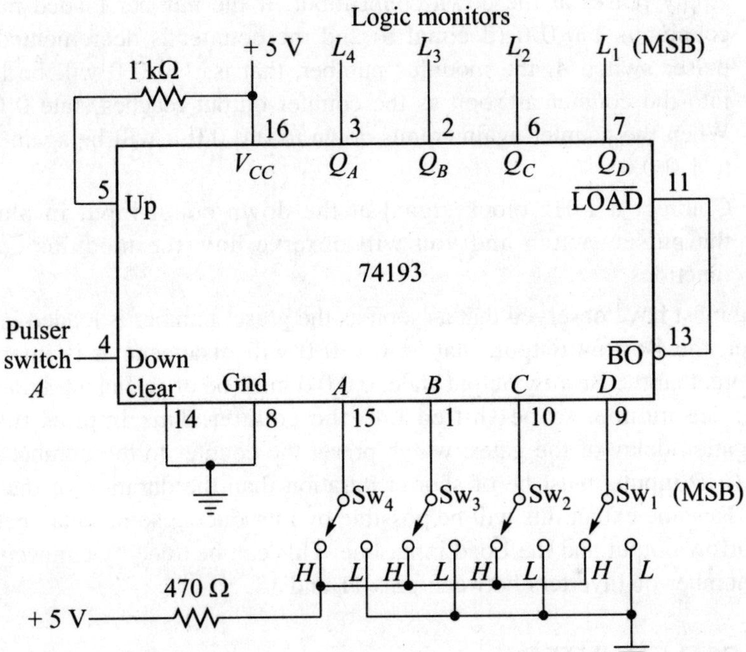

Fig. 12.8. Modulus Counter : IC 74193.

From Fig. 12.8, you will notice that the position of connections shown in the diagram for Experiment 12.6 is not the same as in other previous diagrams using this IC. It is important that you should take this into account while making the connections. Particularly note the change in the position of pin nos. 8, 11 and 12. No connection has been made to pin 12.

Procedure

1. Assemble the circuit given in Fig. 12.8. Particularly note that CLEAR input, pin 14 has been grounded and pin 11 has been connected to pin 13. No connection has been made to pin 12. Pin 5 has been connected to V_{CC} through a 1 k ohm resistor.
2. Connect pin 16 and one end of 1 k ohm and 470 ohm resistors to + 5 V. Make the ground connections and switch on the power supply.

3. The Borrow output, pin 13, detects the state of the Borrow output line when it shows 0 0 0 0. Since it is connected back to the Load input line, the binary number loaded into the *A, B, C* and *D* inputs is shifted into the counter. The decimal number loaded into the counter represents the modulus of the counter.
4. Switch on the power supply and load a number representing the required modulus of the counter into the *A, B, C* and *D* inputs.
5. Apply pulses at the down count input. If the number loaded into the counter is 1 0 0 0 (decimal 8) and the counter is decremented with pulser switch *A*, the modulus number, that is, 1 0 0 0 will be loaded into the counter as soon as the counter output reaches state 0 0 0 0. When the counter again counts down to 0 0 0 0 it will be again preset to 1 0 0 0.
6. Connect a 1 Hz clock signal at the down count input in place of the pulser switch and you will observe how the modulus counter functions.

You must have observed that as soon as the preset number is loaded into the counter, the Borrow output, that is, 0 0 0 0 will disappear. It is important, therefore, that the Borrow output state, 0 0 0 0 must be of sufficient duration to enable the number to be shifted into the counter. This implies that the propagation delay of the gates, which preset the counter to the number at the *A, B, C, D* inputs, must be of shorter duration than the duration of the clock pulse. To some extent this will be possible by introducing some delay between the Borrow output and the Load input line. This can be done by connecting an even number of Inverters between pins 11 and 13.

12.6 BCD COUNTERS

BCD counters, also known as decade counters have ten discrete states which correspond to the decimal numbers 0 through 9 and so they count in tens. These counters which are in common use count in the standard 8421 pure binary code and represent the ten 4-bit codes of this system. Table 12.7 gives the count sequence for a decade counter. This table represents the output of a counter with four flip-flops. You will notice that each input pulse increases the count by 1. On the application of the 10th pulse the counter is reset to 0 0 0 0 and the counting cycle is repeated. Since BCD counters have 10 discrete states, they can divide the input frequency by 10. Notice that Q_D output shows only one output for every 10 input pulses. These counters can be cascaded to increase count capability.

Table 12.7
Counting Sequence of a BCD Counter

Count	*Outputs*			
	Q_D	Q_C	Q_B	Q_A
0	0	0	0	0
1	0	0	0	1
2	0	0	1	0
3	0	0	1	1
4	0	1	0	0
5	0	1	0	1
6	0	1	1	0
7	0	1	1	1
8	1	0	0	0
9	1	0	0	1

EXPERIMENT 12.7 : BCD COUNTER

Objective

To study the performance of IC 7490A as a BCD counter in 2 × 5 configuration.

Materials Required

Logic trainer
Volt-ohm-milliammeter
IC 7490A : Decade counter
Resistor 470 ohm.

A circuit diagram for Experiment 12.7 is given in Fig. 12.9.

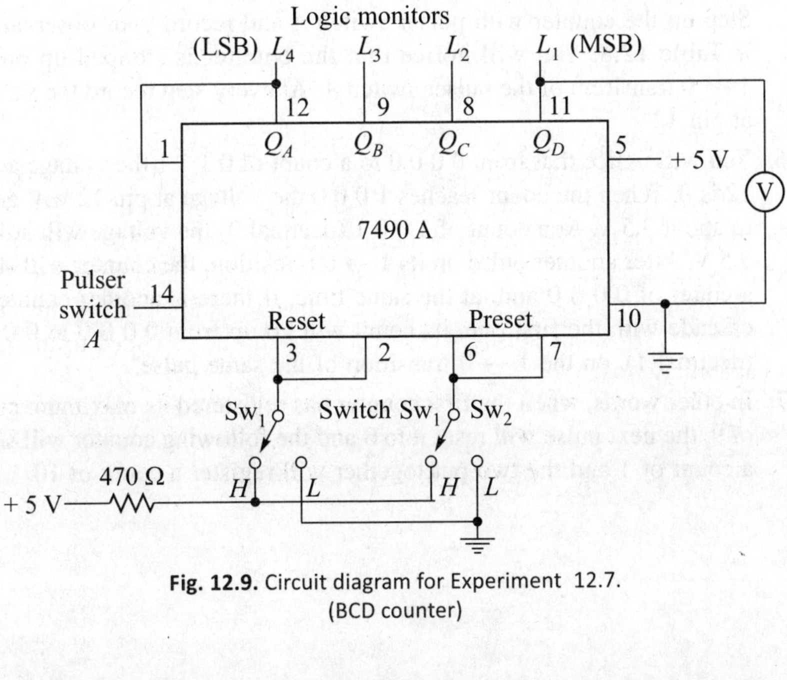

Fig. 12.9. Circuit diagram for Experiment 12.7. (BCD counter)

IC 7490A is an asynchronous counter and it uses four flip-flops. Flip-flop *A* functions as a mod-2 counter and *B, C* and *D* flip-flops form a mod-5 counter. They are not internally connected. They can be connected externally in the 2 × 5 and 5 × 2 configurations. In both cases, they function as decade counters. However only the 5 × 2 configuration produces a symmetrical output.

The counter can be reset with a high input at pin 2 or 3, which makes it possible to reset it from any one of two sources. Normally both are tied together. The counter can be preset to 1 0 0 1 with a high input at pin 6 or 7. Normally they are tied together. This counter can also be used to divide the input frequency by 5, 6 or 7. We will first consider its operation as a decade counter in the 2 × 5 configuration in Experiment 12.7.

Procedure

1. Assemble the circuit given in Fig. 12.9. Notice that the counter is connected in the 2 × 5 configuration as a decade counter. Connect pin 5 and the free end of 470 ohm resistor to + 5 V and make the ground connections marked in the diagram.
2. Set switches Sw_1 and Sw_2 to logic 0 and switch on the power supply. The counter will now come up in one of the states. Apply logic 1 to Sw_1 momentarily and then return it to logic 0. The counter will now be reset to 0 0 0 0.
3. To preset the counter to 1 0 0 1, take switch Sw_2 momentarily to logic 1 and then return it to logic 0. The counter will now be preset to 1 0 0 1 (decimal 9).
4. You will now demonstrate the function of this IC as BCD counter. To do this first put switches Sw_1 and Sw_2 in the logic 0 position to enable the counter to function.
5. Step up the counter with pulser switch *A* and record your observations in Table 12.8. You will notice that the counter is stepped up on the 1 → 0 transition of the pulser switch *A*. At every step record the voltage at pin 12.
6. You will notice that from 0 0 0 0 to a count of 0 1 1 1 the voltage at pin 12 is 0. When the count reaches 1 0 0 0 the voltage at pin 12 will go up to about 3.5 V. At a count of 1 0 0 1 (decimal 9) the voltage will still be 3.5 V. After another pulse on its 1 → 0 transition, the counter will show a count of 0 0 0 0 and, at the same time, if there is another counter in cascade with the first one, its count will go up from 0 0 0 0 to 0 0 0 1 (decimal 1), on the 1 → 0 transition of the same pulse.
7. In other words, when the first counter has registered its maximum count of 9, the next pulse will reset it to 0 and the following counter will show a count of 1 and the two put together will register a count of 10.

Table 12.8

Count	*Outputs*				*Voltage at pin* 12
	Q_D	Q_C	Q_B	Q_A	
0					
1					
2					
3					
4					
5					
6					
7					
8					
9					

8. Your observations should tally with Table 12.7.

12.7 DIVIDE-BY-TEN COUNTER

The BCD counter in Section 12.6 was connected in the 2 × 5 configuration and it counts in the pure binary sequence given in Table 12.7. If you look at the *D* output you will observe that the output is not symmetrical. The output will be symmetrical if it is connected in the 5 × 2 configuration known as the bi-quinary sequence, which is different from the pure binary sequence. In this configuration, output is taken from the Q_A output which is the output of the last flip-flop in the chain. Connections for the bi-quinary sequence have to be made as in Fig. 12.10. This counter will now function as a divide-by-ten scalar having a symmetrical output. Experiment 12.8 describes its operation as a decimal scalar.

EXPERIMENT 12.8 : DIVIDE-BY-TEN COUNTER

Objective

To study the operation of IC 7490A in the 5 × 2 configuration as a divide-by-ten counter.

Materials Required

Logic trainer
IC 7490A : Decade counter
Frequency counter
Resistor 470 ohm
Volt-ohm-milliammcter.

A circuit diagram for Experiment 12.8 is given in Fig. 12.10.

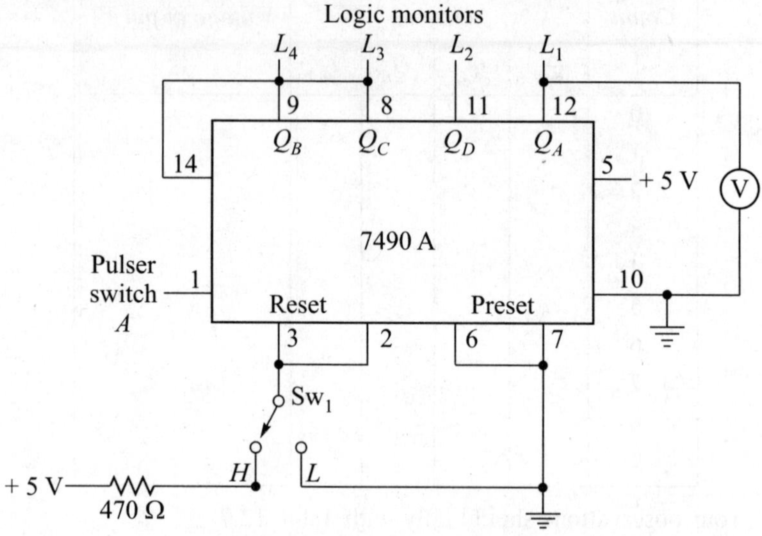

Fig. 12.10. Circuit diagram for Experiment 12.8. (Divide-by-ten counter)

Procedure

1. Assemble the circuit given in Fig. 12.10. Connect a voltmeter as shown in the diagram. The connections are not the same as in Fig. 12.9. Notice that the output is taken from the Q_A output and not from the Q_D output as in Fig. 12.9. Connect pin 5 of the IC and the free end of 470 ohm resistor to + 5 V and pin 10 to ground.
2. Switch on the power supply and reset the counter by momentarily applying a logic 1 input with switch Sw_1. The counter will now be reset to 0 0 0 0.
3. Apply a pulse at input 1 with pulser switch *A*. The counter will now show a count of 0 0 0 1 after the first pulse.
4. After the 5th pulse the counter output will be 1 0 0 0 (not to be confused with decimal 8). You will notice that at the same time the voltage at pin 12 will go up to about 3.5 V.
5. After the 9th pulse the counter output will be 1 1 0 0 and the voltmeter will still show 3.5 V at pin 12. Record your observations in Table 12.9.

Table 12.9

Count	*Outputs*				*Voltage at pin* 12
	Q_A	Q_D	Q_C	Q_B	
0					
1					
2					
3					
4					
5					
6					
7					
8					
9					

6. When you apply the 10th pulse, the counter output will become 0 0 0 0 on the trailing edge ($1 \rightarrow 0$) transition of this pulse, which will also increment the next counter to 0 0 0 1, if another similar counter is connected in cascade.
7. You will notice that for every 10 input pulses at the counter input, there is only 1 output pulse, which shows that the frequency of the train of input pulses has been scaled down by a factor of 10. In other words, if the input frequency is f, the output frequency will be $f/10$.
8. If the input of the counter is connected to a clock and you monitor the output on an oscilloscope you will notice that the output pulse is symmetrical in shape.
9. Your observations in Table 12.9 should tally with Table 12.9 in Appendix 3.
10. Apply a 1 Hz clock signal at the counter input and measure the clock frequency at the output frequency at pin 12 with a frequency counter. You will notice that the output frequency is about 100 Hz, that is one-tenth of the input frequency.

12.8 RING COUNTER

Ring counters provide a sequence of equally spaced timing pulses and, therefore, they find considerable application in logic circuits, which require such pulses for setting in motion a number of operations in a predetermined sequence and at very precise time intervals. Ring counters can be regarded as a variation of shift registers. In our next experiment we will consider IC 7495A which has been connected to perform as a ring counter.

EXPERIMENT 12.9 : RING COUNTER

Objective

To demonstrate the operation of a ring counter using IC 7495A.

Materials Required

Logic trainer

IC 7495A : Shift register

Resistor 470 ohm

A circuit diagram for Experiment 12.9 is given in Fig. 12.11.

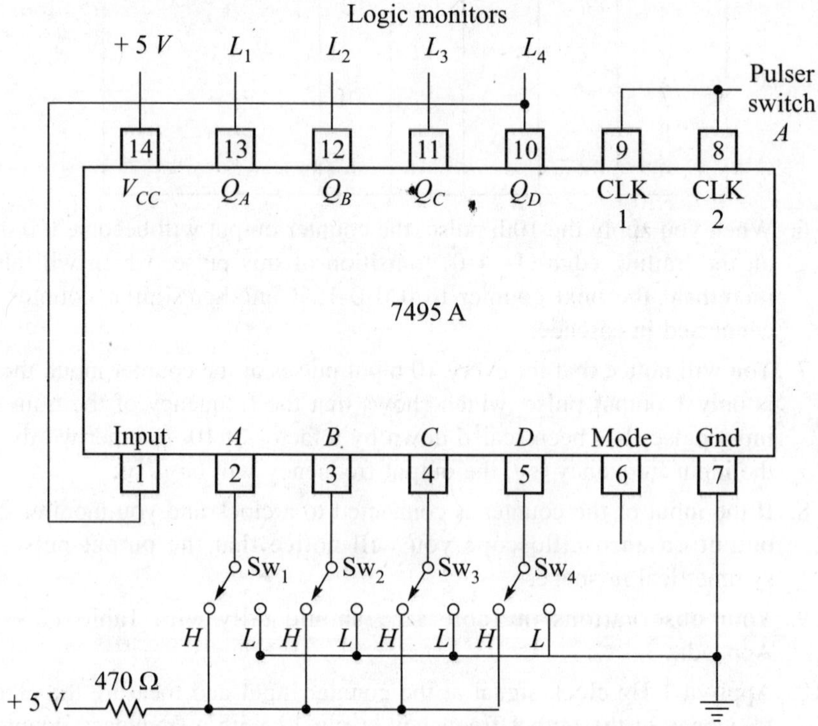

Fig. 12.11. Circuit diagram for Experiment 12.9.

Procedure

1. In Fig. 12.11 a shift register has been connected to function as a ring counter. In the ring counter, the output of the last flip-flop in the chain Q_D has been connected to the serial input, pin 1. Also notice that both the clock inputs 1 and 2 are connected to pulser switch *A* (low output socket).
2. Assemble the circuit given in Fig. 12.11. The four data inputs have been connected to four logic switches. As there are usually not more.

than four logic switches in logic trainers, a high logic level can be applied to the Mode control by connecting it to + 5 V through a 470 ohm resistor and a low logic level can be applied by connecting it to ground.

3. Load binary number 1 0 0 0 into the *A, B, C, D* inputs using logic switches Sw_1 through Sw_4.
4. Apply logic 1 level to the Mode control and apply a single clock pulse with pulser switch *A*. This number is now loaded into the counter.
5. After the number has been loaded into the counter take the mode input line low, which will establish a right shift mode of operation of the counter. It is now ready to function as a ring counter.
6. Apply pulses with pulser switch *A* and record your observations in Table 12.10.

Table 12.10

Pulse count	*Outputs*			
	Q_A	Q_B	Q_C	Q_D
1				
2				
3				
4				
5				

7. Notice that the last pulse recycles it back to 1 0 0 0. Your observations should tally with Table 12.10 in Appendix 3.
8. As long as there is a single 1 in the counter and all the others are 0s, the ring counter will function as expected. If there is more than a single 1, the counter will not function as it should. The only allowed states in the ring counter have a single 1. To ensure that there are no disallowed states, a self-correcting circuit has to be used.

12.9 JOHNSON COUNTER

In the Johnson counter the outputs of the last flip-flop are crossed over and then connected to the *J* and *K* inputs of the first flip-flop as shown in Fig. 12.12. In the ring counter the Q_D and the $\overline{Q}_D$ outputs of the *D* flip-flop are connected respectively to the *J* and *K* inputs of the *A* flip-flop. Because of this cross-connection the Johnson counter is sometimes referred to as a twisted ring counter or a switch tail counter. This counter has a sequence of eight states as given in Table 12.12.

Since this counter uses four flip-flops, the total number of states is 16 out of which 8 states are invalid, which have been listed in Table 12.11. The valid

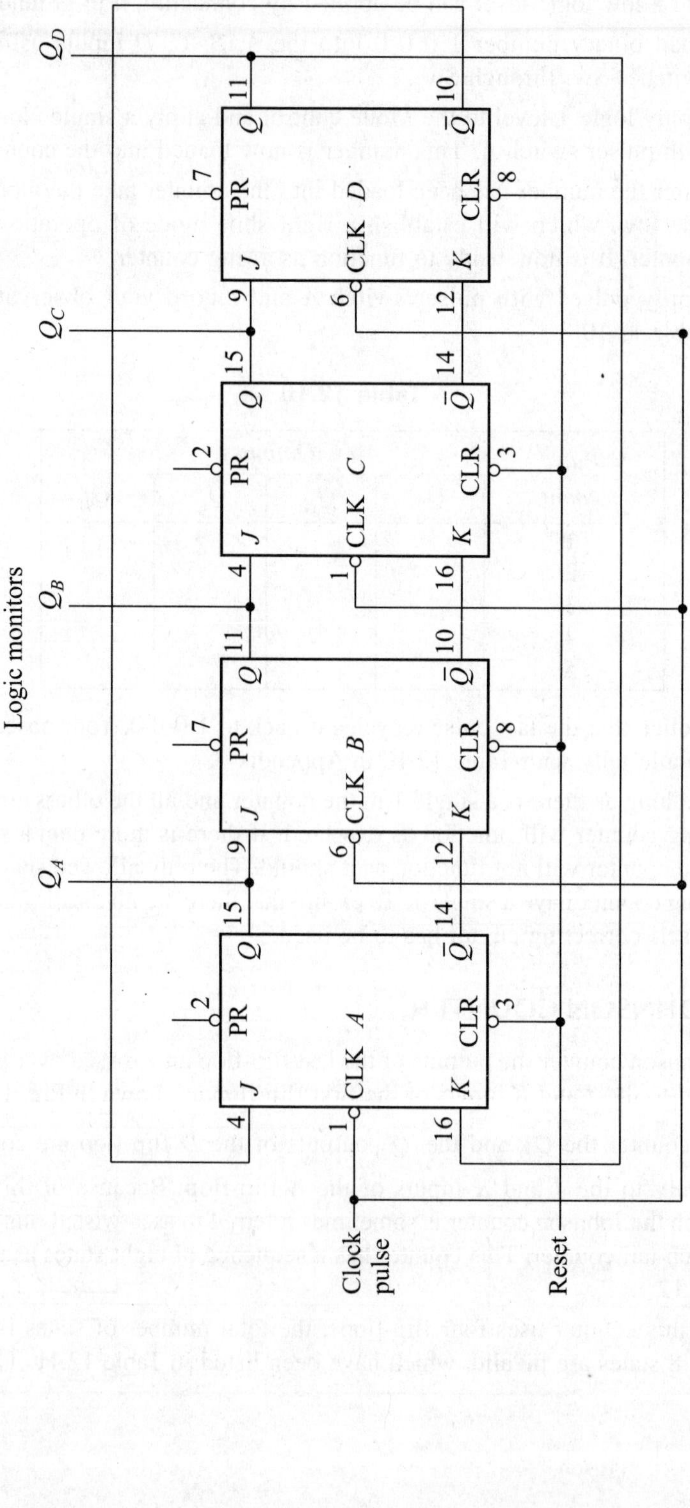

Fig. 12.12. Johnson-counter.

states require decoding, which is different from normal decoding used for standard pure binary count sequence. Notice that state 1 is uniquely defined when the outputs of flip-flops *A* and *D* are low. Thus, a 2-input AND gate with inputs as shown in Table 12.12 can decode state 1. Similarly the other outputs can be decoded by the gates with inputs as shown in this table.

Table 12.11

Q_D	Q_C	Q_B	Q_A	*Binary equivalent*
0	1	0	0	4
1	0	0	1	9
0	0	1	0	2
0	1	0	1	5
1	0	1	1	11
0	1	1	0	6
1	1	0	1	13
1	0	1	0	10

Table 12.12
Valid States

State	Q_D	Q_C	Q_B	Q_A	*Binary equivalent*	*Output decoding*
1	0	0	0	0	0	$\bar{A}$, $\bar{D}$ → $\bar{A}\bar{D}$
2	0	0	0	1	1	A, $\bar{B}$ → $A\bar{B}$
3	0	0	1	1	3	B, $\bar{C}$ → $B\bar{C}$
4	0	1	1	1	7	C, $\bar{D}$ → $C\bar{D}$
5	1	1	1	1	15	A, D → AD
6	1	1	1	0	14	$\bar{A}$, B → $\bar{A}B$
7	1	1	0	0	12	$\bar{B}$, C → $\bar{B}C$
8	1	0	0	0	8	$\bar{C}$, D → $\bar{C}D$

The waveforms for a 4-state Johnson counter are shown in Fig. 12.13.

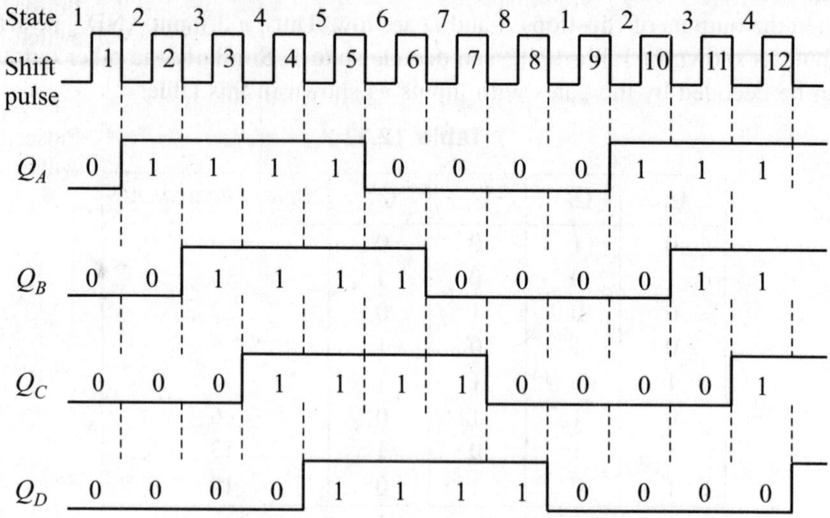

Fig. 12.13. Waveforms for a 4-stage Johnson counter.

If the counter enters a disallowed state it may advance from one disallowed state to another and fail to come out of this situation. To ensure that this does not happen a correcting circuit may be used.

You will notice from the waveform that each of the outputs is square in shape and that the output frequency is one-eighth of the clock frequency. The circuit, therefore, functions as a modulo-eight counter. It follows that by increasing the number of flip-flops, counters with higher modulo numbers can be implemented. Almost any desired count sequence can be produced by decoding the outputs and feeding the results to the input.

A 4-stage counter can also be implemented with similar characteristics by using shift register IC 7495 as demonstrated in Experiment 12.10.

EXPERIMENT 12.10 : JOHNSON COUNTER

Objective

To demonstrate the functioning of a Johnson counter using IC 7495A.

Materials Required

Logic trainer

IC 7495A : Shift register

IC 7404 : Hex Inverter.

A circuit diagram for Experiment 12.10 is given in Fig. 12.14.

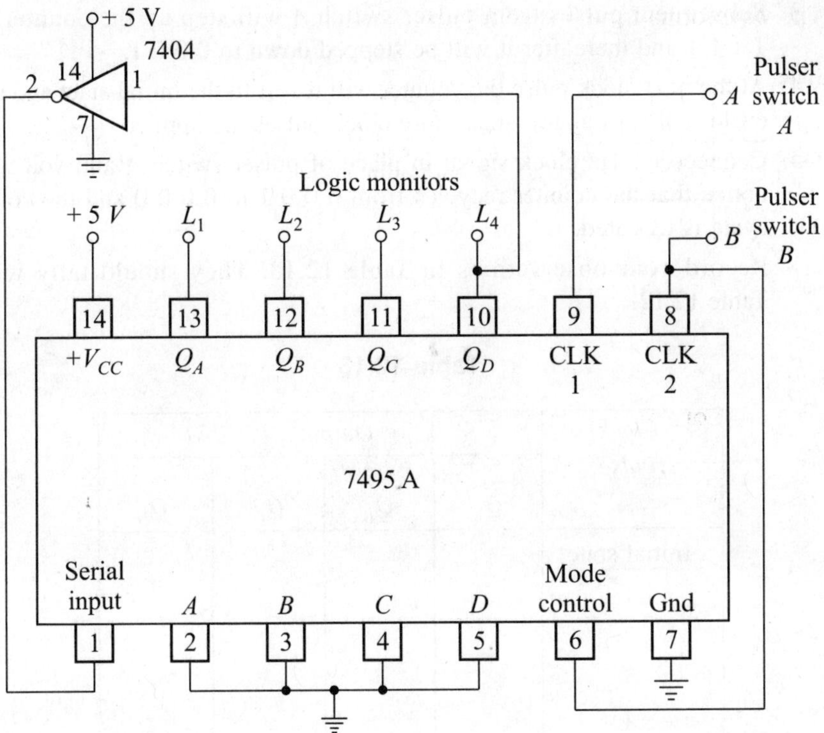

Fig. 12.14. Johnson counter.

You will notice from the circuit diagram for Experiment 12.10 that the normal output Q_D, of the D flip-flop is connected through an Inverter to the Serial input, pin 1. If the counter is initially reset the normal output of the D flip-flop will be 0, which will make the K output of A flip-flop 0 and the J input 1. Since the mode input is connected to pin 8, which goes to pulser switch B (low), the Mode control is normally low. The first clock pulse will make the output of a flip-flop 1 and the outputs of the other flip-flops will remain 0. The next clock pulse will make Q_B, 1 and Q_C and Q_D will remain 0. After some more pulses all flip-flops will be set and thereafter it will recycle back to the initial state as shown in Table 12.12.

Procedure

1. Assemble the circuit given in Fig. 12.14. Connect pin 14 of ICs 7495A and 7404 to + 5 V and pin 7 to ground.

   When you switch on the power supply the counter may come up in any state. Reset the counter to 0 0 0 0 with pulser switch B.

2. When you apply the first clock pulse with pulser switch A the counter output will be 1 0 0 0.

3. Subsequent pulses from pulser switch *A* will step up the counter to 1 1 1 1 and thereafter it will be stepped down to 0 0 0 1.
4. At the next clock pulse the counter will revert to the initial state and the cycle will start again when more clock pulses are applied.
5. Connect a 1 Hz clock signal in place of pulser switch *A* and you will notice that the counter recycles from 0 0 0 1 to 0 0 0 0 and the count cycle is repeated.

   Record your observations in Table 12.13. They should tally with Table 12.12.

Table 12.13

Clock Pulse	*Outputs*			
	Q_A	Q_B	Q_C	Q_D
Initial state				
1				
2				
3				
4				
5				
6				
7				

PROBLEMS

12.1. Draw the waveform diagram for a 3-bit binary ripple down-counter.

12.2. What will be the modulus of IC 74193 in the down-counting mode if the number preset in the counter is 1 0 0 1 ?

12.3. Draw a diagram to build an 8-bit up-down counter using IC 74193.

12.4. Draw a waveform diagram for IC 7490A when used as a BCD counter in 2 × 5 configuration.

12.5. Draw a waveform diagram for IC 7490A when used as a divide-by-ten counter in the 5 × 2 configuration.

12.6. Figure P-12.6 shows IC 7490A connected as a modulo counter. Determine its modulus and explain how it functions.

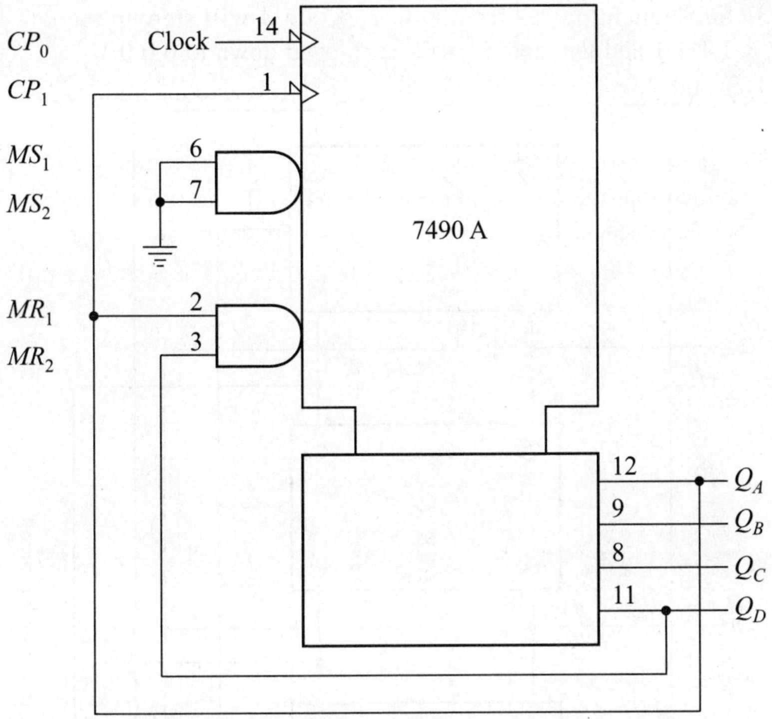

Fig. P-12.6

12.7. Draw a mod-6 counter using IC 7490A.

12.8. Draw a circuit diagram for a decimal scalar which will divide the input frequency by 10, 100, 1000 and 10,000.

12.9. Figure P-12.9 (Page 172) shows a ring counter with a correcting circuit. If any disallowed state occurs in the counter output the ring counter will automatically correct itself after not more than two shift pulses, so that only one of the flip-flops is at binary 1 level. Explain, how the correcting circuit functions.

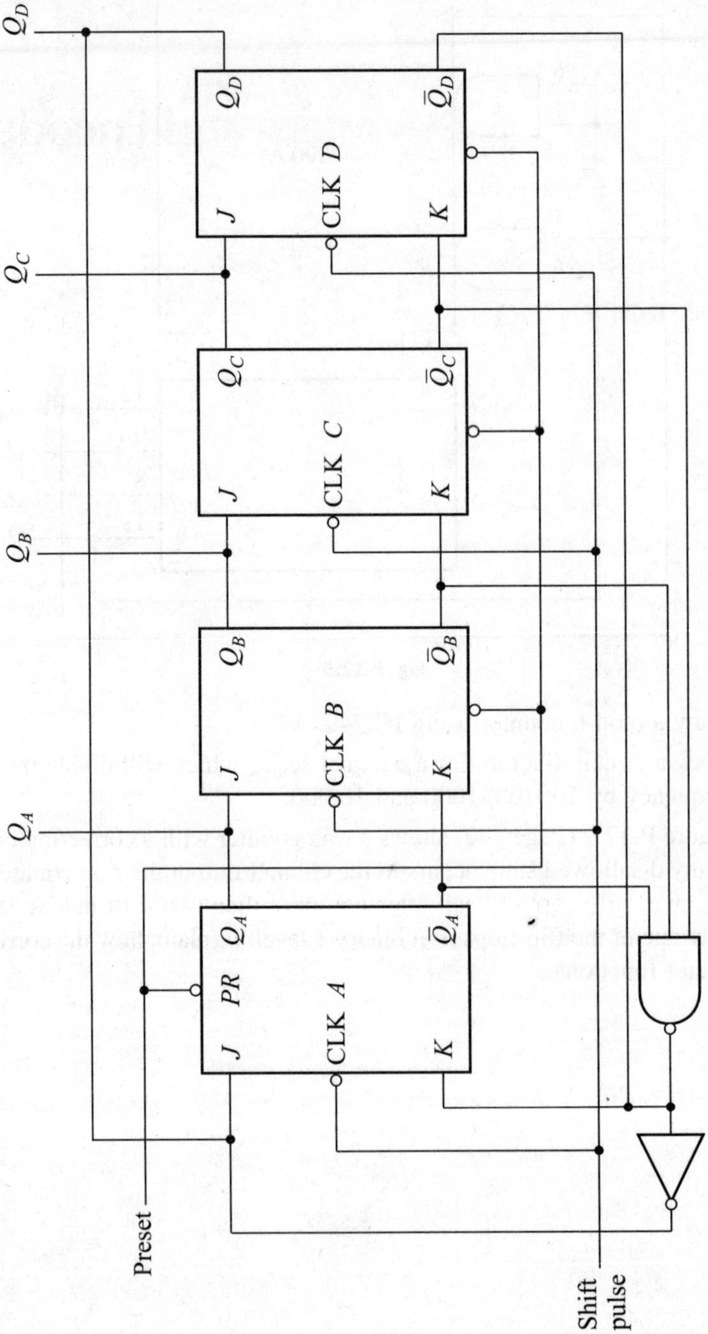

Fig. P-12.9. Ring counter with correcting circuit.

CHAPTER 13

Decoders and Encoders

13.1 INTRODUCTION

In this chapter we will consider Decoders and Encoders, which are classified as combinational logic circuits. They differ from sequential circuits as they possess no memory and the output is the direct result of the state of the inputs. Besides, it bears no relation to prior state of the inputs and the response of the output follows immediately after the inputs. As long as the input values are maintained, the output will undergo no change.

Combinational circuits are provided with a set of inputs, which are connected to a logic network. It operates on the input and produces a set of outputs. Combinational circuits have many configurations; but in this chapter, we are mainly concerned with Decoders and Encoders.

13.2 DECODERS

The input to a decoder is a parallel binary number and it is used to detect the presence of a particular binary number at the input. The output indicates the presence or absence of a specific number at the decoder input.

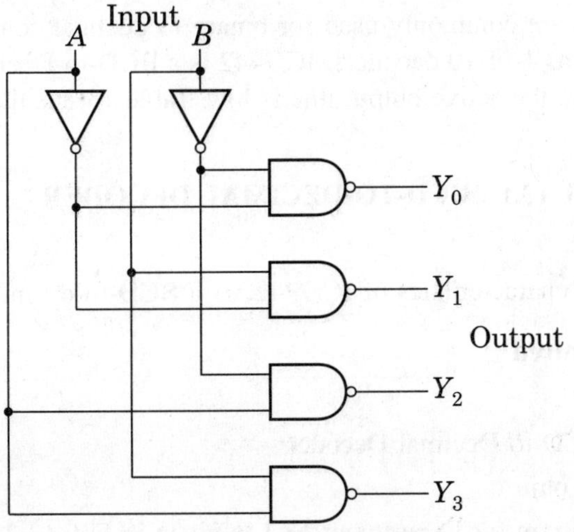

Fig. 13.1. 1-of-4 Decoder.

Let us suppose that a logic network has two inputs *A* and *B*. They will give rise to four states *A*, $\overline{A}$, *B* and $\overline{B}$. The logic circuit can be designed to give you an indication at its output of the decimal value of any set of inputs. If the outputs are connected to logic monitors you can know the decimal value of the inputs by looking at the outputs.

A decoder which will perform this function can be implemented by the circuit shown in Fig. 13.1.

Table 13.1 gives the truth table for this 1-of-4 decoder.

Table 13.1

Inputs		*Outputs*			
A	*B*	Y_0	Y_1	Y_2	Y_3
L	L	L	H	H	H
L	H	H	L	H	H
H	L	H	H	L	H
H	H	H	H	H	L

It is evident from Table 13.1 that for any input combination only one of the outputs is low and all the others are high. A low value at the output represents the state of the input. Also notice that the decoder selects only one of the four outputs at a time. This decoder is, therefore called a 1-of-4 decoder.

Following the same design concept, a 1-of-8, 1-of-10 or 1-of-16 decoder can be implemented.

13.3 BCD-TO-DECIMAL DECODER

These decoders are commonly used for binary to decimal conversion and are also referred to as 1-of-10 decoders. IC 7442 is a BCD-to-Decimal decoder. In this decoder also the active output line is low. Table 13.2 is the truth table for this IC.

EXPERIMENT 13.1 : BCD-TO-DECIMAL DECODER

Objective

To study the characteristics of IC 7442, as a BCD-to-decimal decoder.

Materials Required

Logic trainer

IC 7442 : BCD-to-Decimal Decoder

Resistor 470 ohm.

A circuit diagram for Experiment 13.1 is given in Fig. 13.2.

Table 13.2
Truth Table for IC 7442

	Inputs				*Outputs*									
No.	*D* *MSB*	*C*	*B*	*A* *LSB*	Y_0	Y_1	Y_2	Y_3	Y_4	Y_5	Y_6	Y_7	Y_8	Y_9
0	L	L	L	L	L	H	H	H	H	H	H	H	H	H
1	L	L	L	H	H	L	H	H	H	H	H	H	H	H
2	L	L	H	L	H	H	L	H	H	H	H	H	H	H
3	L	L	H	H	H	H	H	L	H	H	H	H	H	H
4	L	H	L	L	H	H	H	H	L	H	H	H	H	H
5	L	H	L	H	H	H	H	H	H	L	H	H	H	H
6	L	H	H	L	H	H	H	H	H	H	L	H	H	H
7	L	H	H	H	H	H	H	H	H	H	H	L	H	H
8	H	L	L	L	H	H	H	H	H	H	H	H	L	H
9	H	L	L	H	H	H	H	H	H	H	H	H	H	L
Invalid	H	L	H	H	H	H	H	H	H	H	H	H	H	H
	H	L	H	H	H	H	H	H	H	H	H	H	H	H
	H	H	L	L	H	H	H	H	H	H	H	H	H	H
	H	H	H	L	H	H	H	H	H	H	H	H	H	H
	H	H	H	H	H	H	H	H	H	H	H	H	H	H

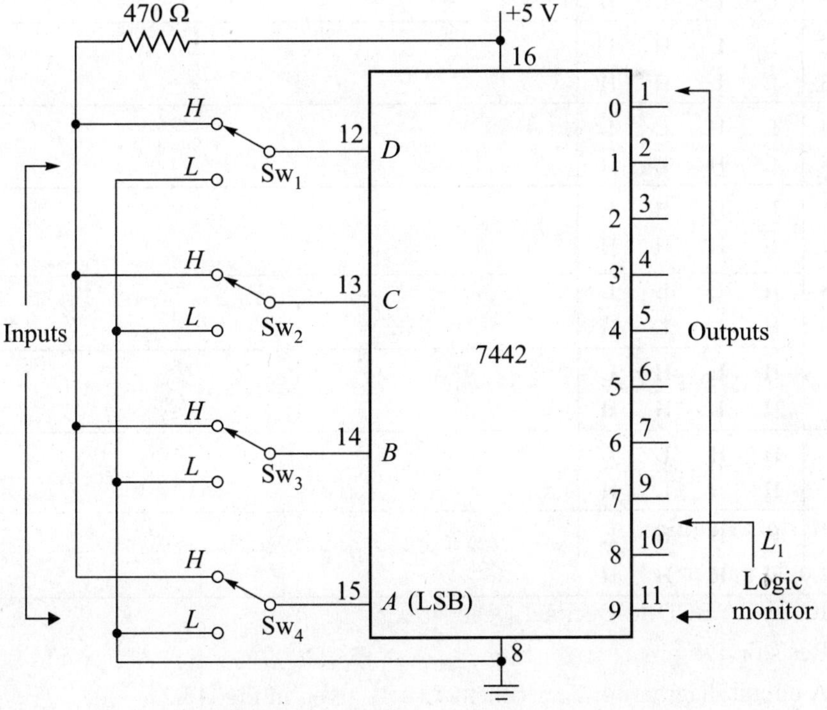

Fig. 13.2. Circuit diagram for Experiment 13.1.

Notice that in Fig. 13.2 the output values in decimal numbers are shown within the IC and the pin numbers have been indicated on the outside. Also note that *A* is the LSB and *D* is the MSB of the input number. Only one logic monitor has been used and the output logic values are determined by touching the appropriate pin number with the logic probe.

Procedure

1. Assemble the circuit given in Fig. 13.2. Connect pin 16 of the IC to + 5 V and pin 8 to ground. Switch on the power supply.
2. Apply the inputs given in Table 13.3 and with the logic probe ascertain the output value at each of the outputs from 0 to 9. Enter your observations in Table 13.3. Your observations should tally with Table 13.2.

Table 13.3

	Inputs				*Outputs*									
	MSB			*LSB*										
No.	*D*	*C*	*B*	*A*	Y_0	Y_1	Y_2	Y_3	Y_4	Y_5	Y_6	Y_7	Y_8	Y_9
0	L	L	L	L										
1	L	L	L	H										
2	L	L	H	L										
3	L	L	H	H										
4	L	H	L	L										
5	L	H	L	H										
6	L	H	H	L										
7	L	H	H	H										
8	H	L	L	L										
9	H	L	L	H										
	H	L	H	L										
	H	L	H	H										
	H	H	L	L										
	H	H	L	H										
	H	H	H	L										
	H	H	H	H										

During your observations you must have noticed that only the output corresponding to the selected input is low and all the other outputs are high. This is because in this decoder the active output line is low. You must also have noticed that for all inputs from *H L H L* to *H H H H* all outputs are high and not a single output is low. The reason is that this is a 8 4 2 1 BCD 1-of-10 decoder and it ignores all inputs from 1 0 1 0 to 1 1 1 1.

13.4 ENCODERS

Encoders are also combinational logic circuits and they are exactly the opposite of decoders. Encoders accept one or more inputs and generate a multibit output code. Thus, it provides binary coded outputs from an input selected from a given number of inputs. We will consider a very simple encoder shown in Fig. 13.3. The encoder has pushbutton inputs which are labelled 1, 2 and 3. The encoder outputs are taken from NAND gates 1 and 2. These gates provide output in a 2-bit binary code.

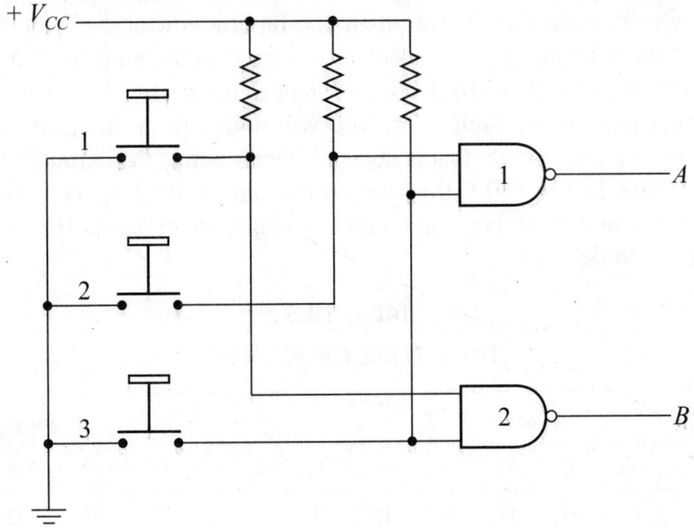

Fig. 13.3

We have assigned decimal numbers 1, 2 and 3 to three switches. When switch 1 is depressed we conclude that the input to the encoder is decimal 1. When none of the switches is depressed the input is 0 and, therefore, the input to the NAND gates is high. Consequently, the outputs *A* and *B* are low, 0 0. When switch 1 is depressed both inputs to gate 1 are high and so its output is low, whereas only one input to gate 2 is high and the other is low. Therefore, the output of the encoder is 0 1, which tallies with the input. Encoders are used for translating decimal keyboard, as in a calculator, into a binary or BCD output code. For the encoder we have just discussed the truth table given in Table 13.4.

Table 13.4

Decimal inputs	*Binary outputs*	
	A	*B*
0	0	0
1	0	1
2	1	0
3	1	1

13.5 DECIMAL PRIORITY ENCODER : IC 74147

A logic symbol for this encoder is given in Fig. 13.4 (Experiment 13.2). The diagram shows that the inputs and outputs are active-low. The inputs are marked X_1 to X_9 . The subscripts indicate the decimal number of the inputs. The BCD outputs are marked *D* (MSB), *C, B, A.* As has been mentioned it is a priority encoder which means that if more than one input is low at the same time, the highest of these inputs gets encoded in the output. For instance, if X_4 and X_6 are low, the output will be 0 1 1 0, equivalent to decimal 6. The truth table for this encoder is given in Table 13.5. You will notice from this truth table that when all the inputs are high, that is none of them is active the output is *H H H H* which is equivalent to 0 0 0 0 as the outputs are active-low. For instance, if input X_5 is low and all higher order inputs are high, the output is *H L H L* which is equivalent to decimal 5.

Table 13.5
Truth Table for IC 74147

Inputs									*Outputs*			
X_1	X_2	X_3	X_4	X_5	X_6	X_7	X_8	X_9	*D*	*C*	*B*	*A*
H	H	H	H	H	H	H	H	H	H	H	H	H
X	X	X	X	X	X	X	X	L	L	H	H	L
X	X	X	X	X	X	X	L	H	L	H	H	H
X	X	X	X	X	X	L	H	H	H	L	L	L
X	X	X	X	X	L	H	H	H	H	L	L	H
X	X	X	X	L	H	H	H	H	H	L	H	L
X	X	X	L	H	H	H	H	H	H	L	H	H
X	X	L	H	H	H	H	H	H	H	H	L	L
X	L	H	H	H	H	H	H	H	H	H	L	H
L	H	H	H	H	H	H	H	H	H	H	H	L

H = HIGH logic level; L = LOW logic level; X = Irrelevant

EXPERIMENT 13.2 : DECIMAL PRIORITY ENCODER : IC 74147

Objective

To demonstrate the operation of IC 74147 as a decimal priority encoder.

Materials Required

Logic trainer

IC 74147 : Decimal priority encoder

IC 7404 : Hex Inverter

A circuit diagram for Experiment 13.2 is given in Fig. 13.4.

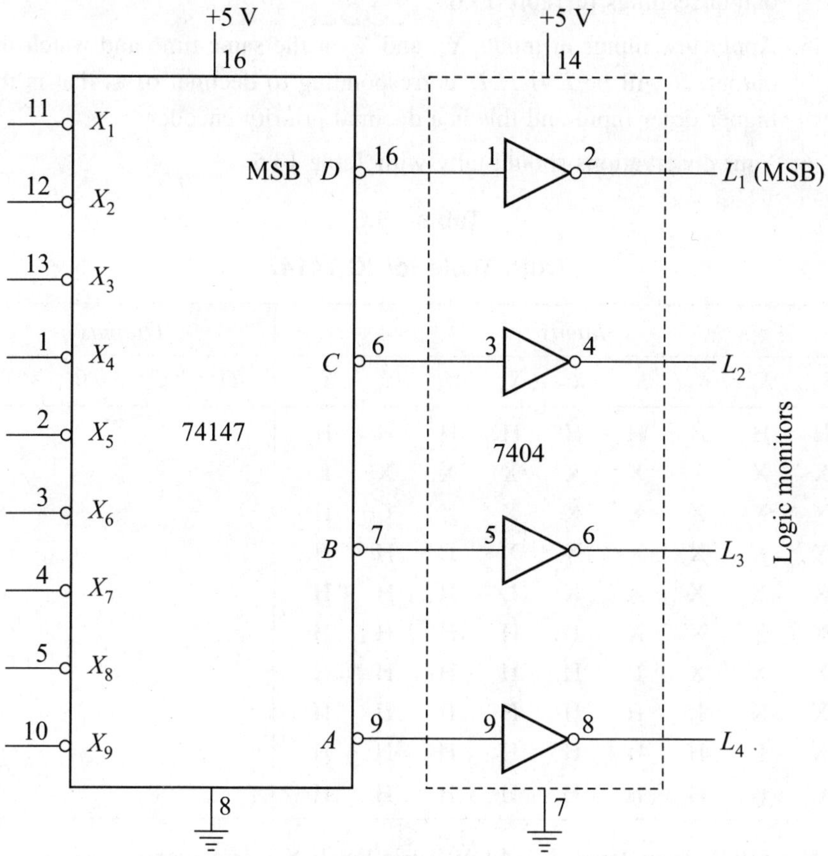

Fig. 13.4. Circuit diagram for Experiment 13.2. (Decimal priority encoder)

Procedure

1. Assemble the circuit given in Fig. 13.4. Since the inputs of the decoder are active-low, the required input can be applied by connecting the pins to ground. As the outputs are also low, Inverters have been used at the outputs to convert low output to a high output.

2. Connect pin 16 of IC 74147 and pin 14 of IC 7404 to + 5 V. Also connect pin 8 of IC 74147 and pin 7 of IC 7404 to ground. Switch on the power supply.
3. Apply a low logic level at X_1 (pin 11). The input pins which have not been connected to ground will automatically be at high level. Although Table 13.5 shows an output of *H H H L*, the logic monitor will show an output of *L L L H* as the Inverter inverts the output.
4. Leave all the inputs open and notice the output. It will be *L L L L*. Now apply low input to pins from X_2 to X_9, one at a time and record your output readings in Table 13.6.
5. Apply low inputs at inputs X_4, and X_6 at the same time and watch the output. It will be *L H H L*, corresponding to decimal 6, as that is the higher order input and this is a decimal priority encoder.
6. Your observations should tally with Table 13.5.

Table 13.6

Truth Table for IC 74147

Inputs									*Outputs*			
X_1	X_2	X_3	X_4	X_5	X_6	X_7	X_8	X_9	*D*	*C*	*B*	*A*
H	H	H	H	H	H	H	H	H				
X	X	X	X	X	X	X	X	L				
X	X	X	X	X	X	X	L	H				
X	X	X	X	X	X	L	H	H				
X	X	X	X	X	L	H	H	H				
X	X	X	X	L	H	H	H	H				
X	X	X	L	H	H	H	H	H				
X	X	L	H	H	H	H	H	H				
X	L	H	H	H	H	H	H	H				
L	H	H	H	H	H	H	H	H				

H = HIGH logic level; L = LOW logic level; X = Irrelevant

PROBLEMS

13.1. For IC 7442 decoder, what will be the decimal equivalent of the output if the inputs are as follows?

(*a*) *L L H H*
(*b*) *L H H H*
(*c*) *H L L H*
(*d*) *L L L L*

13.2. What is the reason for invalid outputs in IC 7442 when it is used as a BCD-to-decimal decoder?

13.3. How will you use IC 7442, BCD-to-decimal decoder, so that it functions as an Octal decoder?

13.4. IC 7442 can also be used to operate a lighted decimal display. How will you design this circuit? Look up the data sheets for a BCD-to-decimal decoder which gives a higher output current than is obtainable from IC 7442.

13.5. Determine the state of the output of IC 74147 if the inputs are:

(*a*) X_4
(*b*) X_7
(*c*) X_2

13.6. What will be the inputs to the priority encoder IC 74147 if the outputs are as follows?

(*a*) *H H H H*
(*b*) *H H H L*
(*c*) *H H L L*
(*d*) *H L L L*

CHAPTER 14

Multiplexers and Demultiplexers

14.1 INTRODUCTION

Multiplexers and Demultiplexers are some of the most useful logic devices and they can be used for numerous applications. Multiplexers can select any one of a number of inputs and route them to a single output. Demultiplexers have a single input and many outputs. The input of a demultiplexer can be routed to any of the output channels. For this reason a demultiplexer is also known as a data distributor.

14.2 MULTIPLEXERS

To illustrate the concept we will consider a multiplexer with two inputs and a single output as shown in Fig. 14.1.

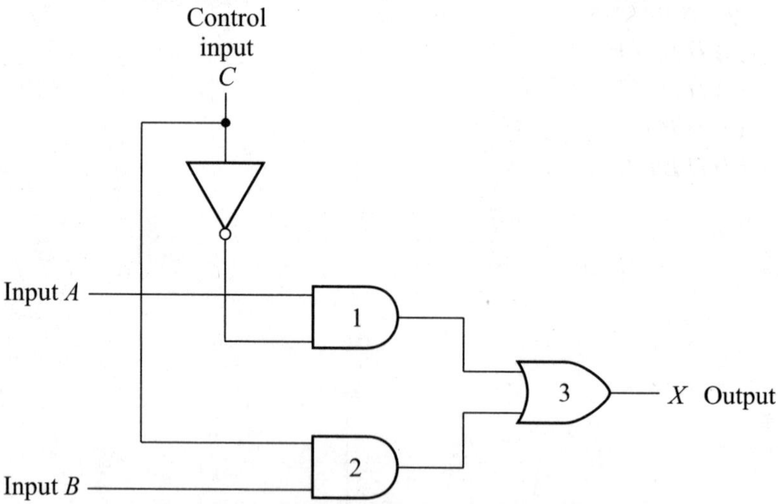

Fig. 14.1. 2-line to 1-line multiplexer.

In Fig. 14.1, A and B are the two inputs : X is the output and C is the control input. If input A is to be routed to the output line, the control input goes low, $C = 0$, which makes both inputs to AND gate 1 high. This enables AND gate 1 and its output passes through the OR gate 3, to the output. When input B is to be routed to the output, the control input, C, is made high, which enables AND gate 2 and its output passes through the OR gate 3 to the output. The output equation in this case is as follows:

$$X = A\overline{C} + BC$$

By using the same basic concept as in the 2-line to 1-line multiplexer, more complex multiplexers can be designed. Incorporating the same design principles a 4-line to 1-line multiplexer is shown in Fig. 14.2.

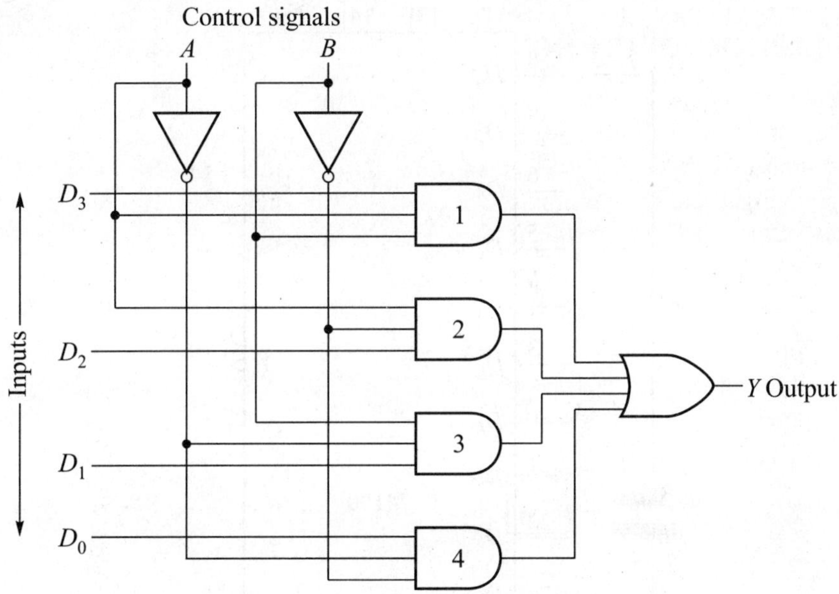

Fig. 14.2. 4-line to 1-line multiplexer.

In Fig. 14.2, the inputs are marked D_0, D_1, D_2 and D_3. The output is Y and the control inputs are A and B. Table 14.1 is the truth table for the multiplexer shown in Fig. 14.2.

Table 14.1

Control inputs		*Output*
A	*B*	*Y*
0	0	D_0
0	1	D_1
1	0	D_2
1	1	D_3

14.3 16-LINE TO 1-LINE MULTIPLEXER : IC 74150

This multiplexer has 16 input lines and a single output. It can select one out of 16 inputs for being routed to a single output. Multiplexers are, therefore, called data selectors. This multiplexer has four control inputs *A, B, C* and *D*. The input data selected depends on the value of the control input. Its logic symbol is given in Fig. 14.3.

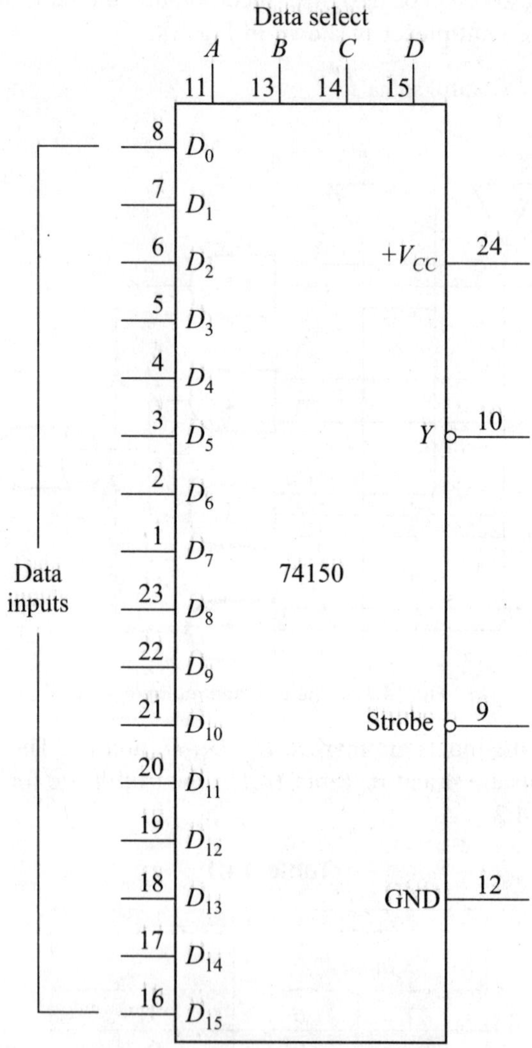

Fig. 14.3. Logic symbol for IC 74150.

The sixteen inputs are from D_0 through D_{15} and a single output marked *Y*. The bubble at the output line indicates that the output is low when the data bit selected is high. In other words, the output is always the complement of the selected data bit. The strobe is active-low. When it is low, the multiplexer is enabled and it is disabled when the strobe is high. Also notice that the subscript of the input line enabled corresponds to the decimal value of the data select inputs. The data select input has four bits and they can generate 16 4-bit words. Each one of the 16 words will enable only one of the 16 input lines, which will be transmitted to the *Y* output. In every case the output will be as follows:

$$Y = \overline{D}_n$$

where '*n*' corresponds to the decimal equivalent of the data select input. Also notice that the output is the complement of the input data. The operation of the multiplexer has been summarized in Table 14.2 which represents its truth table.

Table 14.2

Truth Table for IC 74150

Strobe	*Data select*								*Output*
	MSB			*LSB*	*MSB*			*LSB*	
	A	*B*	*C*	*D*	*A*	*B*	*C*	*D*	*Y*
L	$\overline{A}$	$\overline{B}$	$\overline{C}$	$\overline{D}$	L	L	L	L	$\overline{D}_0$
L	$\overline{A}$	$\overline{B}$	$\overline{C}$	D	L	L	L	H	$\overline{D}_1$
L	$\overline{A}$	$\overline{B}$	C	$\overline{D}$	L	L	H	L	$\overline{D}_2$
L	$\overline{A}$	$\overline{B}$	C	D	L	L	H	H	$\overline{D}_3$
L	$\overline{A}$	B	$\overline{C}$	$\overline{D}$	L	H	L	L	$\overline{D}_4$
L	$\overline{A}$	B	$\overline{C}$	D	L	H	L	H	$\overline{D}_5$
L	$\overline{A}$	B	C	$\overline{D}$	L	H	H	L	$\overline{D}_6$
L	$\overline{A}$	B	C	D	L	H	H	H	$\overline{D}_7$
L	A	$\overline{B}$	$\overline{C}$	$\overline{D}$	H	L	L	L	$\overline{D}_8$
L	A	$\overline{B}$	$\overline{C}$	D	H	L	L	H	$\overline{D}_9$
L	A	$\overline{B}$	C	$\overline{D}$	H	L	H	L	$\overline{D}_{10}$
L	A	$\overline{B}$	C	D	H	L	H	H	$\overline{D}_{11}$
L	A	B	$\overline{C}$	$\overline{D}$	H	H	L	L	$\overline{D}_{12}$
L	A	B	$\overline{C}$	D	H	H	L	H	$\overline{D}_{13}$
L	A	B	C	$\overline{D}$	H	H	H	L	$\overline{D}_{14}$
L	A	B	C	D	H	H	H	H	$\overline{D}_{15}$
H	X	X	X	X	X	X	X	X	H

You will notice from Table 14.2 that when the strobe is low, the output will respond to the data select inputs, and when it is high, as in the last row, the output will be high irrespective of the state of the data select inputs.

EXPERIMENT 14.1 : MULTIPLEXER: 74150

Objective

1. To study the functioning of multiplexer IC 74150.
2. To study the application of multiplexer IC 74150 for the generation of Boolean functions.

Materials Required

Logic trainer

IC 74150 : 16-line to 1-line Multiplexer

IC 74193 : Binary Up-down Counter

Transistor BC 108A (NPN)

LED (Red)

Resistor 330 ohm

Resistor 1 k ohm

Resistor 10 k ohm.

A circuit diagram for Experiment 14.1 is given in Fig. 14.4.

For this experiment, IC 74193 has been used as an up-counter to generate binary numbers from 0 0 0 0 to 1 1 1 1. Particularly notice that in IC 74193 output *D* is the MSB and output *A* is the LSB, whereas in the multiplexer IC 74150A is the MSB and *D* is the LSB.

Apart from the four logic monitors a logic monitor using a transistor has also been used with an LED indicator, which will light up when the multiplexer output at pin 10 is high. You will apply low logic levels at one of the inputs while the other inputs are high so that you can watch the output on the LED indicator.

You will observe the counter output on logic monitors L_1 through L_4. You can reset the counter with pulser switch B and step it up with pulser switch *A* from 0 0 0 0 to 1 1 1 1. Logic switch Sw_1 is connected to the strobe input, pin 9. When a logic 1 is applied with Sw_1, the multiplexer is disabled. It is enabled when a logic 0 is applied to the strobe input with this switch.

Procedure

1. Assemble the circuit given in Fig. 14.4. Connect pin 24 of IC 74150 and pin 16 of IC 74193 to + 5 V and pin 12 of IC 74150 and pin 8 of IC 74193 to ground. Also connect the power supply to the transistor.

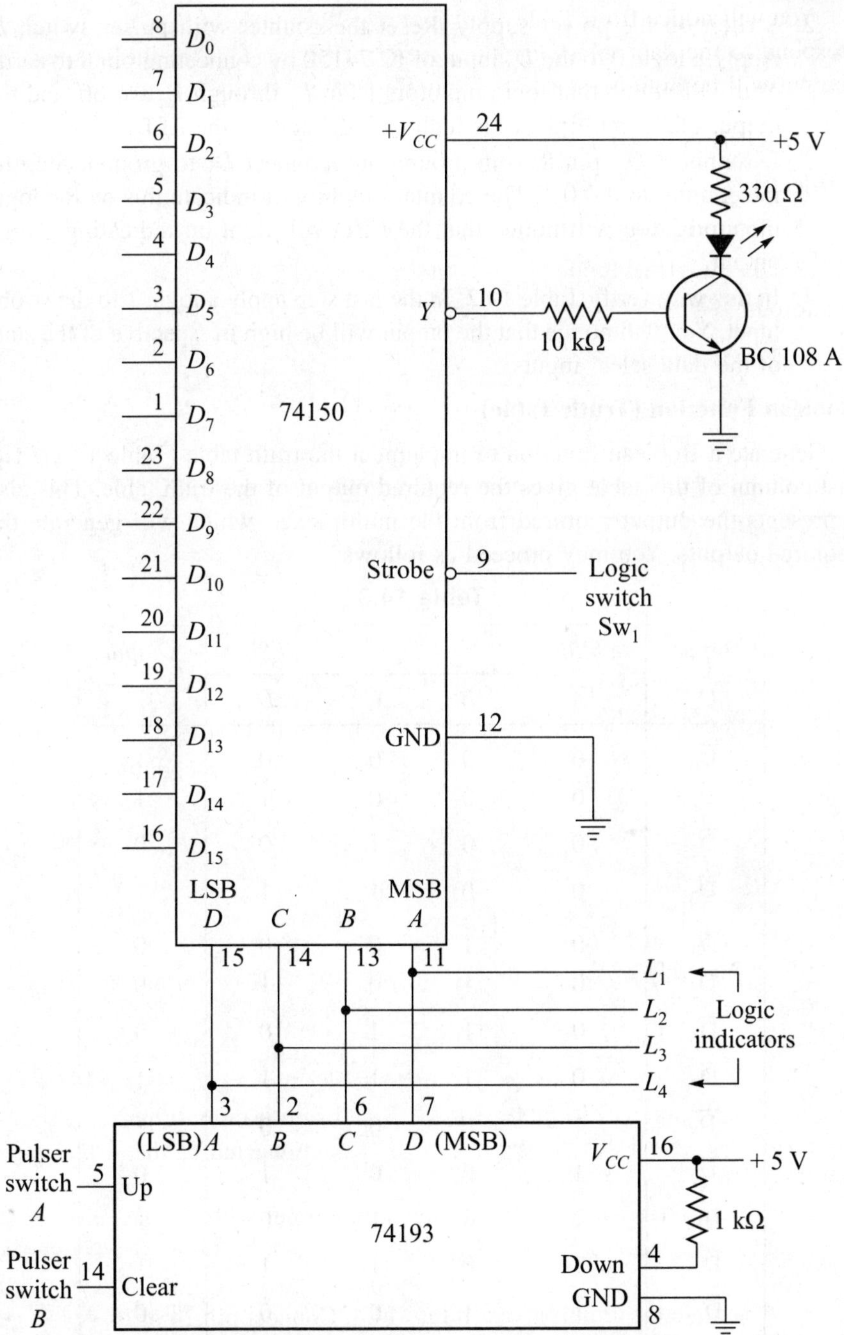

Fig. 14.4. Circuit diagram for Experiment 14.1 (1).

2. Switch on the power supply. Reset the counter with pulser switch *B*. Apply a logic 0 to the D_0 input of IC 74150 by connecting pin 8 to earth. You will notice that logic monitors from L_1 through L_4 are off and the output of the multiplexer is high as indicated by the LED.
3. Disconnect D_0, pin 8 from ground and connect D_1 to ground. Step up the counter to 0 0 0 1. The counter output will indicate this on the logic monitors. You will notice that the LED will light up indicating a high output.
4. In this way verify Table 14.2. In the last step apply a logic 1 to the strobe input. You will notice that the output will be high irrespective of the state of the data select inputs.

Boolean Function (Truth Table)

Generate a Boolean function to implement the truth table (Table 14.3). The last column of this table gives the required output of the truth table. This also represents the output required from the multiplexer, which will generate the required outputs. You may proceed as follows:

Table 14.3

Input	*MSB*			*LSB*	*Output*
D_n	*A*	*B*	*C*	*D*	*Y*
$\overline{D}_0$	0	0	0	0	1
$\overline{D}_1$	0	0	0	1	1
$\overline{D}_2$	0	0	1	0	0
$\overline{D}_3$	0	0	1	1	1
$\overline{D}_4$	0	1	0	0	0
$\overline{D}_5$	0	1	0	1	0
$\overline{D}_6$	0	1	1	0	0
$\overline{D}_7$	0	1	1	1	1
$\overline{D}_8$	1	0	0	0	1
$\overline{D}_9$	1	0	0	1	0
$\overline{D}_{10}$	1	0	1	0	1
$\overline{D}_{11}$	1	0	1	1	1
$\overline{D}_{12}$	1	1	0	0	0
$\overline{D}_{13}$	1	1	0	1	0
$\overline{D}_{14}$	1	1	1	0	0
$\overline{D}_{15}$	1	1	1	1	1

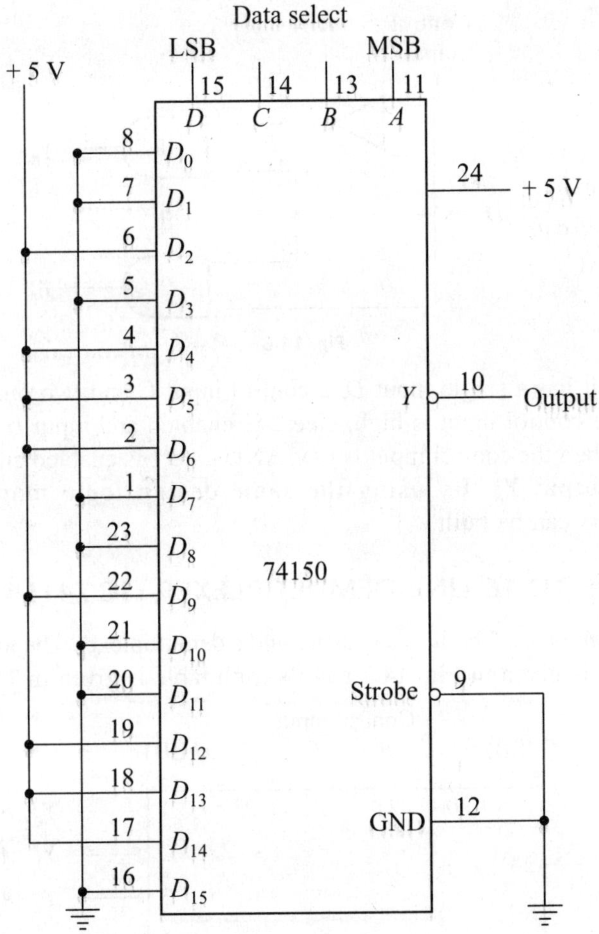

Fig. 14.5. Circuit diagram for Experiment 14.1 (2).

5. Since a low input produces a high output, and a high input produces a low output, all that you have to do to generate the truth table is to connect those inputs to ground against which a high output is required and the remaining inputs to + 5 V as has been shown in Fig. 14.5.
6. Now by sequencing the data select inputs from 0 0 0 0 through 1 1 1 1 the required Boolean function will be generated by the multiplexer in serial form.

14.4 DEMULTIPLEXER

Unlike a multiplexer, the demultiplexer has a single input and many outputs. The input of a demultiplexer can be routed to any of the output lines. Consider the circuit in Fig. 14.6 which shows a 1-line to 2-line demultiplexer.

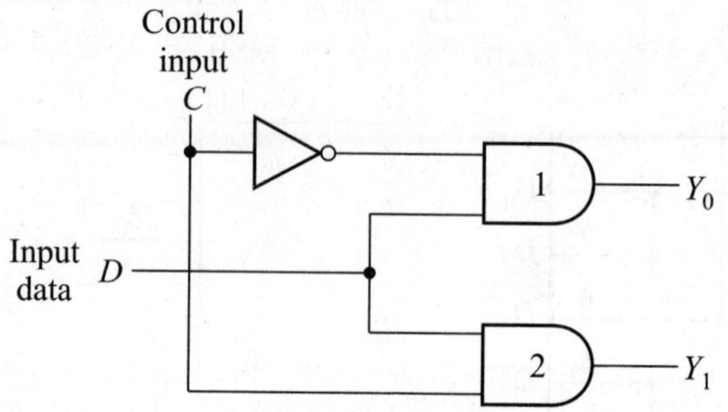

Fig. 14.6

This circuit has a single input D, a control input C and two outputs Y_0 and Y_1. When the control input is high, gale 2 is enabled and input D is routed to output Y_1. When the control input is low, AND gate 1 is enabled and input D is routed to output Y_0. By using the same design logic more complex demultiplexers can be built.

14.5 1-LINE TO 16-LINE DEMULTIPLEXER : IC 74154

This device can be used both as a decoder and a demultiplexer. The logic diagram for IC 74154 is given in Fig. 14.7 and its truth table is given in Table 14.4.

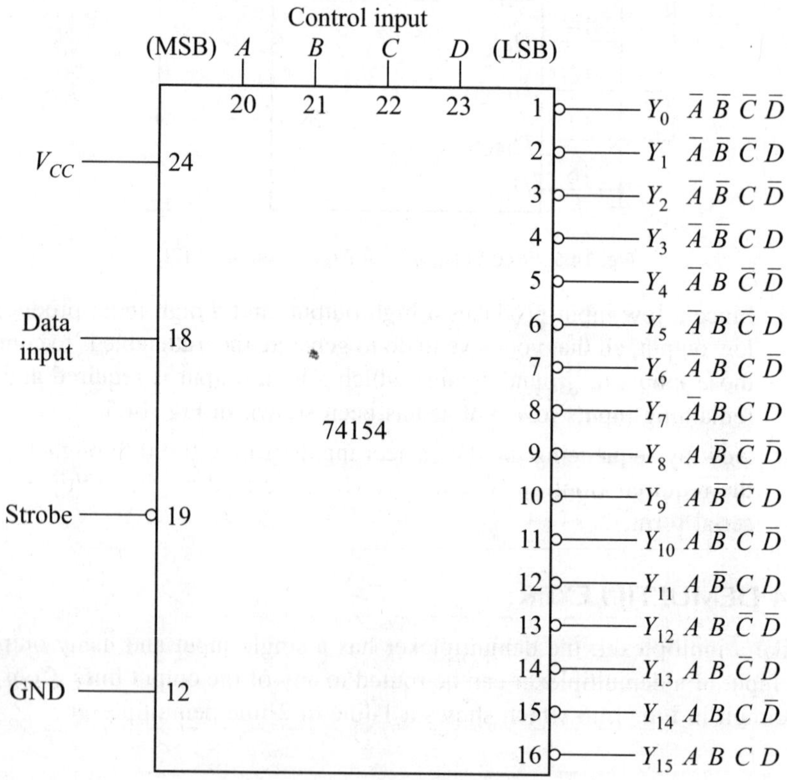

Fig. 14.7. Logic diagram for IC 74154.

Table 14.4

Truth Table for IC 74154

Strobe	*Data Input*	*Control Inputs*				*Outputs*															
		A	B	C	D	Y_0	Y_1	Y_2	Y_3	Y_4	Y_5	Y_6	Y_7	Y_8	Y_9	Y_{10}	Y_{11}	Y_{12}	Y_{13}	Y_{14}	Y_{15}
L	L	L	L	L	L	L	H	H	H	H	H	H	H	H	H	H	H	H	H	H	H
L	L	L	L	L	H	H	L	H	H	H	H	H	H	H	H	H	H	H	H	H	H
L	L	L	L	H	L	H	H	L	H	H	H	H	H	H	H	H	H	H	H	H	H
L	L	L	L	H	H	H	H	H	L	H	H	H	H	H	H	H	H	H	H	H	H
L	L	L	H	L	L	H	H	H	H	L	H	H	H	H	H	H	H	H	H	H	H
L	L	L	H	L	H	H	H	H	H	H	L	H	H	H	H	H	H	H	H	H	H
L	L	L	H	H	L	H	H	H	H	H	H	L	H	H	H	H	H	H	H	H	H
L	L	L	H	H	H	H	H	H	H	H	H	H	L	H	H	H	H	H	H	H	H
L	L	H	L	L	L	H	H	H	H	H	H	H	H	L	H	H	H	H	H	H	H
L	L	H	L	L	H	H	H	H	H	H	H	H	H	H	L	H	H	H	H	H	H
L	L	H	L	H	L	H	H	H	H	H	H	H	H	H	H	L	H	H	H	H	H
L	L	H	L	H	H	H	H	H	H	H	H	H	H	H	H	H	L	H	H	H	H
L	L	H	H	L	L	H	H	H	H	H	H	H	H	H	H	H	H	L	H	H	H
L	L	H	H	L	H	H	H	H	H	H	H	H	H	H	H	H	H	H	L	H	H
L	L	H	H	H	L	H	H	H	H	H	H	H	H	H	H	H	H	H	H	L	H
L	L	H	H	H	H	H	H	H	H	H	H	H	H	H	H	H	H	H	H	H	L
L	H	X	X	X	X	H	H	H	H	H	H	H	H	H	H	H	H	H	H	H	H
H	L	X	X	X	X	H	H	H	H	H	H	H	H	H	H	H	H	H	H	H	H
H	H	X	X	X	X	H	H	H	H	H	H	H	H	H	H	H	H	H	H	H	H

Demultiplexer IC 74154 performs essentially as a distributor, which can distribute the single input data to any one of the output lines. When it is used as a data distributor the input is applied at the data input, and the control inputs *A, B, C* and *D* are used to distribute the data to the desired output line. In this operation the strobe is held low which enables the demultiplexer. If you refer to the truth table, Table 14.4, you will notice that when the data input is low, the corresponding output line also goes low and the other output lines stay high. For instance, if the data input is low and the control input is *L H L H*, corresponding to decimal 5, output line Y_5 is enabled; it goes low while the other output lines stay high.

14.6 MULTIPLEXING

We will consider how multiplexers and demultiplexers find considerable application in multiplexing operations. When a number of digital signals have to be processed in an identical manner or are required to be transmitted over a significant distance, multiplexing technique leads to a reduction in the number of processing circuits and communication channels. We will consider this aspect in Experiment 14.2.

EXPERIMENT 14.2 : MULTIPLEXER-DEMULTIPLEXER DATA TRANSMISSION

Objective

To demonstrate the application of a multiplexer and demultiplexer in data transmission.

Materials Required

Logic trainer
IC 74150 : 16-line to 1-line multiplexer
IC 74154 : 1-line to 16-line demultiplexer
IC 74193 : Binary Up-Down Counter
Transistor : BC 108A (NPN)
LED (Red)
Resistor 330 ohm
Resistor 470 ohm
Resistor 1 k ohm
Resistor 10 k ohm.

A circuit for Experiment 14.2 is given in Fig. 14.8. The data to be transmitted is applied at the multiplexer input. For the purpose of this experiment all the multiplexer inputs are held at a high logic level and for this purpose they are connected to + 5 V through a 470 ohm resistor. The multiplexer output is, therefore, at a low logic level. The data input for the demultiplexer is low which is necessary for the demultiplexer to develop a low logic level at the desired output pin.

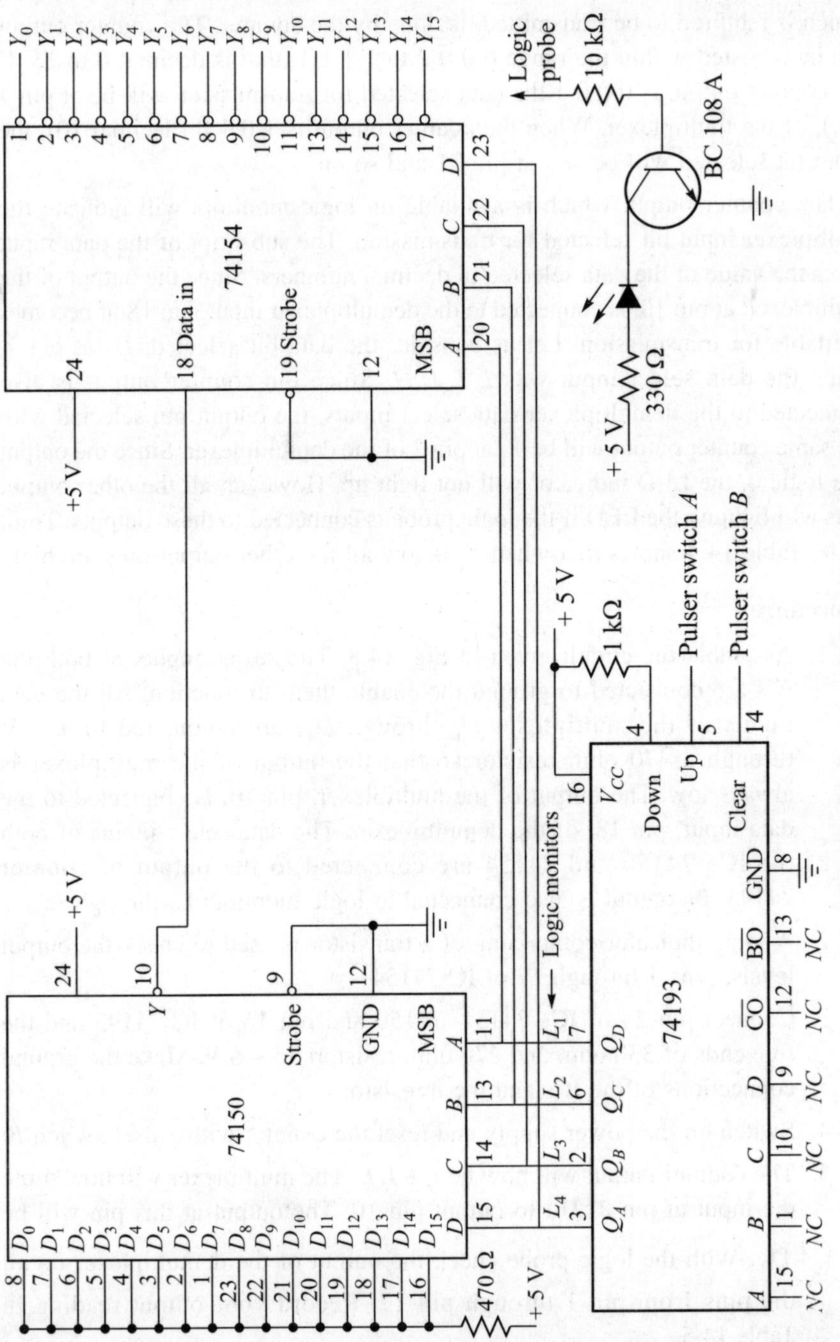

Fig. 14.8. Mutiplexer- Demultiplexer data transmission.

The multiplexer and demultiplexer data select inputs are connected to the output of counter IC 74193. The selection of the data at the multiplexer input, which is required to be transmitted, is done by the counter. The counter output can be adjusted within the range 0 0 0 0 to 1 1 1 1, that is decimal 0 to 15. If the counter output is 0 0 0 1 the data selected for transmission will be at pin 7 (D_1), of the multiplexer. When the counter output is 1 0 1 0 (decimal 10), the input bit selected will be D_{10} at pin 21 and so on.

The counter output which is available on logic monitors will indicate the multiplexer input bit selected for transmission. The subscript of the data input gives the value of the data selected in decimal numbers. Since the output of the multiplexer, at pin 10, is connected to the demultiplexer input, pin 18, it becomes available for transmission. Let us consider the data bit selected, D_1 at pin 7, when the data select input was *L L L H*. Since the counter output is also connected to the demultiplexer data select inputs, the output pin selected with the same counter output will be Y_1 at pin 2 of the demultiplexer. Since the output is a logic 0, the LED indicator will not light up. However, all the other output pins will light up the LED if the logic probe is connected to these outputs. Truth table, Table 14.4 shows that when Y_1 is low all the other output pins are high.

Procedure

1. Assemble the circuit given in Fig. 14.8. The strobe inputs of both the ICs are connected to ground the enable them to function. All the data inputs of the multiplexer D_0 through D_{15} are connected to + 5 V through a 470 ohm resistor so that the output of the multiplexer is always low. The output of the multiplexer, pin 10, is connected to the data input, pin 18, of the demultiplexer. The data select inputs of both the ICs 74150 and 74154 are connected to the output of counter 74193. Its output is also connected to logic monitors L_1 through L_4.
2. A logic indicator comprising of a transistor is used to check the output levels, pins 1 through 17 of IC 74154.
3. Connect pin 24 of ICs 74154, 74150 and pin 16 of IC 74193 and the free ends of 330 ohm and 470 ohm resistors to + 5 V. Make the ground connections of the ICs and the transistor.
4. Switch on the power supply and reset the counter with pulser switch *B*.
5. The counter output will now be *L L L L*. The multiplexer will now route the input at pin 8, D_0, to output pin 10. The output at this pin will be $\overline{D}_0$. With the logic probe check the output of the demultiplexer on all the pins from pin 1 through pin 17. Record your output reading in Table 14.5.

6. Sequence the counter through all the possible states from *L L L L* to *H H H H* and at every stage measure the output level at all the output pins of the demultiplexer and note down the pin numbers which have low output levels.
7. Your observations should tally with Table 14.5 in Appendix 3.

PROBLEMS

14.1. What will be the output of multiplexer IC 74150 when the strobe is high, data inputs are low and are as follows:

(*a*) D_6 ?

(*b*) D_{11} ?

(*c*) D_{14} ?

(*d*) D_0 ?

14.2. In the above example, will it make any difference to the output if the strobe is low and the data inputs are high?

14.3. In Experiment 14.1, a transistor has been used along with an LED. Can you suggest an alternative arrangement so that a transistor is not required?

14.4. Design a multiplexer tree using two ICs 74150 so that it provides for 32 input lines.

14.5. Design a demultiplexer which will satisfy the following truth table (Truth Table AP-14.5)

Truth Table AP-14.5

Control inputs		*Output*
A	*B*	*Y*
0	0	Y_0
0	1	Y_1
1	0	Y_2
1	1	Y_3

14.6. Design a 5-line to 32-line decoder using two ICs 74154.

CHAPTER 15

Seven-Segment LED Display

15.1 INTRODUCTION

Seven-segment displays are widely used in many digital systems, wherever decimal digits are required to be displayed, as in calculators, frequency counters, totalizers etc. These displays are made with LEDs for displaying decimal digits from 0 to 9. A bank of these displays in a row can display very large numbers with several digits. There are some displays which use liquid crystal devices (called LCDs) and they consume very little current. However, we will confine our discussion to LED displays.

15.2 SINGLE DIGIT DISPLAY

A single digit display is shown in Fig. 15.1. You will notice that there are seven segments and they are designated from *a* to *g*. Besides, there is also a decimal point in one corner of the display. In common cathode displays pins 3 and 8 are connected to the negative supply line and in common anode displays these pins are connected to the positive supply line.

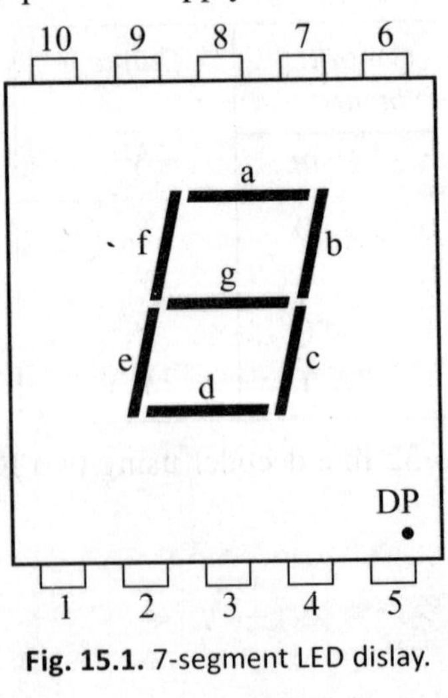

Fig. 15.1. 7-segment LED dislay.

Pin connections for common anode and common cathode displays for some types of displays are given in Table 15.1.

Table 15.1

Pin	*Common anode display FND 507 : FND 567*	*Common cathode display FND 500 : FND 560*
1	Segment *e*	Segment *e*
2	Segment *d*	Segment *d*
3	Common anode	Common cathode
4	Segment *c*	Segment *c*
5	Decimal point	Decimal point
6	Segment *b*	Segment *b*
7	Segment *a*	Segment *a*
8	Common anode	Common cathode
9	Segment *f*	Segment *f*
10	Segment *g*	Segment *g*

Table 15.2 gives the segments required to be illuminated for a given digit. The segments to be illuminated are marked 1 and those not required to be

Table 15.2

BCD Code for 7-segment LED Display

Display	*Decimal digit*	*BCD*		*Code*		*Segment*					*ON*	
		A	*B*	*C*	*D*	*a*	*b*	*c*	*d*	*e*	*f*	*g*
0	0	0	0	0	0	1	1	1	1	1	1	
1	1	0	0	0	1		1	1				
2	2	0	0	1	0	1	1		1	1		1
3	3	0	0	1	1	1	1	1	1		o	1
4	4	0	1	0	0		1	1			1	1
5	5	0	1	0	1	1		1	1		1	1
6	6	0	1	1	0		1	1	1	1	1	
7	7	0	1	1	1	1	1	1				
8	8	1	0	0	0	1	1	1	1	1	1	1
9	9	1	0	0	1	1	1	1			1	1

illuminated are marked 0. The table also gives the decimal numbers displayed as well as the BCD code for each decimal number. In actual practice, a BCD counter is used which is followed by a decoder/driver. The decoder/driver decodes the decimal states of the input from 0 through 9 and develops the seven output signals, one for each segment of the display to operate the LED display.

15.3 BCD-TO-SEVEN-SEGMENT DECODER DRIVER

We will consider two decoder-drivers, ICs 7447A and 7448. These ICs can be used to drive 7-segment LED displays. The pin connections for both these ICs are the same and are given in Fig. 15.2. However, there is some

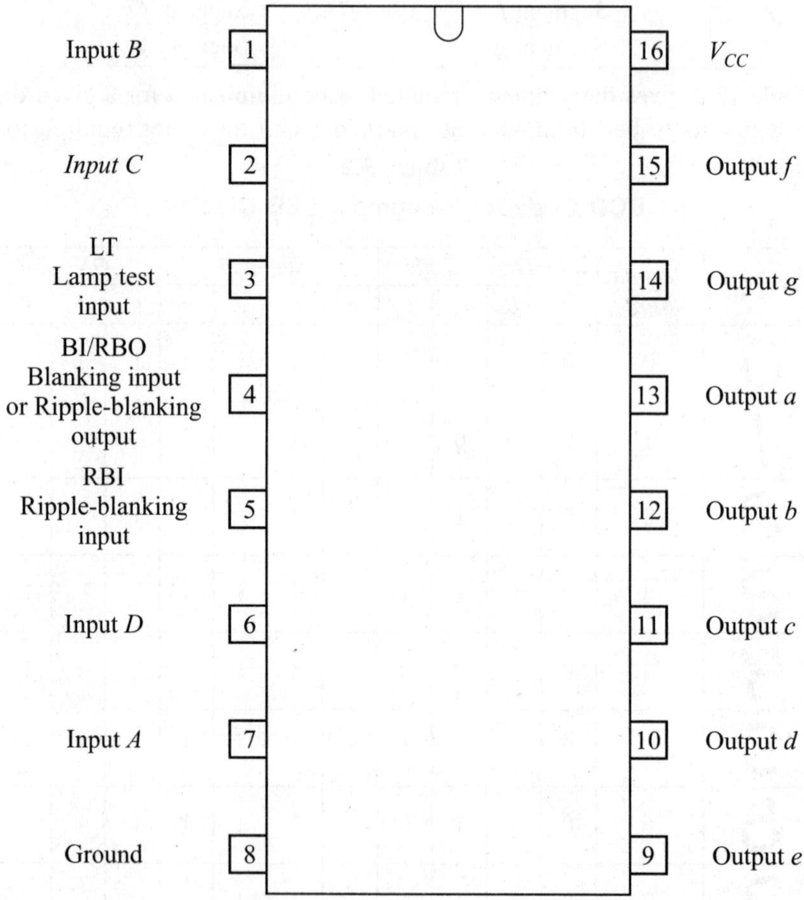

Fig. 15.2. Pin connections (top view) for ICs 7447A and 7448.

difference between these ICs. Whereas IC 7448 incorporates voltage dropping resistors for the LED display, the external resistors are required to be used with IC 7447A. Another significant difference is that IC 7448 is used with common cathode displays while IC 7447A is meant for use with common anode displays. Tables 15.3 and 15.4 give the truth tables for ICs 7447A and 7448.

Table 15.3

Truth Table for IC 7447A

Decimal or Function	*Inputs*						*BI*	*Outputs*							*Note*
	LT	*RBI*	*D*	*C*	*B*	*A*	*RBO*	*a*	*b*	*c*	*d*	*e*	*f*	*g*	
0	H	H	L	L	L	L	H	L	L	L	L	L	L	H	
1	H	X	L	L	L	H	H	H	L	L	H	H	H	H	
2	H	X	L	L	H	L	H	L	L	H	L	L	H	L	
3	H	X	L	L	H	H	H	L	L	L	L	H	H	L	
4	H	X	L	H	L	L	H	H	L	L	H	H	L	L	
5	H	X	L	H	L	H	H	L	H	L	L	H	L	L	
6	H	X	L	H	H	L	H	H	H	L	L	L	L	L	
7	H	X	L	H	H	H	H	L	L	L	H	H	H	H	
8	H	X	H	L	L	L	H	L	L	L	L	L	L	L	
9	H	X	H	L	L	H	H	L	L	L	H	H	L	L	1
10	H	X	H	L	H	L	H	H	H	H	L	L	H	L	
11	H	X	H	L	H	H	H	H	H	L	L	H	H	L	
12	H	X	H	H	L	L	H	H	L	H	H	H	L	L	
13	H	X	H	H	L	H	H	L	H	H	L	H	L	L	
14	H	X	H	H	H	L	H	H	H	H	L	L	L	L	
15	H	X	H	H	H	H	H	H	H	H	H	H	H	H	
BI	X	X	X	X	X	X	L	H	H	H	H	H	H	H	2
RBI	H	L	L	L	L	L	L	H	H	H	H	H	H	H	3
LT	L	X	X	X	X	X	H	L	L	L	L	L	L	L	4

H = Binary 1; L = Binary 0; X = Don't care

Table 15.4

Truth Table for IC 7448

Decimal or Function	Inputs						BI	Outputs							Note
	LT	*RBI*	*D*	*C*	*B*	*A*	*RBO*	*a*	*b*	*c*	*d*	*e*	*f*	*g*	
0	H	H	L	L	L	L	H	H	H	H	H	H	H	L	
1	H	X	L	L	L	H	H	L	H	H	L	L	L	L	
2	H	X	L	L	H	L	H	H	H	L	H	H	L	H	
3	H	X	L	L	H	H	H	H	H	H	H	L	L	H	
4	H	X	L	H	L	L	H	L	H	H	L	L	H	H	
5	H	X	L	H	L	H	H	H	L	H	H	L	H	H	
6	H	X	L	H	H	L	H	L	L	H	H	H	H	H	
7	H	X	L	H	H	H	H	H	H	H	L	L	L	L	
8	H	X	H	L	L	L	H	H	H	H	H	H	H	H	
9	H	X	H	L	L	H	H	H	H	H	L	L	H	H	1
10	H	X	H	L	H	L	H	L	L	L	H	H	L	H	
11	H	X	H	L	H	H	H	L	L	H	H	L	L	H	
12	H	X	H	H	L	L	H	L	H	L	L	L	H	H	
13	H	X	H	H	L	H	H	H	L	L	H	L	H	H	
14	H	X	H	H	H	L	H	L	L	L	H	H	H	H	
15	H	X	H	H	H	H	H	L	L	L	L	L	L	L	
BI	X	X	X	X	X	X	L	L	L	L	L	L	L	L	2
RBI	H	L	L	L	L	L	L	L	L	L	L	L	L	L	3
LT	L	X	X	X	X	X	H	H	H	H	H	H	H	H	4

H = Binary 1; L = Binary 0; X = Don't care

The various functions which can be performed by decoder/drivers ICs 7447A and 7448 which are mentioned in their Truth Tables 15.3 and 15.4 have been discussed below.

1. Inputs BI/RBO represent Blanking input/Ripple blanking output. These inputs are held open or at a high logic level for output functions

from 0 through 15. This input can also be used for controlling the intensity of the display by applying a variable duty cycle pulse signal at this input.

2. If it is desired that decimal 0 is not displayed the Ripple blanking input, RBI, must be held low.
3. When a low logic level is applied to the Blanking input, BI, all segment outputs are off irrespective of the state of the other inputs.
4. When the Ripple blanking input, RBI, and inputs *A, B, C* and *D* are at a low logic level, with the Lamp test input, LT, high, all segment outputs will go off and the Ripple blanking output, RBO, will go to a low level.
5. When the Blanking input/Ripple blanking output, BI/RBO is open or high, and a low is applied to the Lamp test input, all segments will be on. This test can be applied at any stage to check if all segments are in order.
6. Decimal numbers 10 through 14 are displayed by special symbols as shown in Fig. 15.3.

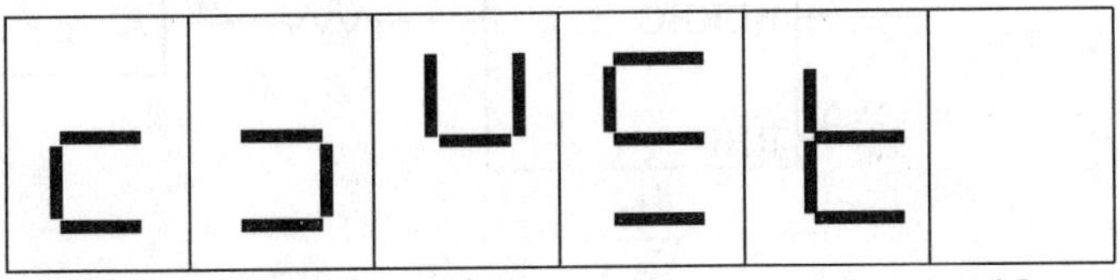

10 11 12 13 14 15

Fig. 15.3

EXPERIMENT 15.1 : SEVEN-SEGMENT LED DISPLAY WITH IC 7447A

Objective

To demonstrate the display function performed by Decoder/Driver IC 7447A when used with 7-segment LED display FND 567.

Materials Required

Logic trainer

IC 7447A : Decoder/Driver

7-segment LED display : FND 567

Resistors 330 ohm (7)

Resistor 470 ohm.

A circuit diagram for Experiment 15.1 is given in Fig. 15.4.

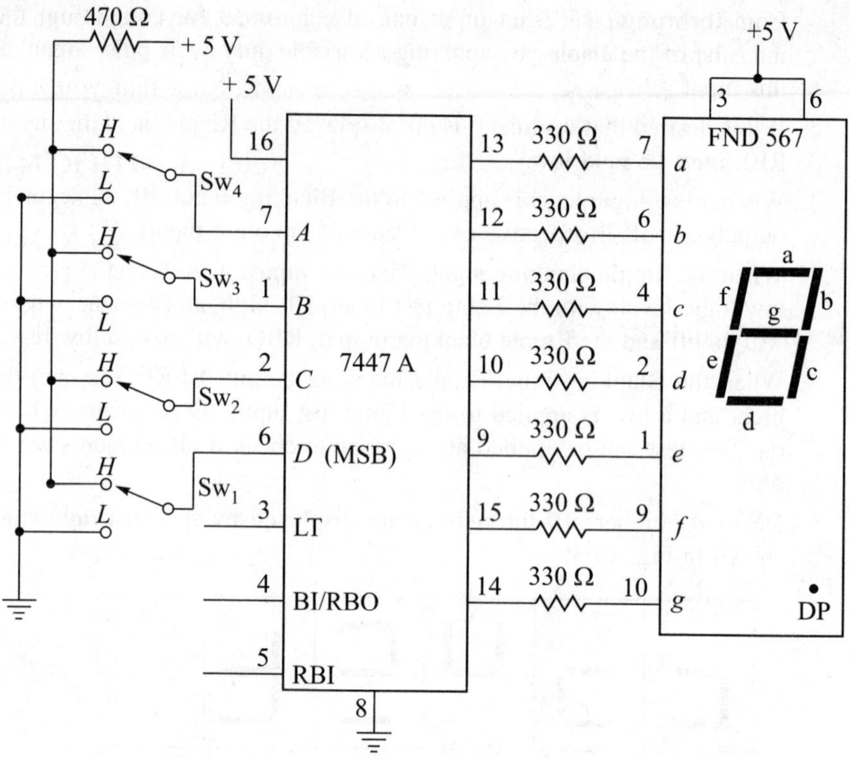

Fig. 15.4. Circuit diagram for Experiment 15.1.

In Experiment 15.1, we will consider the basic display functions of IC 7447A, that is we will pay attention only to the display of decimal digits from 0 through 9. The display of decimal numbers from 10 through 15 will be accomplished by special symbols. The other functions listed in Truth Table 15.3 will be considered in Experiment 15.2 when we deal with decoder/ driver IC 7448.

Procedure

1. Assemble the circuit given in Fig. 15.4. Connect pin 16 of IC 7447A, pins 3 and 8 of the LED display and the free end of 470 Ohm resistor to + 5 V. Also connect pin 8 of the IC 7447A to ground.
2. Connect pin 4, BI/RBO to a high logic level (that is through the 470 ohm resistor to + 5 V). Also connect pins 3 and 5 to high logic level.
3. Switch on the power supply. Apply input 0 0 0 0 at the *A, B, C* and *D* inputs. A zero (0) will now be displayed.
4. Apply *B, C* and *D* inputs corresponding to functions 0 through 9 as given in Table 15.3. Digits 0 through 9 will now be displayed.

5. Now apply *B, C* and *D* inputs corresponding to functions 10 through 15. The display will now be as given in Fig. 15.3.
6. Keep a record of your observations to make sure that you have understood the various functions discussed here.

EXPERIMENT 15.2 : SEVEN-SEGMENT LED DISPLAY WITH IC 7448

Objective

To demonstrate the various functions performed by Decoder/Driver IC 7448 when used with 7-segment LED display FND 500/560.

Materials Required

Logic trainer
IC 7448 : Decoder/Driver
7-segment LED display : FND 500/560
Resistor 470 ohm.

A circuit diagram for Experiment 15.2 is given in Fig. 15.5.

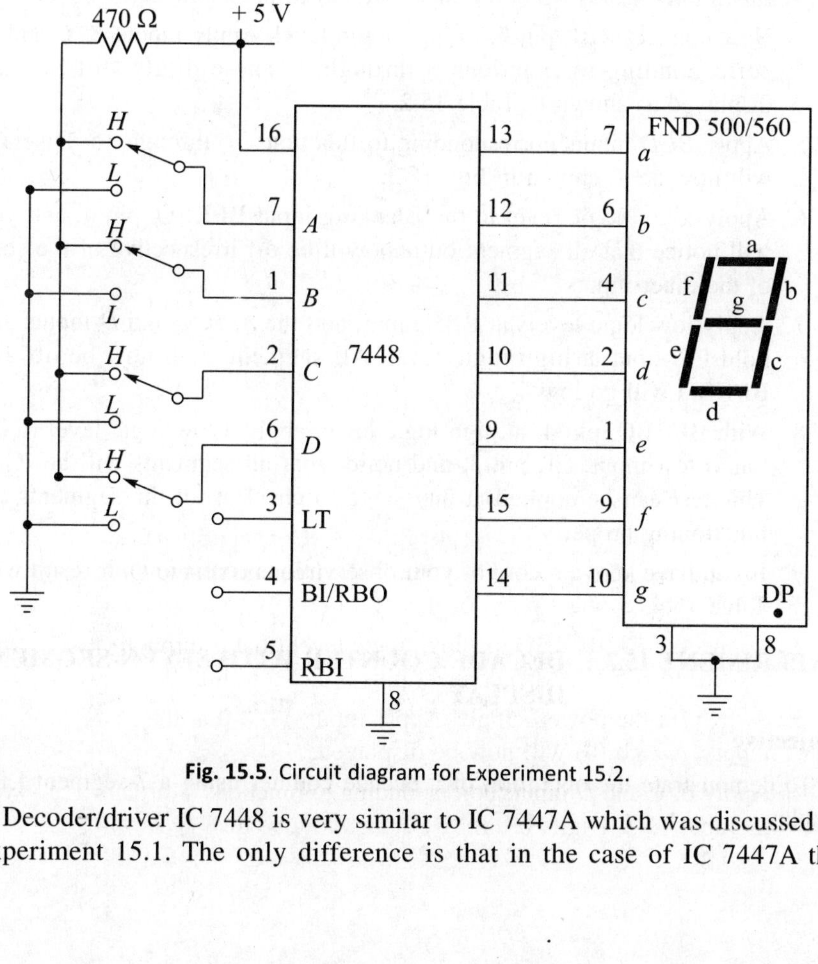

Fig. 15.5. Circuit diagram for Experiment 15.2.

Decoder/driver IC 7448 is very similar to IC 7447A which was discussed in Experiment 15.1. The only difference is that in the case of IC 7447A the segments which are on have low output, whereas in the case of IC 7448 the on segments have high output. It is for this reason that common anode displays are used with IC 7447A and common cathode displays are used with IC 7448. As IC 7447A does not incorporate voltage dropping resistors, they have been used externally to limit the current to a safe level. Resistors of 330 ohm were used to limit the current to about 10 mA. IC 7448 does not require external resistors.

Procedure

1. Assemble the circuit given in Fig. 15.5. Connect pin 16 of IC 7448 to + 5 V and pin 8 of the IC and pins 3 and 8 of the LED display to ground.
2. Switch on the power supply. Refer to Truth Table 15.4 for IC 7448 and first of all verify the functions marked 0 through 9. Apply high inputs at pins 3, 4 and 5 (that is, pins marked LT, RBI and BI/RBO). Apply a 0 0 0 0 input at pins marked *A, B, C* and *D.* A 0 will now be displayed.
3. When you connect RBI pin 5, to ground the 0 will disappear. This function is useful when 0 is not required to be displayed.
4. Now connect RBI, pin 5, to high logic level. Apply inputs *B, C* and *D* corresponding to functions 0 through 9. These digits will now be displayed as shown in Table 15.2.
5. Apply *BCD* inputs corresponding to functions 10 through 15. Displays will now be as shown in Fig. 15.3.
6. Apply a low logic level to the Blanking input BI/RBO, pin 4, and you will notice that all segment outputs will be off irrespective of the state of the other inputs.
7. Apply low logic levels at RBI, pin 5, and the *A, B, C* and *D* inputs and hold LT input at high logic level. All segments will now be off and BI/RBO will go low.
8. With BI/RBO, pin 4, at high logic level, apply a low logic level to the Lamp test input, LT, pin 3, and notice that all segments will be 'On'. This test can be applied at any stage to check if all the segments are functioning properly.
9. If you have kept a record of your observations compare your result with Table 15.4.

EXPERIMENT 15.3 : DECADE COUNTER WITH SEVEN-SEGMENT DISPLAY

Objective

To demonstrate the operation of a decade counter using a 7-segment LED display.

Materials Required

Logic trainer
IC 7448 : Decoder/Driver
IC 7490A : Decade counter
7-segment display : FND 500/560.

A circuit diagram for Experiment 15.3 is given in Fig. 15.6.

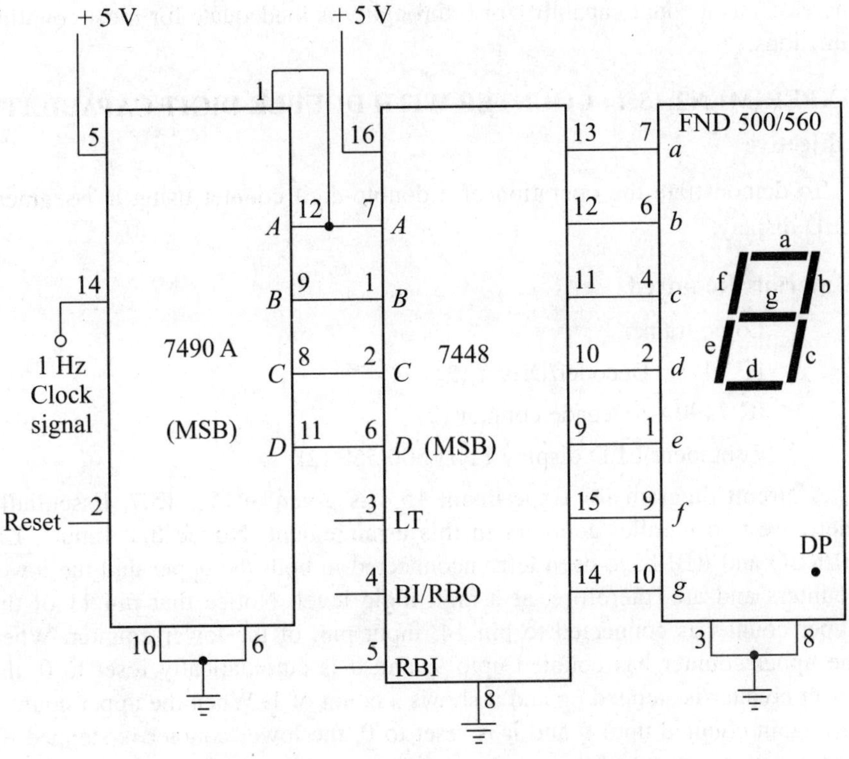

Fig. 15.6. Circuit diagram for Experiment 15.3.

The circuit for Experiment 15.3, which is given in Fig. 15.5, incorporates a BCD counter which was not there in previous circuits. This enables the circuit to count the input pulses in the standard 8421 BCD code. The output of the counter is fed to decoder/driver IC 7448 which drives the display.

Procedure

1. Assemble the circuit given in Fig. 15.6. Notice that pin 1 of IC 7490A is connected to pin 12, which enables it to function as a decade counter. Connect pin 5 of IC 7490A and pin 16 of IC 7448 to + 5 V and make the ground connections as shown in Fig. 15.6.
2. Connect a 1 Hz clock signal to pin 14 of IC 7490A and switch on the power supply.

3. Reset the counter by taking the reset, pin 2, high momentarily. The circuit will start counting from 0 through 9. When it has reached the count of 9 the counter will be reset to 0 and the same counting cycle will be repeated.
4. You will notice that inputs, LT, BI/RBO and RBI are left unconnected and therefore at a high logic level as required for this circuit.

We will consider in Experiment 15.4 how the count capability can be increased as a count capability of 0 through 9 is inadequate for most counting functions.

EXPERIMENT 15.4 : COUNTER WITH DOUBLE-DIGIT CAPABILITY

Objective

To demonstrate the operation of a double-digit counter using a 7-segment LED display.

Materials Required

Logic trainer

IC 7448 : Decoder/Driver (2)

IC 7490A : Decade counter (2)

7-segment LED display FND 500/560 (2).

A circuit diagram for Experiment 15.4 is given in Fig. 15.7. Essentially there are two parallel counters in this arrangement. Notice that inputs, LT, BI/RBO and RBI have been left unconnected in both the upper and the lower counters and are, therefore, at a high logic level. Notice that pin 11 of the upper counter is connected to pin 14, input pin, of the lower counter. When the upper counter has counted upto 9, and it is automatically reset to 0, the lower counter is stepped up and it shows a count of 1. When the upper counter has again counted upto 9 and it is reset to 0, the lower counter is stepped up and it shows a count of 2.

In this way when the circuit has counted upto 99, both counters will show a count of 0 0 at the next input pulse.

Notice that the reset inputs of both the upper and lower counters are connected together. Before starting the counting operation the reset input is momentarily taken high, which resets both the counters. The count capability can be increased by adding more counting circuits.

Procedure

1. Assemble the circuit given in Fig. 15.7. Leave pins 3, 4 and 5 of ICs 7448 unconnected. Notice that both ICs 7490A have their pin 1 connected to pin 12 and, therefore, they will function as decade counters.

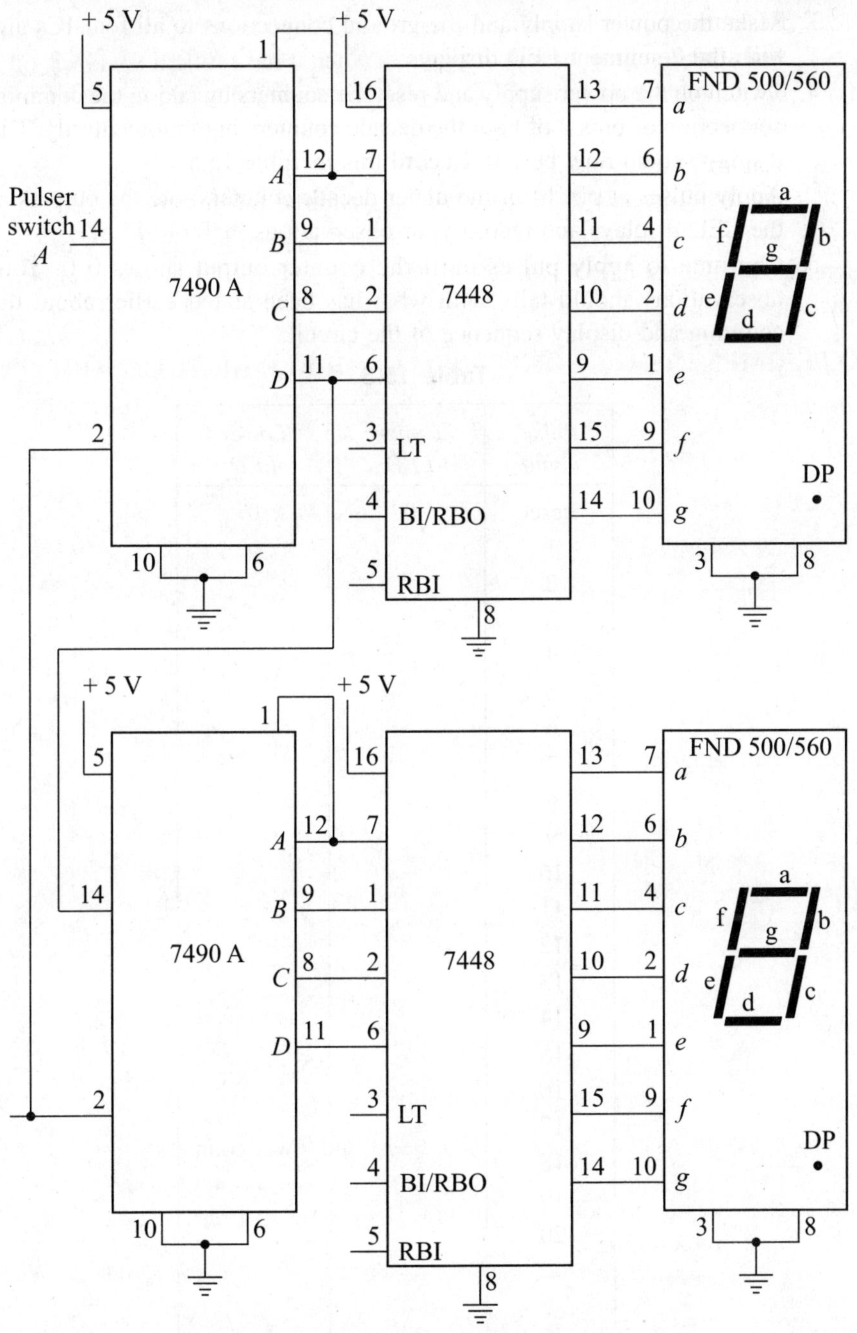

Fig. 15.7. Circuit diagram for Experiment 15.4.

2. Connect pin 14 of upper IC 7490A to pulse switch *A* and also connect pin 11 of the upper IC 7490A to pin 14 of the lower IC 7490A.

3. Make the power supply and the ground connections to all four ICs and both the 7-segment LED displays.
4. Switch on the power supply and reset the counters by taking the common connection of pins 2 of both the decade counters high momentarily. The display should now be 0 0. Record this in Table 15.5.
   Apply pulses at pin 14 of the upper decade counter, note the outputs in the LED displays and record your observations in Table 15.5.
5. Continue to apply pulses until the counter output shows 0 0. Your observations should tally with what has been stated earlier about the counting and display sequence of the circuit.

Table 15.5

Pulse count	*Upper LED*	*Lower LED*
Reset	0	0
1		
2		
3		
4		
5		
6		
7		
8		
9		
10		
11		
12		
13		
14		
15		
16		
17		
18		
19		
20		
.		
.		
.		
.		
99		
100		

15.4 DIGITAL DISPLAY WITH MEMORY

In Experiment 15.4, you had assembled a counter using a BCD counter 7490A, a decoder/driver 7448 and a 7-segment LED display. In this circuit as soon as a clock pulse was applied at the counter input, the counter output was displayed by the LED display. In the experiment that you will now perform a memory register has been incorporated between the BCD counter and the decoder/driver. The BCD counter output is stored in the memory register and it is displayed only when the memory contents are transferred to the decoder/driver. The memory register used in this experiment is IC 7495A. This register has been used in the parallel-in/parallel-out mode. You have studied the operation of this register in Experiment 11.1.

The input, pin 14, of IC 7490A is connected to pulser switch *A* which is used to step up the counter. This counter is reset by applying a momentary high input at reset pin 2. In normal operation the reset pin is held low by connecting it to ground.

The BCD counter output will appear at the input of the memory register IC 7495A. When a pulse is applied at the clock input, pin 8, of the register, the contents at the input of the register are transferred to its output and are displayed after processing by the decoder/driver.

EXPERIMENT 15.5 : COUNTER DISPLAY USING MEMORY REGISTER

Objective

To study the operation of a digital display with memory.

Materials Required

Logic trainer

IC 7490A : Decade counter

IC 7495A : Shift register

IC 7448 : BCD-to-7-segment Decoder/Driver

7-segment LED display : FND 500/560

Resistor 470 ohm

A circuit diagram for Experiment 15.5 is given in Fig. 15.8.

Procedure

1. Assemble the circuit given in Fig. 15.8. Connect pulser switch *A* to pin 14 of IC 7490A and pulser switch *B* to pin 8 of IC 7495A. Make the power supply and ground connections shown in Fig. 15.8. Also remember to connect pin 6 (Mode control) of IC 7495A to + 5 V through a 470 ohm resistor. Make the remaining connections shown in the diagram.

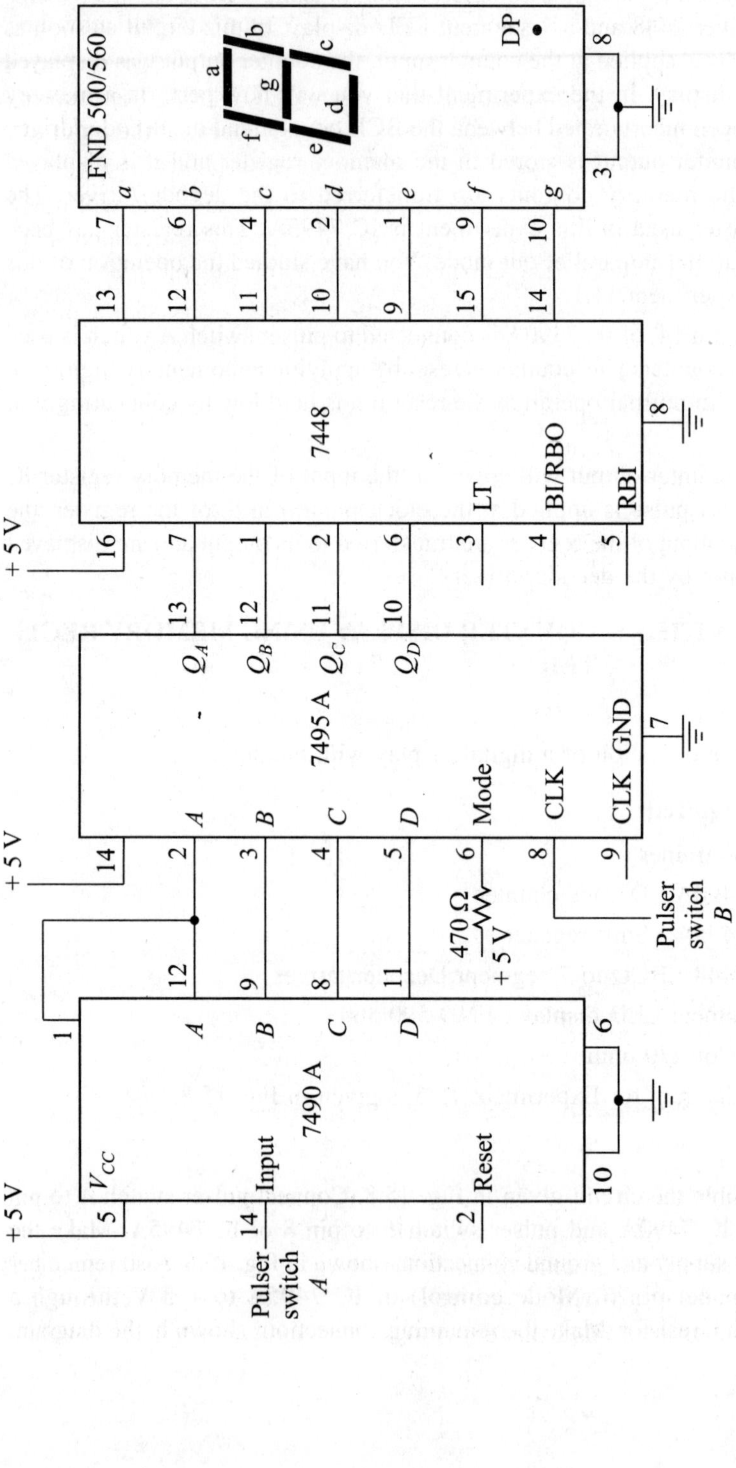

Fig. 15.8. Circuit diagram for Experiment 15.5.

2. Switch on the power supply and reset the counter by applying a momentary high input at reset pin 2 of IC 7490A.
3. Apply a single pulse at pin 8 of IC 7495A. A zero (0) will now be displayed by the LED display.
4. Apply a single clock pulse at the input of counter IC 7490A. The counter output will now be 0 0 0 1. This output will also appear at the input of the shift register IC 7495A; but it will not yet be displayed.
5. Apply a pulse at pin 8 of IC 7495A. The contents at its input will now appear at its output and will be displayed after being decoded and processed by IC 7448. The display will now show a count of 1.
6. Step up the counter by one step at a time and after every step apply a pulse at pin 8 of IC 7495A. You will notice the display changing after every step from 1 through 9.

CHAPTER 16

Semiconductor Memory

16.1 INTRODUCTION

Memories are an essential requirement of most digital systems, especially computers, microprocessors, calculators and you can think of many more devices. Memories are used to store data and to make it available for use as and when required. Where a large amount of data is to be stored, bulk storage devices like tapes and discs are commonly used. When the storage capacity required is not so large semiconductor memories are employed.

If data is not required to be retrieved from memory for whatever purpose, read only memories called ROMs serve the purpose. These memories are dedicated to perform a particular task.

There are situations when data storage is of a temporary nature. Devices which provide this facility are known as RAMs, random access read-write memories. In these devices, data can be written into or read there from. Semiconductor memories consist of an array of memory cells arranged in a rectangular pattern and are fabricated on a silicon wafer using bipolar or MOS technology. Each memory cell stores one bit of data, a binary 0 or 1. Logic circuits are used for reading the contents from memory. The outputs are buffered, which ensures that the memory contents are not lost during the read operation.

There are other memory devices which employ more recent technologies; but we are confining our discussion to RAMs and we have chosen RAM 7489, which is very representative of this class of devices, called static RAMs. If you get a good grasp of this device you will be in a good position to make use of RAMs with larger capabilities.

16.2 RANDOM ACCESS MEMORY (RAM) : IC 7489

A block diagram of RAM 7489 is given in Fig. 16.1 and pin connections and logic symbol in Fig. 16.2 and Fig. 16.3. This RAM has 64 memory cells arranged in 16 rows and 4 columns. There are 4 address data inputs, which are used to select one of the sixteen 4-bit words in the memory. The words are numbered from 0 to 15. There are four data input lines from D_1 through D_4 into which data is fed for being written into the memory. There are four outputs from $\overline{O}_1$ through $\overline{O}_4$. You must remember that data at the output is the complement of the data in the memory.

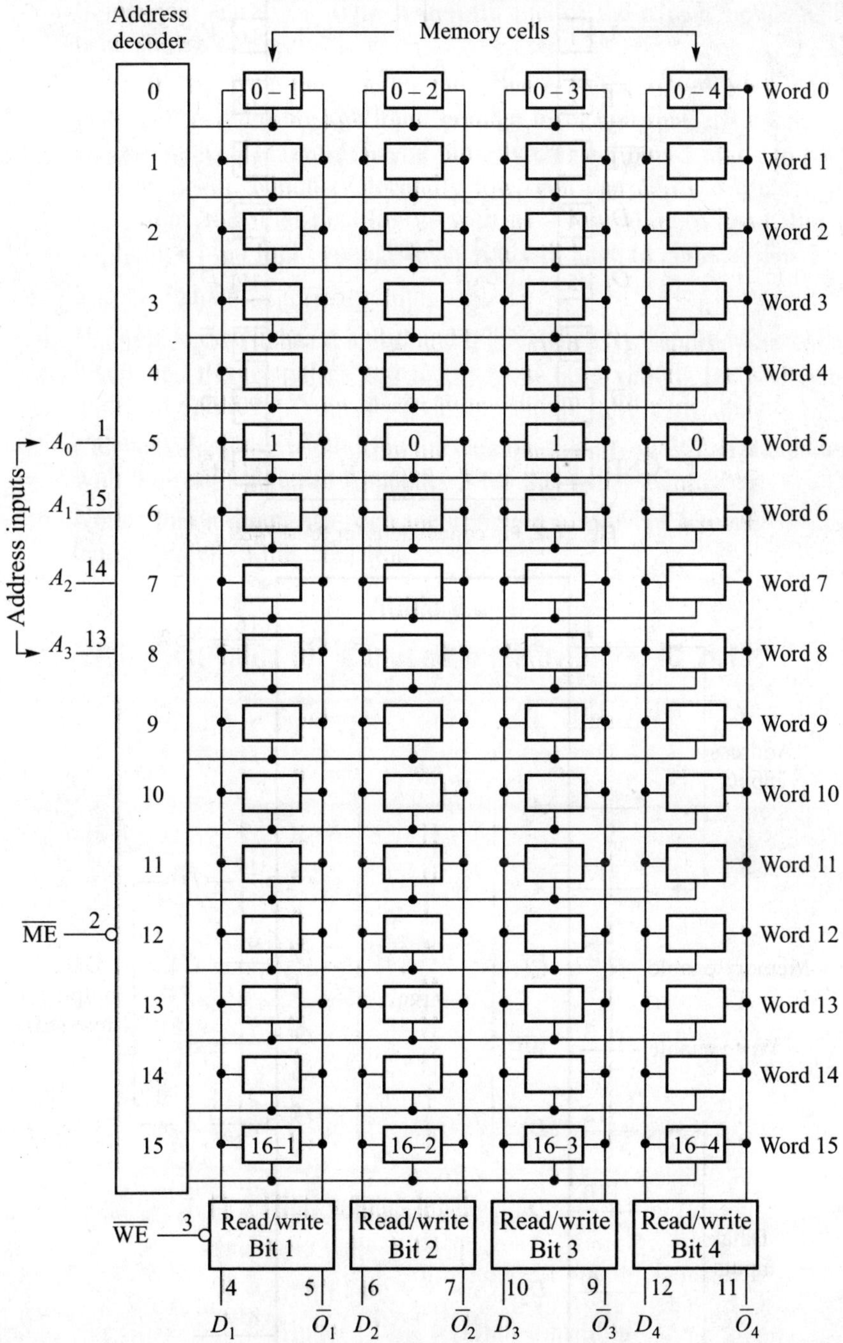
Address decoder
Memory cells
0 – 1
0 – 2
0 – 3
0 – 4
Word 0
Word 1
Word 2
Word 3
Word 4
Word 5
Word 6
Word 7
Word 8
Word 9
Word 10
Word 11
Word 12
Word 13
Word 14
Word 15
16–1
16–2
16–3
16–4
Address inputs
A_0
A_1
A_2
A_3
$\overline{ME}$
$\overline{WE}$
Read/write Bit 1
Read/write Bit 2
Read/write Bit 3
Read/write Bit 4
D_1
$\overline{O}_1$
D_2
$\overline{O}_2$
D_3
$\overline{O}_3$
D_4
$\overline{O}_4$

Fig. 16.1

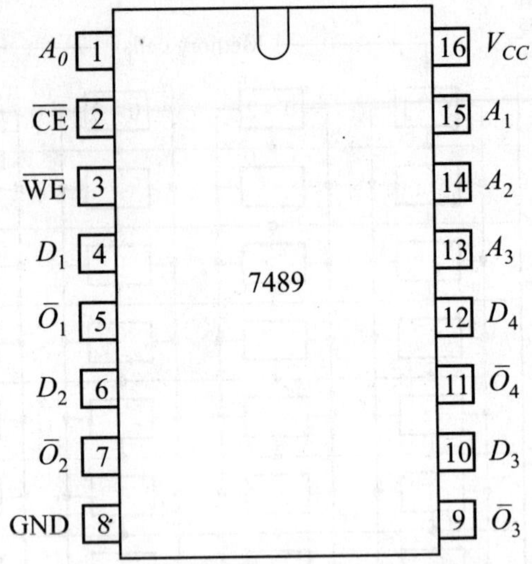

Fig. 16.2. Pin connections for RAM 7489.

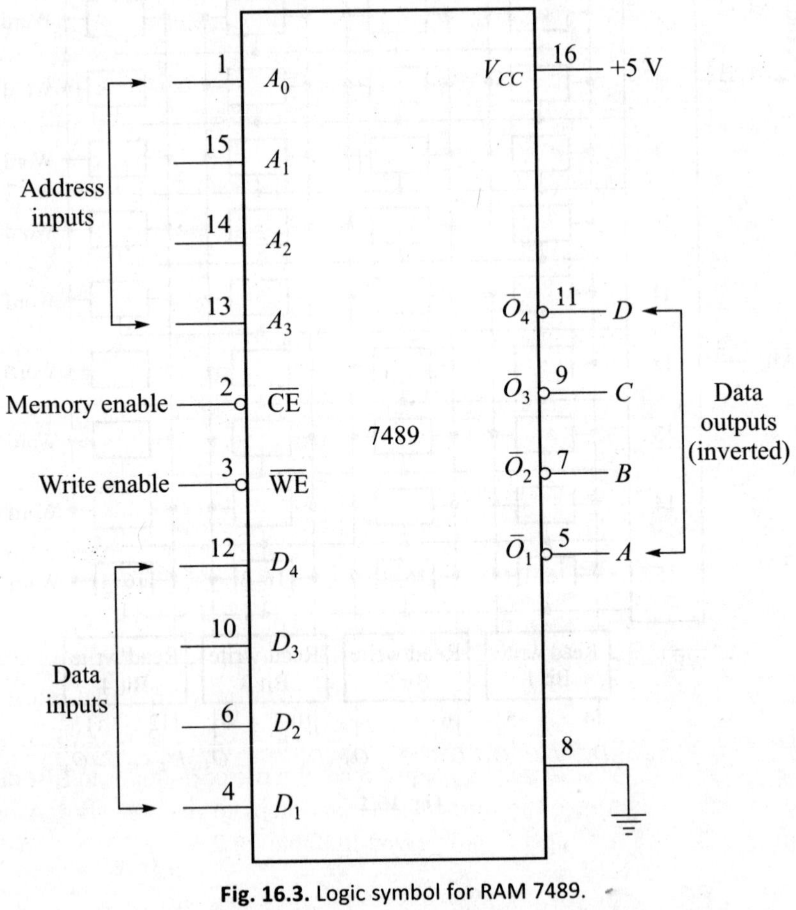

Fig. 16.3. Logic symbol for RAM 7489.

There are two enable inputs which are labelled Memory Enable, $\overline{ME}$ and the other Write Enable, $\overline{WE}$. Memory Enable, $\overline{ME}$ is held low both for reading data from the memory, as well as for writing data into the memory. Moreover, in either case, the address input is used to select one of the sixteen rows from which the word in the memory is to be read or written into.

For the read operation, $\overline{ME}$ is held low and $\overline{WE}$ high. The 4-bit data word from the row selected by the address input can now be read from the output. However, the data at the output will be the complement of the data in the memory. For instance, if the word in the memory is 1 1 0 0, the word at the output will be 0 0 1 1.

As in the read operation, when you have to write into the memory, select the appropriate input address to select the row, apply the data to be written into the memory at the data inputs, and hold both $\overline{ME}$ and $\overline{WE}$ low. The data will go into the memory at the selected address and it will also appear at the output in a complemented form.

Since the output of the RAM has open-collector transistors, pull-up resistors will be required to be connected to V_{CC} from each output. If you want the output to be available in an uncomplemented form, Inverters will have to be used at each output. If you refer to Experiment 16.1 on this RAM, you will see how Inverters have to be connected.

A function table for RAM 7489 gives a summary of what has been discussed above about the operational requirements of this RAM. Table 16.1 is the function table for this RAM.

Table 16.1

Function Table for RAM IC 7489

Inputs		*Operation*	*State of outputs*
$\overline{ME}$	$\overline{WE}$		
L	L	Write	Complement of data inputs
L	H	Read	Complement of selected word
H	L	Inhibit entry	Undetermined
H	H	Hold	High

You will also notice from the function table, Table 16.1 that when both $\overline{ME}$ and $\overline{WE}$ go high, the outputs go high, which indicates that the RAM is disabled. In fact, the right way to begin a read or write operation is to first disable the RAM and then take the further necessary steps to read or write. A proper sequence for the read operation is given in Table 16.2.

Table 16.2
Read Sequence for IC 7489

Sequence of operations	*Control inputs*		*State of outputs*
	$\overline{ME}$	$\overline{WE}$	
1. Disable RAM	H	H	All outputs go high
2. Select address	H	H	All outputs are still high
3. Read data	L	H	Data at the selected address in the memory available at the outputs
4. Disable chip	H	H	All outputs go high

For reading data at other addresses you will have to repeat steps 2 to 4 after every read operation.

While following the read operation in Table 16.2, also refer to Fig. 16.4, which gives the switching characteristics of the RAM in the read cycle. As a first step you should disable the RAM by holding both $\overline{\text{ME}}$ and $\overline{\text{WE}}$ high. The next step is to select the address if the word in the memory is to be read. You will notice from Fig. 16.4 that the Memory Enable is brought low at the same time. However, it takes some time for the data to appear at the output. This propagation delay, t_{PHL}, is the period of time it takes for stable data to appear at the output after the transition of $\overline{\text{ME}}$ from high to low level. The maximum time delay is 50 n and it is typically 33 ns. You will also notice from Fig. 16.4 that the $\overline{\text{ME}}$ and address input lines are held stable during the entire read operation. Figure 16.4 also shows that the output is the complement of the data word stored in the memory.

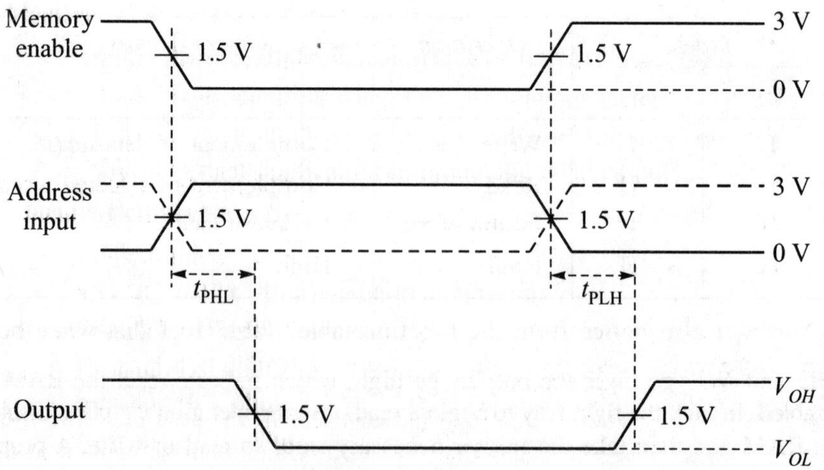

Fig. 16.4. Switching characteristics of RAM 7489 in the Read cycle.

There is another significant time delay, t_{PHL}, which indicates the time it takes for the outputs to return to the high state after the transition of $\overline{ME}$ from low to high level. The RAM is now disabled as both $\overline{ME}$ and $\overline{WE}$ are high and you can now read the next word. This delay is typically 26 ns although the maximum listed value is 50 ns.

The proper sequence to be followed for the write operation for RAM IC 7489 is given in Table 16.3.

Table 16.3

Write Sequence for IC 7489

Sequence of operations	*Control inputs*		*State of outputs*
	$\overline{ME}$	$\overline{WE}$	
1. Disable RAM	H	H	All outputs go high
2. Select address	H	H	All outputs are still high
3. Apply at the data inputs the word to be stored in the memory.	H	H	All outputs are still high
4. Write data into the memory at the selected address	L	L	Data is written into the memory and it also appears at the output
5. Disable chip	H	H	All outputs go high

While following the write sequence given in Table 16.3 also refer to Fig. 16.5 which gives the switching characteristics of RAM 7489 in the write cycle.

The sequence to be followed for writing a word into the memory of the RAM is as follows:

1. Disable the RAM by taking both $\overline{ME}$ and $\overline{WE}$ high.
2. Select the address in the memory where the word is to be stored.
3. Apply at the data inputs the word to be stored. During these operations the output will remain high.
4. Now hold $\overline{ME}$ low.
5. The address input takes time t_1 to get stabilized.
6. Take $\overline{WE}$ low after this time has elapsed. In the case of RAM 7489 t_1 is 0.0 ns, which means that the address input can be applied at the

same time or before $\overline{WE}$ goes low. It is important that $\overline{WE}$ should be held low for a minimum period of time t_w, so that the data can be written into the memory.

7. Another important requirement is that the address input should remain stable for a time t_2, called the address hold time, after $\overline{WE}$ goes high.
8. Finally the data input should remain stable for a period t_{su}, called the data set up time before $\overline{WE}$ goes high, and for a period, t_h, called the data hold time, after $\overline{WE}$ goes high.
9. Lastly, after the write operation has been completed, and $\overline{ME}$ goes high, it will take time t_{sr} called the sense recovery time, for the outputs to return to the high state.

Explanation of the several time parameters mentioned in Fig. 16.5 is as follows:

t_1 **Address set up time:**

This is the minimum period after the address gets stabilized and before $\overline{WE}$ goes low. For RAM 7489 the minimum time required is 0.0 ns, that is the two events may occur at the same time.

t_2 **Address hold time:**

This is the minimum period of time for the address input to remain stable after $\overline{WE}$ goes high. The minimum time for this RAM is 5 ns.

t_w **Write enable low state:**

This is the minimum period of time for which $\overline{WE}$ should be held low for the data to be stored in the memory. According to data sheets it is 40 ns.

t_{su} **Data set up time:**

It is measured from the point when the data becomes stable upto the point when $\overline{WE}$ goes high. (The data input should get stabilized for a minimum period of time prior to $\overline{WE}$ going low. In this case t_{su} is the same as t_w.

t_h **Data hold time:**

It is the time for which the data must remain stable for a period after $\overline{WE}$ goes high. It is about 5 ns minimum.

t_{sr} **Sense recovery time:**

This is the time after the write operation is completed, that is when $\overline{WE}$ goes high, and the outputs return to the high state. The data sheets specify it as 70 ns.

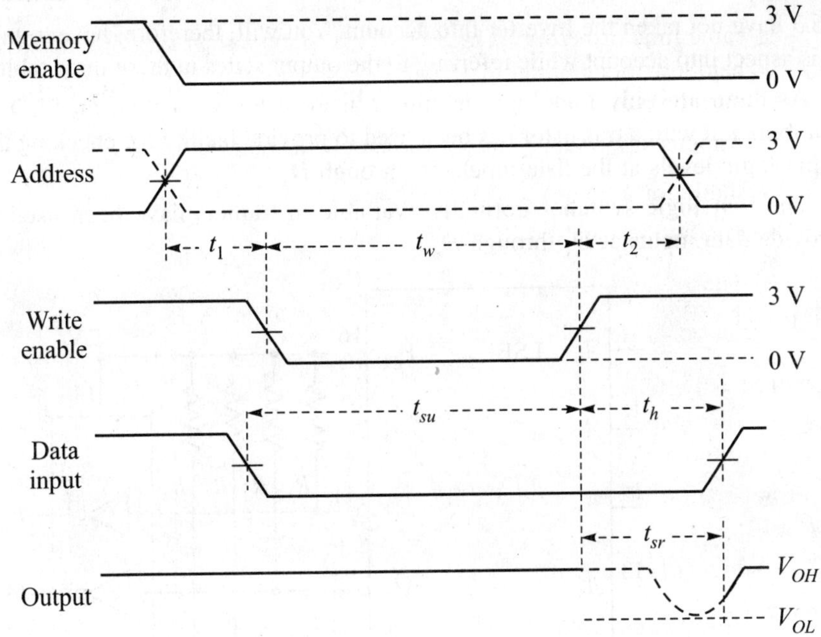

Fig. 16.5. Switching characteristics of RAM 7489 in the Write cycle.

EXPERIMENT 16.1 : RANDOM ACCESS MEMORY (RAM) : IC 7489

Objective

To study the operating procedures for:

1. writing 4-bit words into the memory of RAM 7489;
2. reading words recorded in the memory of this RAM.

Materials Required

Logic trainer

IC 7489 : Random Access Memory (RAM)

IC 7404 : Hex Inverter

BC 108A : NPN Transistor

Resistor 330 ohm

Resistor 470 ohm

Resistors 1 k ohm (4)

Resistor 10 k ohm

LED (Red).

A circuit diagram for Experiment 16.1 is given in Fig. 16.6. As the outputs of RAM 7489 are active-low, we have used Inverters at all the four outputs so that the outputs will be high. The outputs in Tables 16.1, 16.2 and

16.3 have not taken the Inverter into account. You will, therefore, have to take this aspect into account while referring to the output states in these three tables.

As there are only four logic monitors in most logic trainers, an LED in combination with a transistor has been used to provide facility for checking the input logic levels at the data inputs D_1 through D_4.

The four logic switches normally available in trainers have been used to provide data inputs at D_1 through D_4.

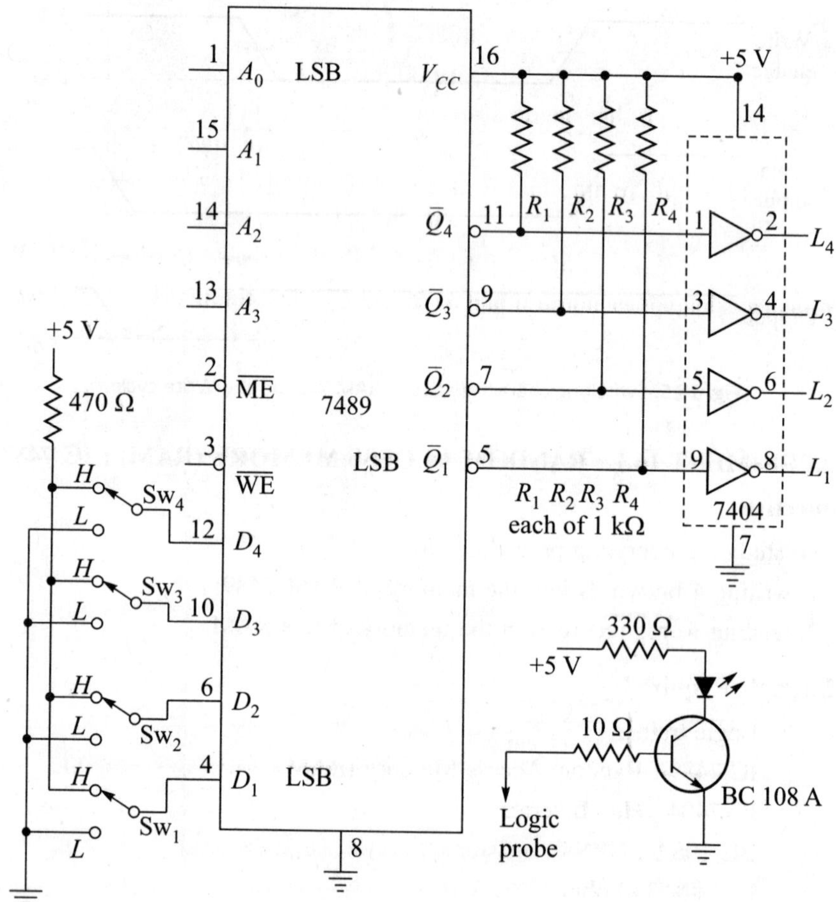

Fig. 16.6. Circuit diagam for Experiment 16.1.

For providing the required logic levels at the address select inputs A_0 through A_3 and the Memory Enable and Write Enable inputs, the pins will have to be connected to ground for low logic levels and to + 5 V through a 470 ohm resistor for high logic levels.

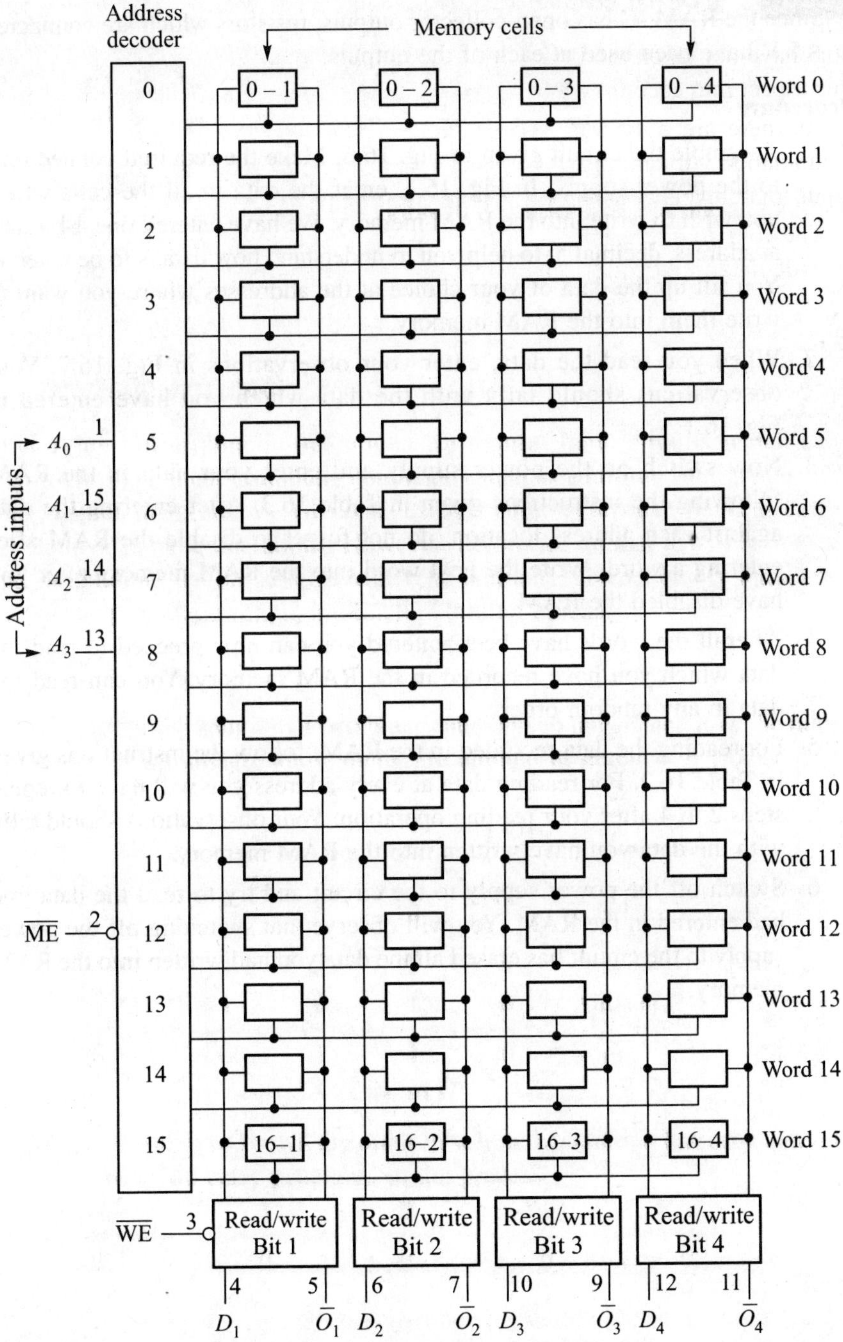
Address decoder
Memory cells
0 – 1
0 – 2
0 – 3
0 – 4
16–1
16–2
16–3
16–4
Word 0
Word 1
Word 2
Word 3
Word 4
Word 5
Word 6
Word 7
Word 8
Word 9
Word 10
Word 11
Word 12
Word 13
Word 14
Word 15
Address inputs
A_0 1
A_1 15
A_2 14
A_3 13
$\overline{ME}$ 2
$\overline{WE}$ 3
Read/write Bit 1
Read/write Bit 2
Read/write Bit 3
Read/write Bit 4
D_1 4
$\overline{O}_1$ 5
D_2 6
$\overline{O}_2$ 7
D_3 10
$\overline{O}_3$ 9
D_4 12
$\overline{O}_4$ 11

Fig. 16.7

Since the RAM IC has open-collector outputs, resistors which are connected to + 5 V have been used at each of the outputs.

Procedure

1. Assemble the circuit given in Fig. 16.6. Make the required connections to the power supply. In Fig. 16.1, enter the data in all the cells which you wish to write into the RAM memory. We have entered one 4-bit data at address, decimal 5, to help you to understand how data is to be entered. You fill up the data of your choice at the addresses where you want to write them into the RAM memory.
2. When you read the data, enter your observations in Fig. 16.7. Your observations should tally with the data which you have entered in Fig. 16.1.
3. Now switch on the power supply and enter your data in the RAM following the instructions given in Table 16.3. After entering the data against each address location, do not forget to disable the RAM after entering a word. Write the next word into the RAM memory after you have disabled the RAM.
4. Afterall the words have been entered you can now proceed to read the data which you have recorded in the RAM memory. You can read the data in any random order.
5. For reading the data recorded in the RAM, follow the instructions given in Table 16.2. For reading data at every address you will have to repeat steps 2 to 4 after your reading operation. Your observations should tally with the data you have written into the RAM memory.
6. Switch off the power supply to the circuit and try to read the data you had entered in the RAM. You will observe that switching off the power supply to the circuit has erased all the data you had written into the RAM memory.

CHAPTER 17

Analog/Digital and Digital/Analog Conversion

17.1 INTRODUCTION

Although most signals which we come across are analog in nature, processing of signals is more convenient and efficient if they are first converted to their equivalent digital form. Even music is now digitally recorded and later converted to the analog form for playback. Therefore, for processing some types of signals we have to convert them first from analog to digital form and thereafter from digital to analog form. To accomplish this we would require A/D and D/A converters. In this chapter, we will concern ourselves with simple techniques to achieve this.

17.2 SUMMING NETWORK D/A CONVERTER

We will consider a 4-bit input and devise a circuit which will weigh the four binary digits with weights of 8, 4, 2 and 1 and sum up the binary currents or voltages to produce an analog output. This can be accomplished by using a resistor network so that the resistors are weighted inversely in proportion to the binary digit weights. A summation circuit is shown in Fig. 17.1 which we will consider in Experiment 17.1.

In this circuit, (Fig. 17.1) since R_3 is 10 k ohm, the MSB will contribute a current of 0.5/10 k or 0.05 mA, the input voltage being 0.5 V. The other three bits will contribute currents of 0.025 mA, 0.125 mA and 0.006 mA which are inversely in proportion to their digit weights.

If you take the output voltage into consideration instead of the input current, the expression for the output voltage will be as follows:

$$\begin{aligned} V_{out} &= -V\left(\frac{R_f}{R_3}+\frac{R_f}{R_2}+\frac{R_f}{R_1}+\frac{R_f}{R_0}\right) \\ &= -0.5\left(\frac{80\text{ k}}{10\text{ k}}+\frac{80\text{ k}}{20\text{ k}}+\frac{80\text{ k}}{40\text{ k}}+\frac{80\text{ k}}{80\text{ k}}\right) \\ &= -0.5(8+4+2+1) \\ &= -0.5(15) \\ &= -7.5\ V \end{aligned}$$

This resistor network for digital to analog conversion has some serious drawbacks. This requires very high precision resistors of many different values. The accuracy of the analog output depends a great deal on resistor tolerance. Besides, the resistor values are also limited by the input and output impedance of the Op-amp.

In Experiment 17.1, we will see how it actually works out in practice before we consider a better method to overcome the limitations of the method we have just described.

EXPERIMENT 17.1 : DIGITAL/ANALOG CONVERTER

Objective

To evaluate the performance of a D/A converter based on a summing network.

Materials Required

Logic trainer
Digital Voltmeter
IC 741 : Operational amplifier
Power supply + 12 V, – 12 V
Potentiometer 10 k ohm
Resistor 1.5 k ohm
Resistor 10 k ohm
Resistor 20 k ohm
Resistor 40 k omh
Resistors 80 k ohm (2).

A cicuit diagram for Experiment 17.1 is given in Fig. 17.1.

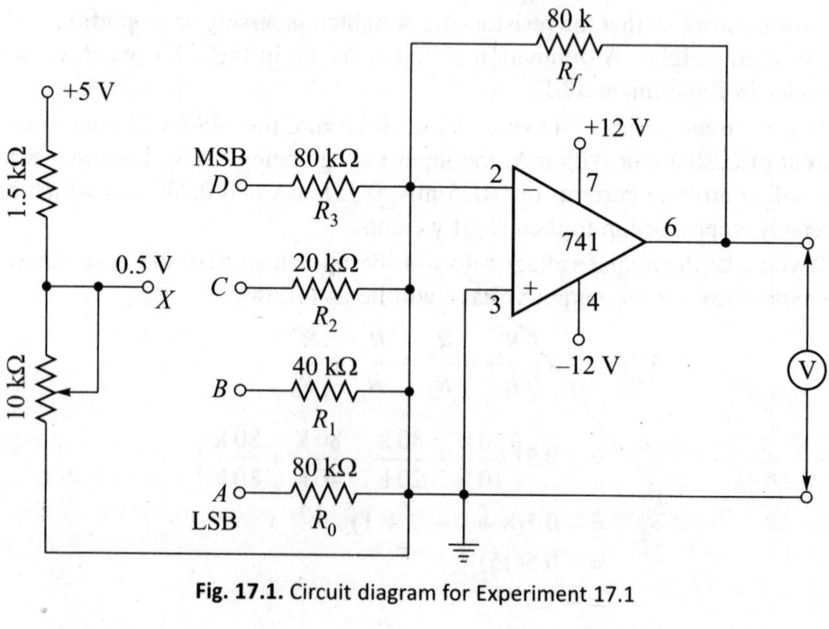

Fig. 17.1. Circuit diagram for Experiment 17.1

In the circuit diagram given for this experiment in Fig. 17.1, for high logic levels inputs *D, C, B* and *A* are connected to terminal *X* which delivers 0.5 V and for low logic levels these inputs are connected to ground. The high logic level voltage is obtained from a 5 V power supply through a voltage divider. The voltage can be adjusted to deliver exactly 0.5 V with the help of the 0.10 k ohm potentiometer. If resistors of the values indicated are not available you may use resistors of the following preferred values, 10 k ohm, 22 k ohm, 39 k ohm and 82 k ohm.

You may now proceed with the experiment as follows:

Procedure

1. Assemble the circuit given in Fig. 17.1. Connect the 5 V supply as shown in the diagram and with the help of the potentiometer adjust the voltage at X to 0.5 V.
2. Connect the + 12 V supply to pin 7 of the Op-amp and the – 12 V supply to pin 4 of the Op-amp. Connect pin 3 of the IC to ground and the voltmeter between pin 6 of the IC and ground.
3. Apply the logic inputs as shown in Table 17.1 and record the output voltage in each case of this table. Your observations should, by and large, tally with Table 17.1 given in Appendix 3.

The voltage output which you have recorded represents the digital input to the Op-amp.

Table 17.1

Inputs				*Output voltage V*
MSB			*LSB*	
D	*C*	*B*	*A*	
0	0	0	0	
0	0	0	1	
0	0	1	0	
0	0	1	1	
0	1	0	0	
0	1	0	1	
0	1	1	0	
0	1	1	1	
1	0	0	0	
1	0	0	1	
1	0	1	0	
1	0	1	1	
1	1	0	0	
1	1	0	1	
1	1	1	0	
1	1	1	1	

17.3 LADDER NETWORK (D/A CONVERTER)

This ladder network is commonly known as the *R/2R* ladder network for the simple reason that it uses resistors of only two values, *R* and *2R*. A 4-bit D/A converter using this network is given in Fig. 17.2. The upper end of the ladder network is connected to the inverting input of the Op-amp and the lower end is chosen to give the required gain. The beauty of this network is that the input current provided to the Op-amp by the four inputs is in direct proportion to the digit weights of the input. Thus, the MSB provides eight times as much current as the LSB. This is due to the fact that looking up or down from any one of the four nodes 1, 2, 3 and 4 the resistance is *2R* ohm.

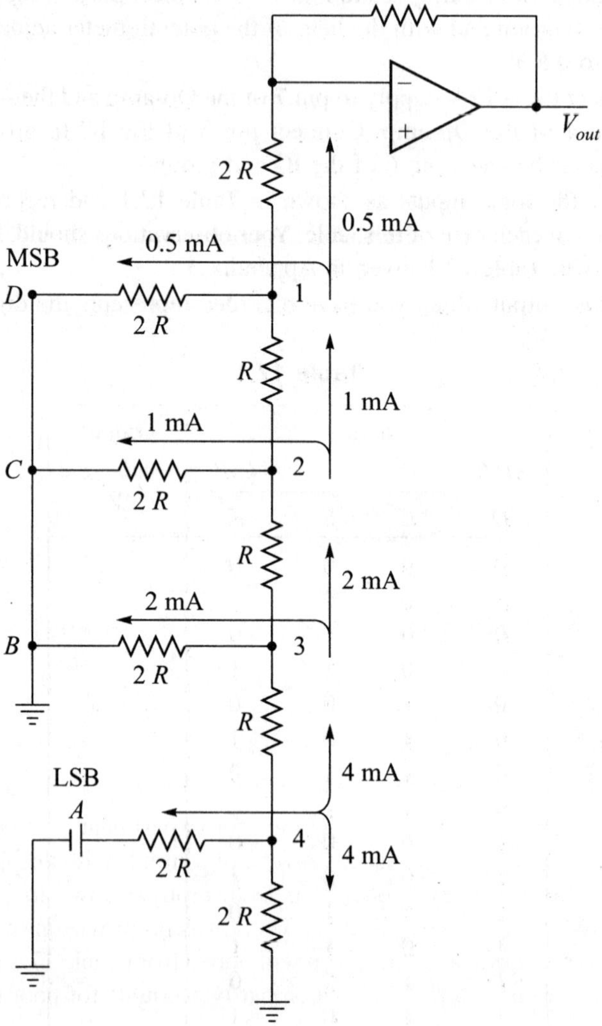

Fig. 17.2. Curent distribution in R/2R Ladder network.

Let us begin with the *D* input and apply a logic 1 level to it. The *D* input will see a resistance of *2R* ohm looking down and the same resistance looking up. If the current flowing through the *D* input is 8 mA, only 4 mA will flow into the Op-amp and the remaining 4 mA will travel down. Analyzing in the same way, the current division from the other input bits has been shown in Fig. 17.2. You will notice that the logic 1 input at the LSB gives rise to 4 mA current flowing up and the same current flows down. In the upward movement this current of 4 mA from the LSB gets divided at every node with the result that it contributes a current of only 0.5 mA to the Op-amp, which is l/8th of the current contributed by the MSB. The contribution of the current flow into the Op-amp will be as follows:

Input *D* (MSB)	4 mA
Input *C*	2 mA
Input *B*	1 mA
Input *A* (LSB)	0.5 mA

You will notice that these currents are in direct proportion to the digit weights.

EXPERIMENT 17.2 : LADDER NETWORK (D/A CONVERTER)

Objective

To study the operation of a D/A converter based on a *R/2R* network.

Materials Required

Logic trainer
Digital Voltmeter
IC 741 : Operational amplifier
Potentiometer 4.7 k ohm
Potentiometers 10 k ohm (2)
Resistor 1.5 k ohm
Resistor 5.6 k ohm
Resistors 10 k ohm (3)
Resistors 22 k ohm (5).

A circuit diagram for Experiment 17.2 is given in Fig. 17.3.

You will observe from the circuit diagram for Experiment 17.2 that provision has been made to null the d.c. offset voltage at the output. With this improvement the offset voltage in this experiment has been brought down to 0.1 mV as is mentioned in Table 17.2 in Appendix 3. There was no provision in Experiment 17.1 for reducing the offset voltage. You will notice from Table 17.1 in Appendix 3 that the offset voltage was 2.0 V, which partly accounts for poor results from that experiment.

You may now proceed with Experiment 17.2.

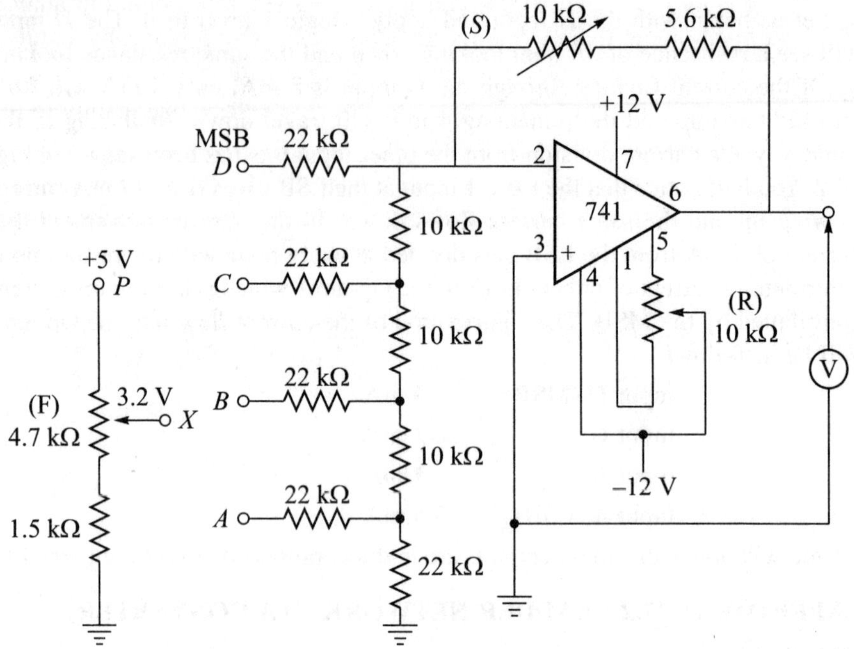

Fig. 17.3. Circuit diagram for Experiment.

Procedure

1. Assemble the circuit given in Fig. 17.3. Connect + 12 V to pin 7 of the Op-amp and pin 4 to – 12 V. Connect the voltmeter between pin 6 of the Op-amp and ground. Also connect pin 3 of the Op-amp to ground. Make the remaining ground connections and also connect the 5 V power supply at terminal *P* and adjust the voltage at *X* to 3.2 V with potentiometer *F*.
2. Switch on the power supply to the Op-amp and connect inputs *D, C, B* and *A* to ground. Adjust potentiometer *R* so that the d.c. offset voltage as shown by the voltmeter is as small as possible.
3. Connect input *D* to point *X,* while inputs *C, B* and *A* should remain connected to ground. Adjust potentiometer *S* so that the voltmeter shows exactly 1.6 V. Potentiometer *S* should not be readjusted thereafter.
4. Connect input *D* also to ground. Inputs *C, B* and *A* are already connected to ground.
5. Now apply the logic inputs at points *D, C, B* and *A* as given in Table 17.2. For high logic level connect the inputs to point *X* and for low logic level connect the inputs to ground. Apply the logic inputs in the order given in Table 17.2 and every time record the output voltage in Table 17.2.
6. Your observations should more or less tally with Table 17.2 in Appendix 3.

7. The output voltage which you have recorded represents the analog equivalent of the digital input. Check the compliance by manual calculation to see how the results agree with your calculation.
8. You may notice some discrepancies; but that will be due mainly to the resistors used in the experiment.

You must have noticed that the output voltage rises by 0.2 V for an increase of the binary input count by 1. The resolution of a D/A converter can be defined by the number of input bits. It can also be stated as a percentage as follows:

$$\text{Per cent resolution} = \frac{1}{2^n - 1} \times 100 = 6.7\% = (n = 4)$$

The D/A converter we have just considered has 16 input combinations as it is a 4-bit converter and its resolution is, therefore 6.7%, which works out to about 0.21 V as the input voltage is 3.2 V.

Table 17.2

Inputs				*Output voltage*
D	*C*	*B*	*A*	
0	0	0	0	
0	0	0	1	
0	0	1	0	
0	0	1	1	
0	1	0	0	
0	1	0	1	
0	1	1	0	
0	1	1	1	
1	0	0	0	
1	0	0	1	
1	0	1	0	
1	0	1	1	
1	1	0	0	
1	1	0	1	
1	1	1	0	
1	1	1	1	

17.4 ANALOG/DIGITAL CONVERTER : ADC 0804

This is an 8-bit successive approximation A/D converter. In this converter the analog input is first sampled and digitized. Thereafter it generates an 8-bit binary output which is directly proportional to the analog input voltage. The binary output has a range of 0 0 0 0 0 0 0 0 to 1 1 1 1 1 1 1 1 for an input voltage of 5 V. The inputs and outputs are TTL compatible and it is designed to operate with the 8080A microprocessor. The conversion time for this converter is 100 μs. The only external components for this converter are a resistor and a capacitor. It also has a built-in clock generator.

The input voltage of 5 V is applied across a 10 k ohm potentiometer. Since this is an 8-bit converter its resolution is as follows:

$$\frac{1}{2^8 - 1} = \frac{1}{255} \text{ of the full scale voltage.}$$

For an input voltage of 5 V it works out to 0.02 V, for an increase of the binary output count by 1.

In this converter the conversion process is started by applying a high to low (1 → 0) transition at the $\overline{\text{WR}}$ input. An interrupt, INTR, has been provided which can signal the microprocessor when the A/D conversion has been completed.

EXPERIMENT 17.3 : A/D CONVERTER : ADC 0804

Objective

To study the operation of successive approximation A/D converter ADC 0804.

Materials Required

Logic trainer

Digital Voltmeter

IC ADC 0804 : A/D Converter

Transistor : BC 108A (NPN)

Resistor 220 ohm

Resistor 6.8 k ohm

Resistor 10 k ohm

10 k ohm Potentiometer

Capacitor 150 pF.

An experimental set up for Experiment 17.3 is given in Fig. 17.4. Since there are eight outputs in the converter it will require eight logic monitors. As most logic trainers have only 4 logic monitors, an extra logic monitor using a transistor has been shown in the diagram. The logic levels of those outputs which are not connected to regular logic monitors can be checked with this extra logic monitor.

Procedure

1. Assemble the circuit given in Fig. 17.4. Connect pin 20 of the IC to + 5 V as well as one end of the potentiometer as shown in the diagram. Also connect the power supply to the logic monitor. Make the ground connections shown in the diagram. Connect a digital voltmeter at the analog input terminals.

2. To convert the analog input to digital form all that you have to do is to apply a 1 → 0 transition at the $\overline{WR}$ input. Carry out the conversions mentioned in Table 17.3 and check your results with Table 17.3 in Appendix 3.

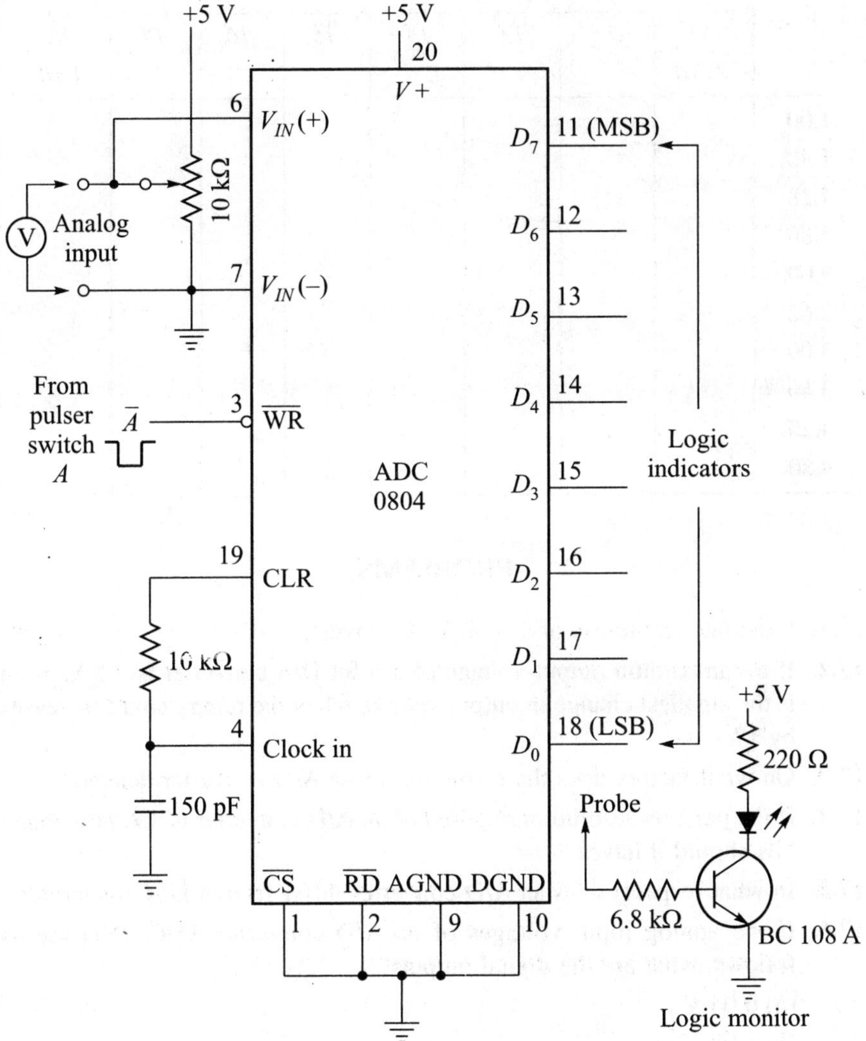

Fig. 17.4. Circuit diagram for Experiment 17.3.

Since this A/D converter has a resolution of 0.02 V, an increase of 0.02 V in the analog input will increase the binary output count by 1. Hence, if the binary output is, for instance, 1 0 0 0 1 1 0, which is equivalent to decimal 70 the equivalent input voltage would be 70 × 0.02 V or 1.40 V which is the input voltage in the table against this digital output.

Table 17.3

Analog input voltage	*Digital output*							
	IC pin Nos.							
	11 MSB	*12*	*13*	*14*	*15*	*16*	*17*	*18 LSB*
1.00								
1.40								
1.28								
1.80								
2.00								
2.62								
3.00								
3.60								
4.20								
4.80								

PROBLEMS

17.1. Calculate resolution of a 6-bit D/A converter.

17.2. If the maximum output voltage of a 6-bit D/A converter is 7.5 V, what is the smallest change in output voltage when the binary count increases by 1?

17.3. On what factors does the resolution of an A/D converter depend?

17.4. If the per cent resolution required of an A/D converter is 3.2, how many bits should it have?

17.5. In what respects does an A/D converter differ from a D/A converter?

17.6. If the analog input voltages of an A/D converter ADC 0804 are as follows, what are the digital outputs?

(*a*) 0.04 V

(*b*) 1.08 V

(*c*) 3.29 V

(*d*) 4.07 V

The circuit is arranged as in Fig. 17.4.

Appendices

APPENDIX 1

Pin Connections for Integrated Circuits

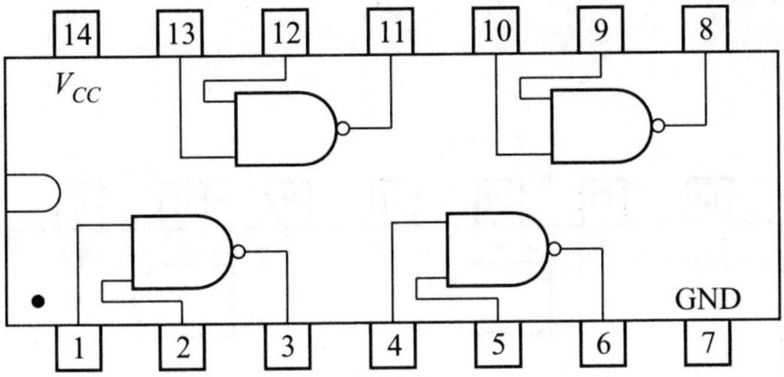

7400

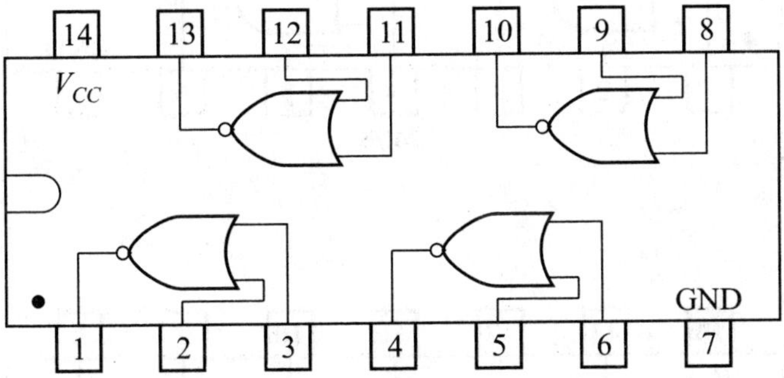

7402

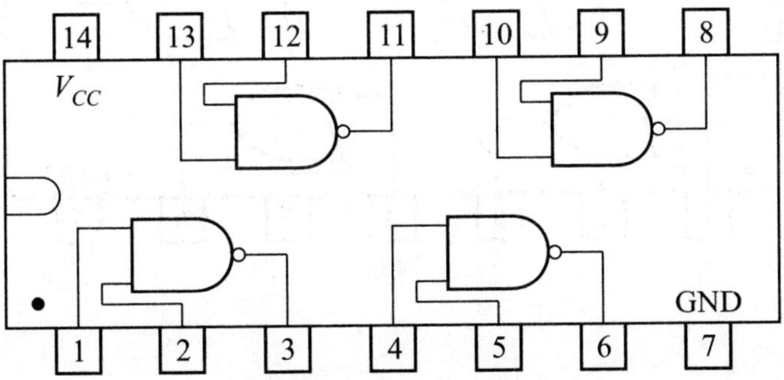

7403 (Open Collector)

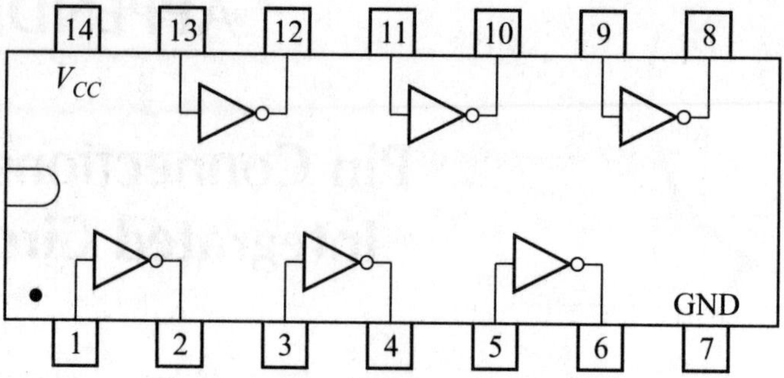

7404

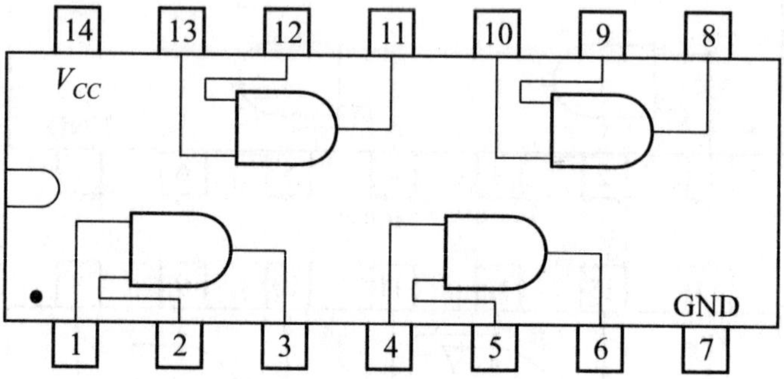

7408

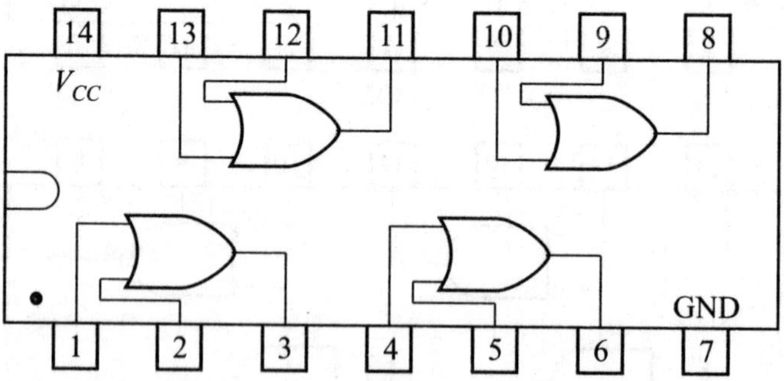

7432

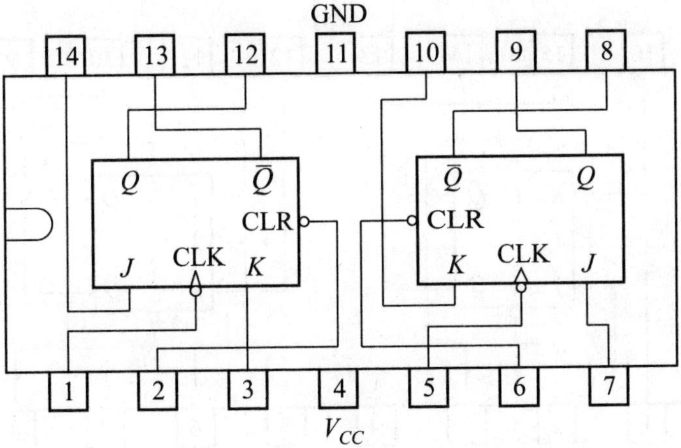

7473

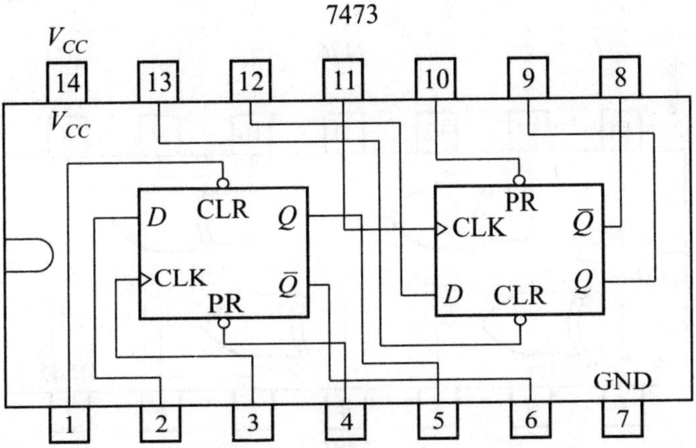

7474

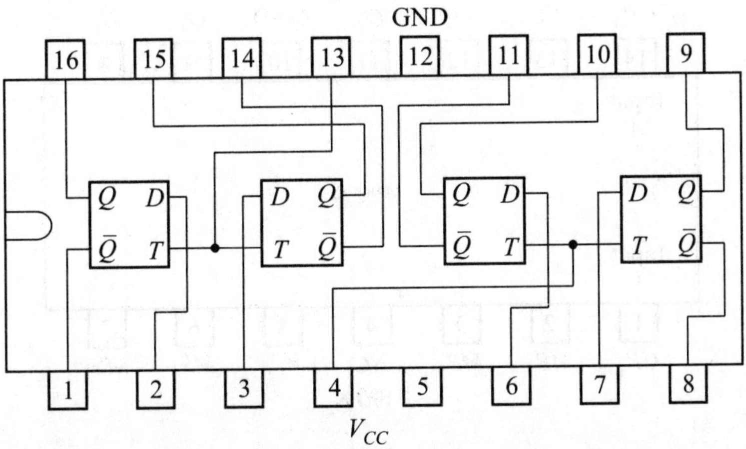

7475

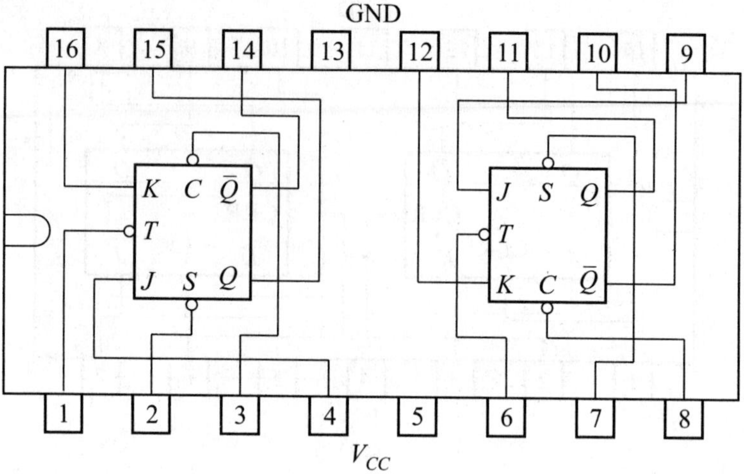

7476

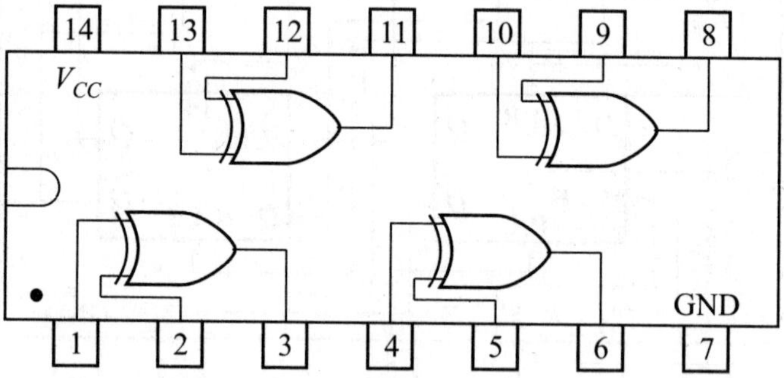

7486

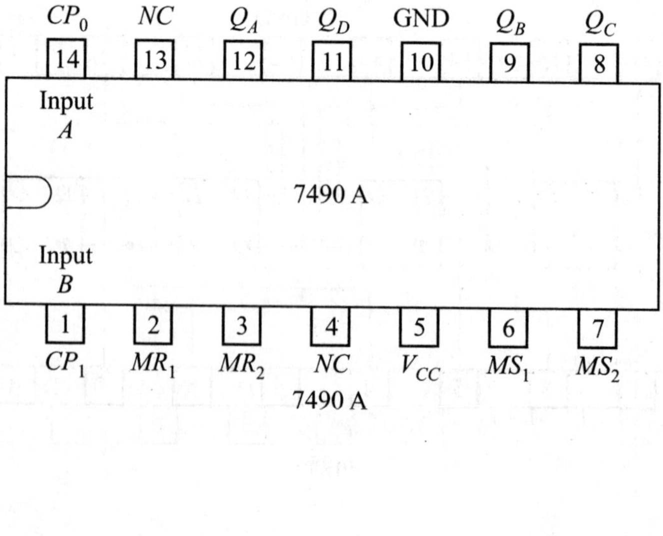

7490 A

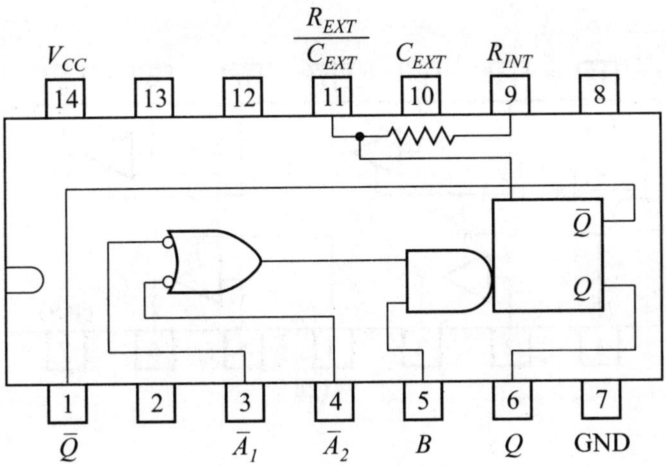

74121

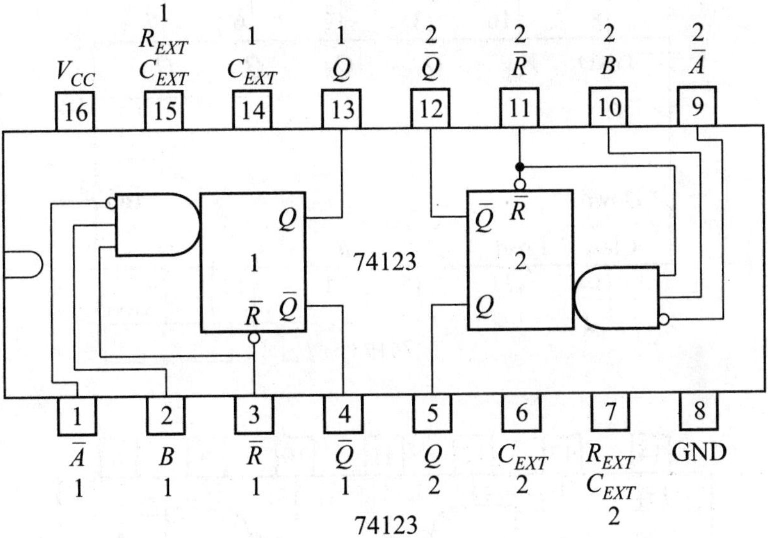

74123

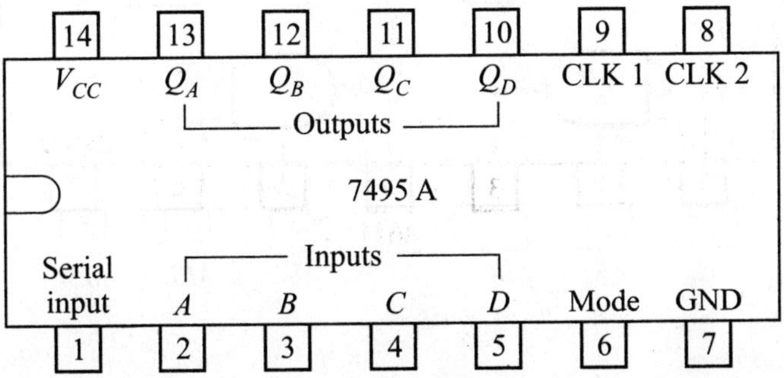

7495 A

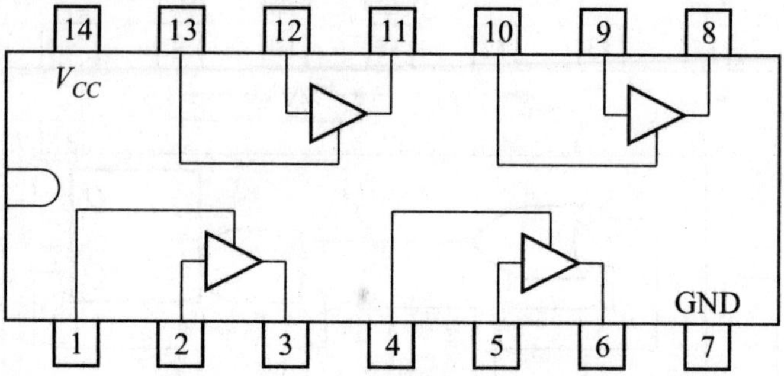

74126

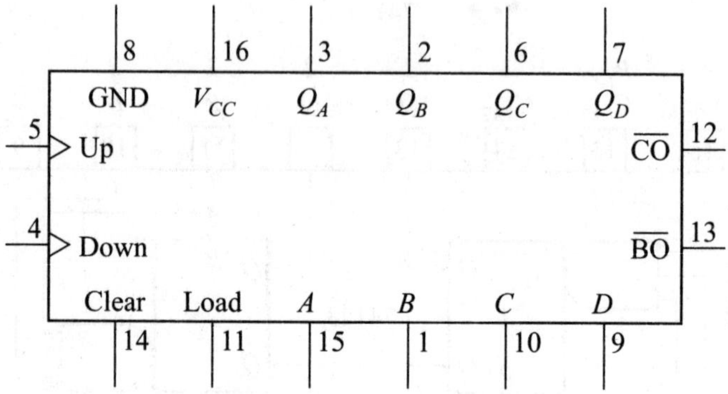

74193

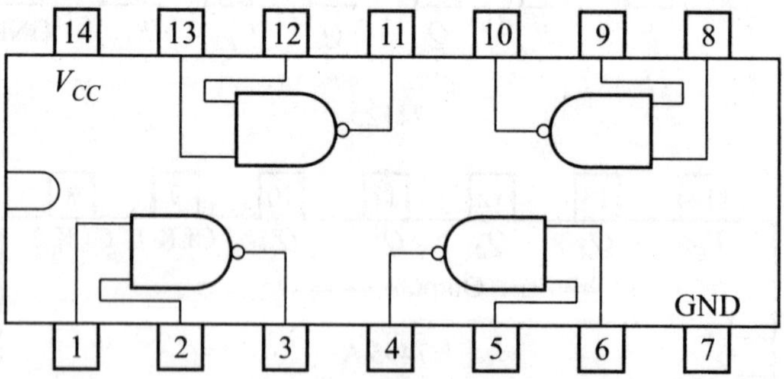

4011

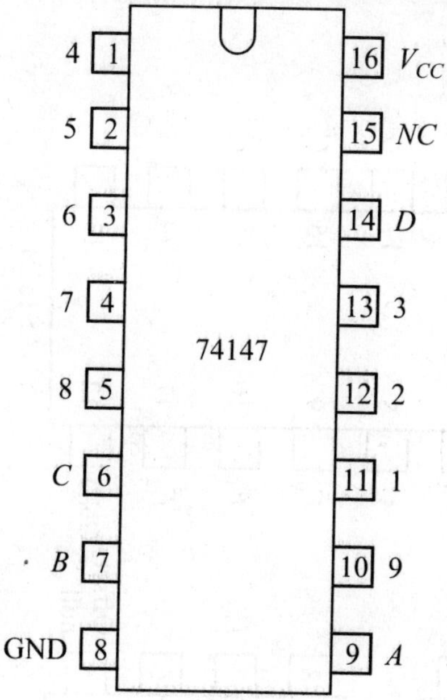

74147

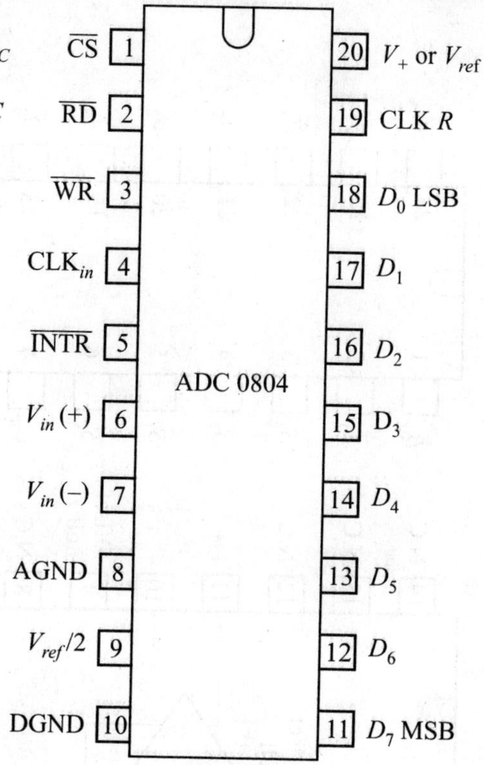

ADC 0804

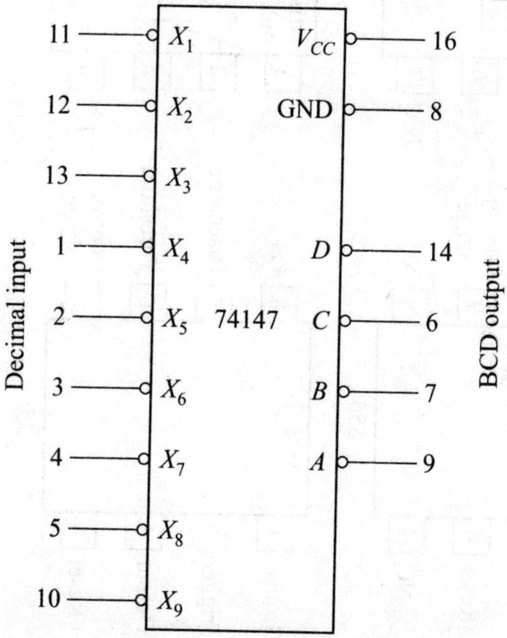

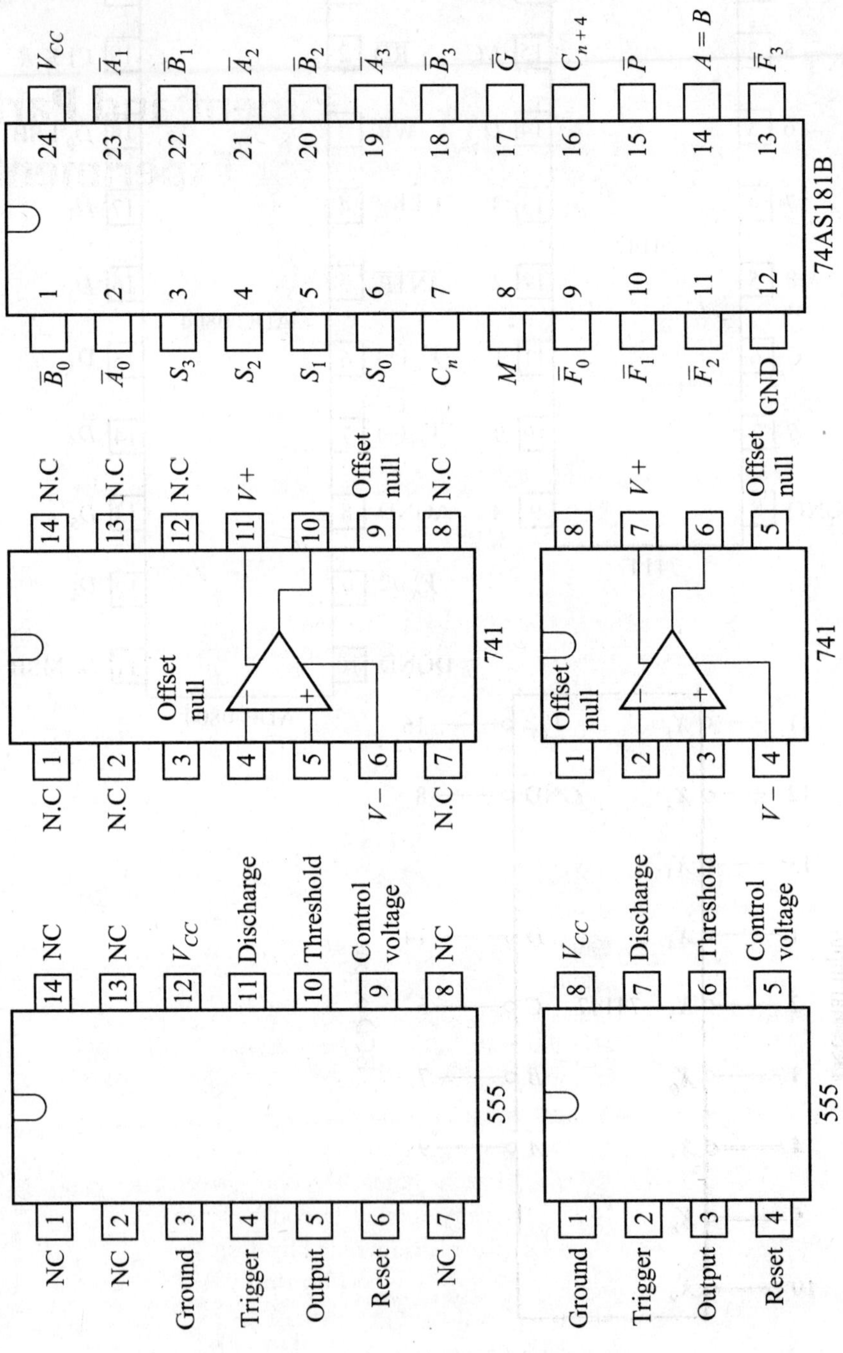
74AS181B
$\bar{B}_0$ 1
$\bar{A}_0$ 2
S_3 3
S_2 4
S_1 5
S_0 6
C_n 7
M 8
$\bar{F}_0$ 9
$\bar{F}_1$ 10
$\bar{F}_2$ 11
GND 12
13 $\bar{F}_3$
14 $A = B$
15 $\bar{P}$
16 C_{n+4}
17 $\bar{G}$
18 $\bar{B}_3$
19 $\bar{A}_3$
20 $\bar{B}_2$
21 $\bar{A}_2$
22 $\bar{B}_1$
23 $\bar{A}_1$
24 V_{CC}
741
N.C 1
N.C 2
Offset null 3
4
5
$V-$ 6
N.C 7
8 N.C
9 Offset null
10
11 $V+$
12 N.C
13 N.C
14 N.C
741
Offset null 1
2
3
$V-$ 4
5 Offset null
6
7 $V+$
8
555
NC 1
NC 2
Ground 3
Trigger 4
Output 5
Reset 6
NC 7
8 NC
9 Control voltage
10 Threshold
11 Discharge
12 V_{CC}
13 NC
14 NC
555
Ground 1
Trigger 2
Output 3
Reset 4
5 Control voltage
6 Threshold
7 Discharge
8 V_{CC}

APPENDIX 2

Equipment and Parts Required for Experiments

Equipment

Logic trainer
Volt-ohm-milliammeter
Frequency counter
Dual-trace oscilloscope
Digital voltmeter.

Diodes

Switching Silicon Diodes (2)
LEDs (Red) (2)
Seven-segment LED display : FND 567
Seven-segment LED display : FND 500/560 (2)
Lamp : 6 V, 60 mA.

Transistors

CK 100 (PNP)
CL 100 (NPN)
BC 108A (NPN) (2)
Quartz Crystal : 1 MHz.

Integrated Circuits

7400	:	Quad 2-input NAND gates
7402	:	Quad 2-input NOR gates
7403	:	Quad 2-input open-collector NAND gates
7404	:	Hex Inverter
7408	:	Quad 2-input AND gates
7432	:	Quad 2-input OR gates
7442	:	BCD-to-Decimal Decoder
7447A	:	BCD to seven-segment Decoder-driver
7448	:	BCD to seven-segment Decoder-driver (2)
7473	:	Dual JK (Level-triggered) Flip-flop
7474	:	Dual D Edge-triggered Flip-flop
7475	:	Quad Latch (Level-sensitive)
7476	:	Dual JK master-slave Flip-flop
7483	:	4-bit Full Adder

7486	:	Quad Exclusive-OR gates
7489	:	Random Access Memory (RAM)
7490A	:	Decade counter (2) (÷ 10)
7495A	:	4-bit Shift Register
74121	:	Non-retriggerable Monostable
74123	:	Retriggerable Monostable
74126	:	Tri-state Quad Buffer
74147	:	Decimal Priority Encoder
74150	:	16-line to 1-line Multiplexer
74154	:	1-line to 16-line Demultiplexer
74181	:	Arithmetic Logic Unit
74193	:	Binary Up-Down Counter
555	:	Timer
741	:	Operational Amplifier
4011	:	Quad CMOS NAND gates
ADC 0804	:	Analog-Digital Converter.

Resistors

150 Ohm (2)
220 Ohm
330 Ohm (7)
470 Ohm
560 Ohm
1 k Ohm (4)
1.2 k Ohm
1.5 k Ohm
2 k Ohm
4.7 k Ohm
5.6 k Ohm
6.8 k Ohm (2)
10 k Ohm (3)
18 k Ohm
20 k Ohm
22 k Ohm (5)
40 k Ohm
80 k Ohm (2)
100 k Ohm.

Capacitors

100 pF (2)
150 pF
.01 μF (2)
10 μF Electrolytic
470 μF Electrolytic

Potentiometers

4.7 k Ohm
10 k Ohm (2).

APPENDIX 3

Completed Experimental Tables

Experiment 5.1

Table 5.2

Truth Table for 3-Input NAND Gate

Inputs			*Output*
A	*B*	*C*	*Y*
0	0	0	1
0	0	1	1
0	1	0	1
0	1	1	1
1	0	0	1
1	0	1	1
1	1	0	1
1	1	1	0

Experiment 5.2

Table 5.7

Truth Table for 2-Input NOR Gate

Inputs		*Output*
A	*B*	*Y*
0	0	1
0	1	0
1	0	0
1	1	0

Experiment 5.3

Table 5.8

Inputs		*Output*	*Voltage*
A	*B*	*Y*	*at pin 3*
0	0	1	5 V
0	1	0	0
1	0	0	0
1	1	0	0.1 V

Experiment 5.4

Table 5.9

Inputs				*Output*	*Voltage*
A	*B*	*C*	*D*	*Y*	*at pin 3*
0	0	0	0	1	5 V
0	0	0	1	1	
0	0	1	0	1	
0	0	1	1	0	
0	1	0	0	1	
0	1	0	1	1	
0	1	1	0	1	
0	1	1	1	0	
1	0	0	0	1	
1	0	0	1	1	
1	0	1	0	1	
1	0	1	1	0	
1	1	0	0	0	
1	1	0	1	0	
1	1	1	0	0	
1	1	1	1	0	0.1 V

Experiment 6.2

Table 6.4

Truth Table and Parity for 3-Input XOR Gate

Inputs			*Output*	*Parity*
A	*B*	*C*	*Y*	
0	0	0	0	Even
0	0	1	1	Odd
0	1	0	1	Odd
0	1	1	0	Even
1	0	0	1	Odd
1	0	1	0	Even
1	1	0	0	Even
1	1	1	1	Odd

Experiment 7.1

Table 7.1

TTL NAND gate		*CMOS NAND gate*
Input	*Output*	*Output*
High	0.12 V	5 V
Low	5 V	0.01 V

Experiment 7.2

Table 7.2

Fig. 7.5	*Fig. 7.6*
V_1 3.16 V	V_2 0.12 V
I_1 4.5 mA	I_2 5 mA

Experiment 7.3

Table 7.3

Fig. 7.6	*Fig. 7.7*
V_1 0.1 V	V_3 3.2 V
V_2 4.8 V	V_4 75 mV
I_1 3.3 mA	I_3 2.2 mA
I_2 55 mA	I_4 54 mA

Experiment 7.4

Table 7.4

V_1 3.7 V	V_2 0.3 V
I_1 0.3 mA	I_2 14 mA

Experiment 7.5

Table 7.5

V_1 2.8 V	V_2 57 mV
I_1 3.8 mA	I_2 54 mA

Experiment 10.2

Table 10.6

S. No.	*Augend*				*Addend*				*Sum*				*Carry*
	A_3	A_2	A_1	A_0	B_3	B_2	B_1	B_0	S_3 L_4	S_2 L_3	S_1 L_2	S_0 L_1	L_5
1	0	1	0	1	1	0	0	0	1	1	0	1	
2	1	0	0	1	0	1	0	0	1	1	0	1	
3	1	0	0	0	0	1	1	0	1	1	1	0	
4	1	1	0	0	0	0	1	0	1	1	1	0	
5	0	0	0	1	1	1	0	0	1	1	0	1	
6	1	0	1	1	0	1	1	0	0	0	0	1	1

Experiment 10.3

Table 10.9

S. No.	*Minuend*				*Subtrahend*				*Sum*				*Carry*
	A_3	A_2	A_1	A_0	B_3	B_2	B_1	B_0	S_3 L_4	S_2 L_3	S_1 L_2	S_0 L_1	L_5
1	0	1	1	1	0	0	1	1	0	1	0	0	
2	0	1	0	1	0	1	0	0	0	0	0	1	
3	0	1	1	0	0	0	0	1	0	1	0	1	
4	0	0	1	0	0	0	0	1	0	0	0	1	
5	0	1	0	0	0	1	0	1	1	1	1	1	
6	0	1	0	0	0	1	1	0	1	1	1	0	

Experiment 10.4

S. No.	*Select inputs*				*M*	C_n	*Word A*				*Word B*				*Output*			
	S_3	S_2	S_1	S_0			A_3	A_2	A_1	A_0	B_3	B_2	B_1	B_0	F_3	F_2	F_1	F_0

Additions

Table 10.11

1	1	0	0	1	0		0	0	1	1	0	0	0	1	0	1	1	1
2							0	0	0	1	0	0	1	0	0	0	1	1
3	0	1	0	0			0	1	0	0	0	0	1	0	0	1	1	0

Subtractions

Table 10.12

1	0	1	1	0	0	1	1	0	0	1	0	0	1	0	0	1	1	1
2							1	1	0	1	0	1	1	1	0	1	1	0
3							0	1	1	0	0	0	1	1	0	0	1	1

Magnitude Comparison

Table 10.13

						C_{n+4} Output	$A = B$ Output	
1	0 1 1 0	1	0	1 0 1 0	0 1 0 0	*H*	*L*	*A* > *B*
2				0 0 1 0	1 0 0 0	*L*		*B* > *A*
3				1 0 0 1	1 0 0 1	*H*	*H*	A = *B*

Experiment 12.1

Table 12.1

Input	*Outputs*			
Pulse count *n*	Q_D L_1	Q_C L_2	Q_B L_3	Q_A L_4
0 (Clear)	0	0	0	0
1	0	0	0	1
2	0	0	1	0
3	0	0	1	1
4	0	1	0	0
5	0	1	0	1
6	0	1	1	0
7	0	1	1	1
8	1	0	0	0
9	1	0	0	1
10	1	0	1	0
11	1	0	1	1
12	1	1	0	0
13	1	1	0	1
14	1	1	1	0
15	1	1	1	1
16 (0)	0	0	0	0

Recycle (from 15 back to 0)

Experiment 12.2

Table 12.2

Input	Outputs			
Pulse count n	Q_D L_1	Q_C L_2	Q_B L_3	Q_A L_4
0 (Clear)	0	0	0	0
1	1	1	1	1
2	1	1	1	0
3	1	1	0	1
4	1	1	0	0
5	1	0	1	1
6	1	0	1	0
7	1	0	0	1
8	1	0	0	0
9	0	1	1	1
10	0	1	1	0
11	0	1	0	1
12	0	1	0	0
13	0	0	1	1
14	0	0	1	0
15	0	0	0	1
16 (0)	0	0	0	0

Recycle

Experiment 12.8

Table 12.9

Count	Outputs				Voltage at pin 12
	Q_A	Q_D	Q_C	Q_B	
0	0	0	0	0	0
1	0	0	0	1	0
2	0	0	1	0	0
3	0	0	1	1	0
4	0	1	0	0	0
5	1	0	0	0	3.5 V
6	1	0	0	1	3.5 V
7	1	0	1	0	3.5 V
8	1	0	1	1	3.5 V
9	1	1	0	0	3.5 V
10	0	0	0	0	0

Experiment 12.9

Table 12.10

Pulse count	*Outputs*			
	Q_A	Q_B	Q_C	Q_D
1	1	0	0	0
2	0	1	0	0
3	0	0	1	0
4	0	0	0	1
5	1	0	0	0

Experiment 14.2

Table 14.5

IC 74150	*Data Select*				*IC 74154*	
Input	*MSB*		*Input*		*Outputs*	
	A	*B*	*C*	*D*	*Pin No.*	*Data*
D_0	L	L	L	L	1	$\overline{D}_0$
D_1	L	L	L	H	2	$\overline{D}_1$
D_2	L	L	H	L	3	$\overline{D}_2$
D_3	L	L	H	H	4	$\overline{D}_3$
D_4	L	H	L	L	5	$\overline{D}_4$
D_5	L	H	L	H	6	$\overline{D}_5$
D_6	L	H	H	L	7	$\overline{D}_6$
D_7	L	H	H	H	8	$\overline{D}_7$
D_8	H	L	L	L	9	$\overline{D}_8$
D_9	H	L	L	H	10	$\overline{D}_9$
D_{10}	H	L	H	L	11	$\overline{D}_{10}$
D_{11}	H	L	H	H	13	$\overline{D}_{11}$
D_{12}	H	H	L	L	14	$\overline{D}_{12}$
D_{13}	H	H	L	H	15	$\overline{D}_{13}$
D_{14}	H	H	H	L	16	$\overline{D}_{14}$
D_{15}	H	H	H	H	17	$\overline{D}_{15}$

Experiment 17.1

Table 17.1

Inputs				*Output voltage*
D	*C*	*B*	*A*	*V*
0	0	0	0	0.2 V
0	0	0	1	0.51 V
0	0	1	0	1.0 V
0	0	1	1	1.58 V
0	1	0	0	1.9 V
0	1	0	1	2.4 V
0	1	1	0	2.9 V
0	1	1	1	3.3 V
1	0	0	0	4.1 V
1	0	0	1	4.54 V
1	0	1	0	5.0 V
1	0	1	1	5.43 V
1	1	0	0	5.7 V
1	1	0	1	6.12 V
1	1	1	0	6.55 V
1	1	1	1	6.95 V

Experiment 17.2

Table 17.2

Inputs				*Output voltage*
D	*C*	*B*	*A*	
0	0	0	0	0.1 mV
0	0	0	1	0.23 V
0	0	1	0	0.45 V
0	0	1	1	0.68 V
0	1	0	0	0.86 V
0	1	0	1	1.09 V
0	1	1	0	1.30 V
0	1	1	1	1.53 V
1	0	0	0	1.60 V
1	0	0	1	1.84 V
1	0	1	0	2.1 V
1	0	1	1	2.35 V
1	1	0	0	2.54 V
1	1	0	1	2.73 V
1	1	1	0	2.92 V
1	1	1	1	3.18 V

Experiment 17.3

Table 17.3

Analog input voltage	*Digital output*							
	IC Pin Nos.							
	11 MSB	*12*	*13*	*14*	*15*	*16*	*17*	*18 LSB*
1.00	0	0	1	1	0	0	1	0
1.40	0	1	0	0	0	1	1	0
1.28	0	1	0	0	0	0	0	0
1.80	0	1	0	1	1	0	1	0
2.00	0	1	1	0	0	1	0	0
2.62	1	0	0	0	0	0	1	1
3.00	1	0	0	1	0	1	1	0
3.60	1	0	1	1	0	1	0	0
4.20	1	1	0	1	0	0	1	0
4.80	1	1	1	1	0	0	0	0

APPENDIX 4

Answers to Selected Problems

Chapter 3

3.3. Inverters can be used for the following applications:

(*a*) Complementation

(*b*) Buffering

(*c*) Adding propagation delay of 10 ns per gate with or without inversion to signal

(*d*) Converting AND gates to NAND gates and NAND gates to AND gates etc.

(*e*) As clock oscillators by feeding the output of the second Inverter in tandem to the input of the first Inverter with appropriate circuit for biasing the Inverters.

Chapter 5

5.11

Truth Table for a 2-Input NOR Gate using Negative Logic

Inputs		*Output*
A	*B*	*Y*
1	1	0
1	0	1
0	1	1
0	0	1

It functions as a NAND gate with positive logic.

Chapter 6

6.4

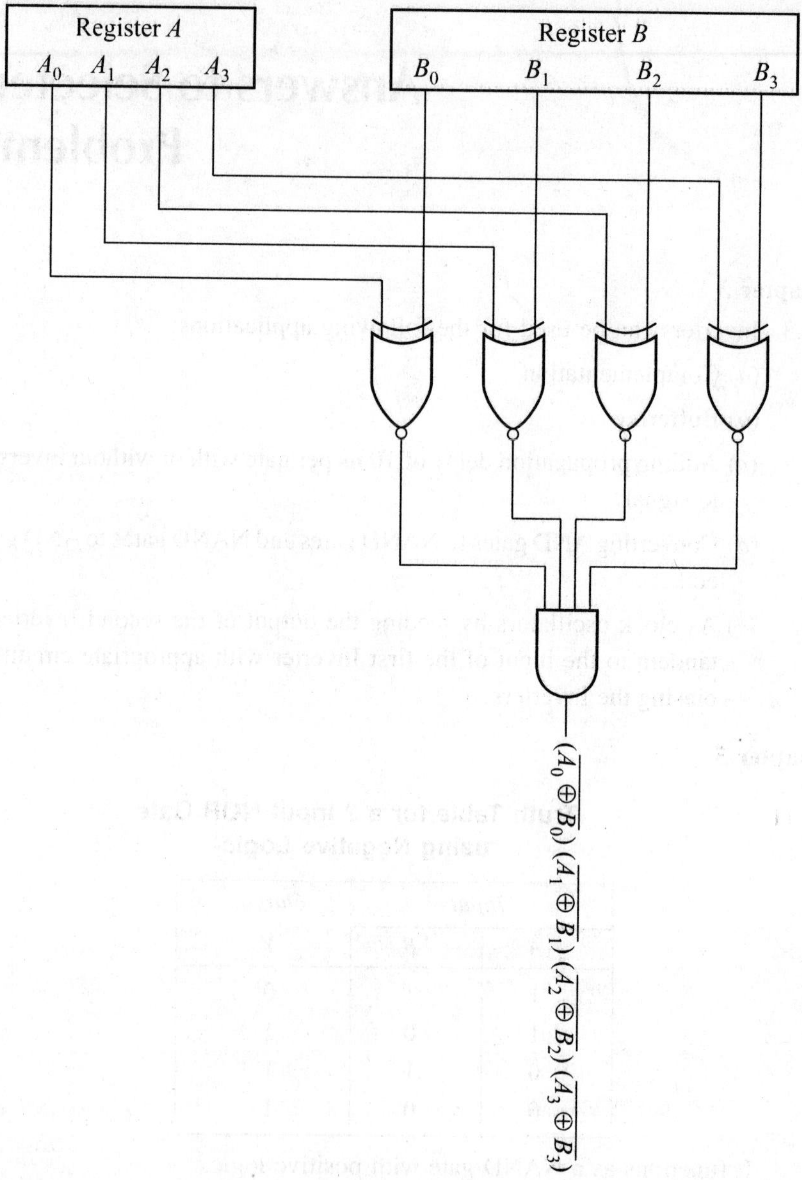

Fig. A-6.4

Chapter 8

8.7 30 kHz square wave

8.9 29

8.10 8

Chapter 9

9.3 $R_B = 3750$ ohm

$C = 115$ pF

9.5 $C = 14.4\ \mu F$

9.7 $P_W = 24\ \mu S$

9.9 Ground pins 1 and 9 and connect pin 2 to pin 12.

Chapter 12

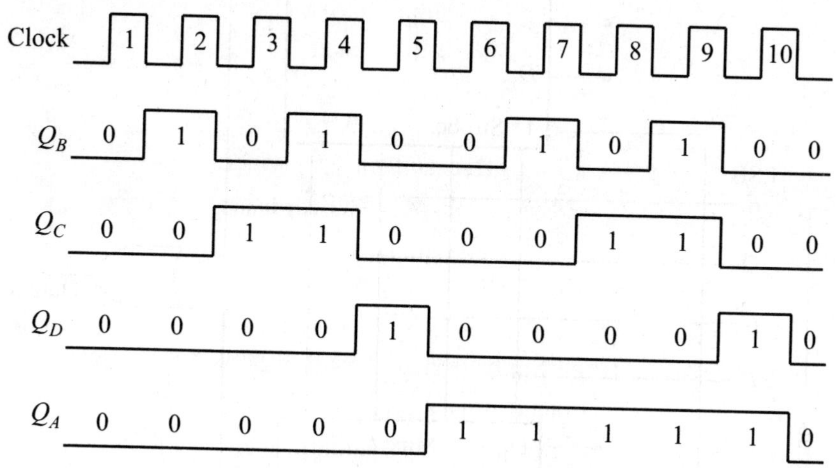

Fig. A-12.5. Waveforms in bi-quinary sequence.

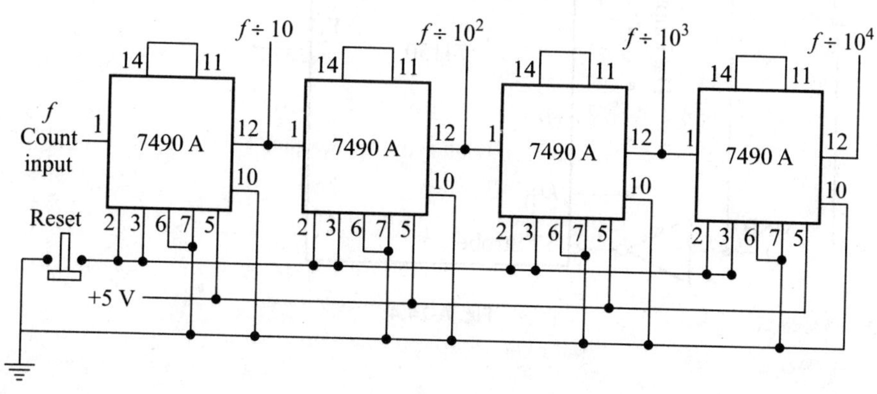

Fig. A-12.8. Decimal scalar.

Chapter 14

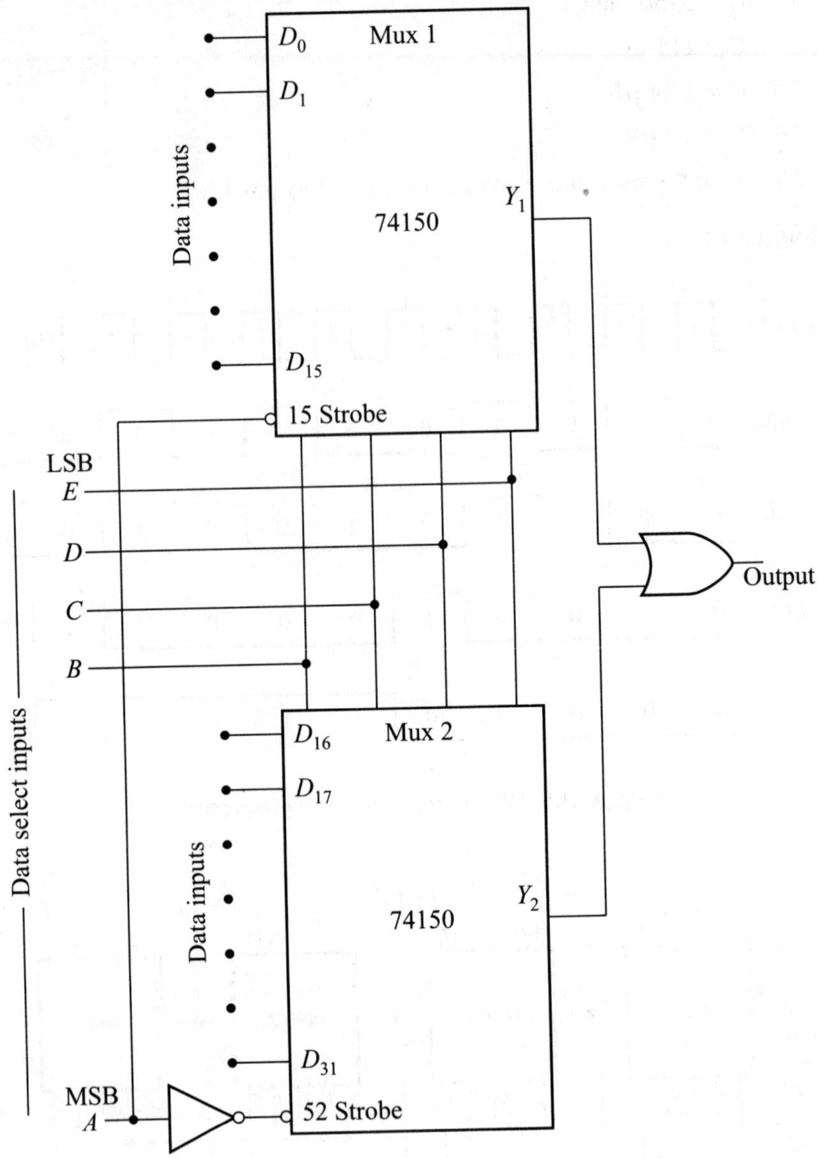

Fig. A-14.4

Chapter 14

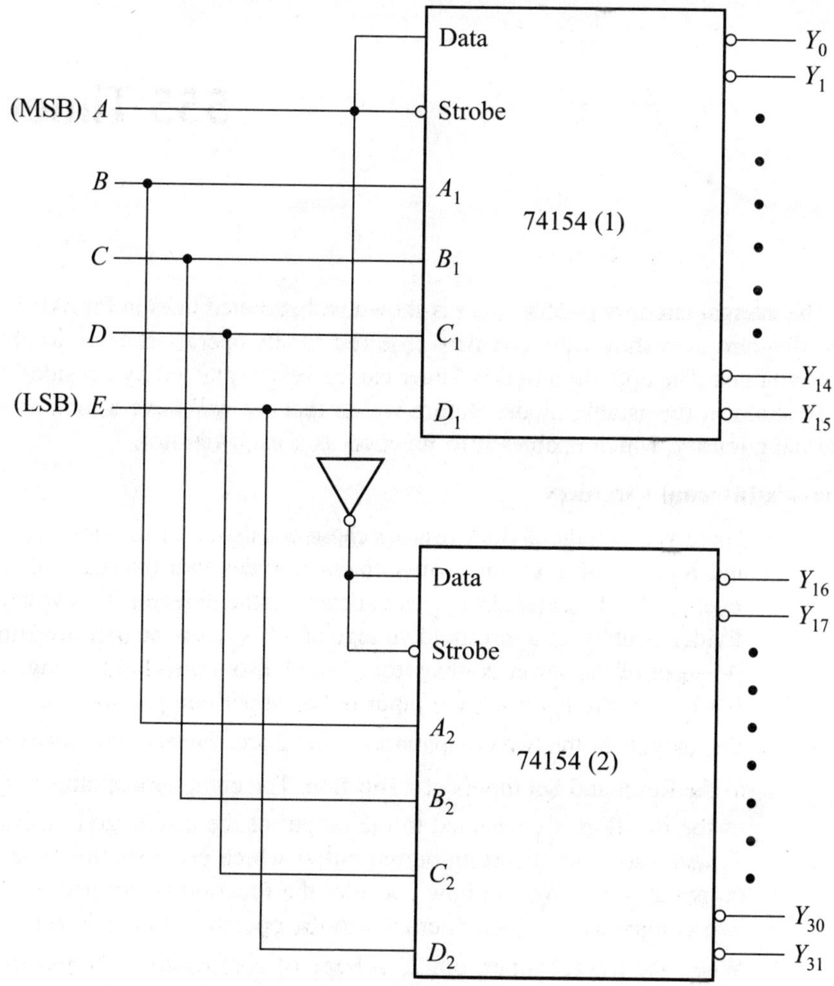

Fig. A-14.6. 5-line to 32-line decoder.

Chapter 17

17.6 (*a*) 0 0 0 0 0 0 1 0

(*b*) 0 0 1 1 0 1 1 0

(*c*) 1 0 1 0 0 1 0 0

(*d*) 1 1 0 0 1 0 1 1

APPENDIX 5

555 Timer

The internal circuitry of 555 Timer is shown within dotted lines in Fig. AP 5.1. The diagram also shows the circuitry required for its operation as an astable multivibrator. The operation of this Timer can be best explained by considering its function in the astable mode. Before we do that we will have a look at its internal circuitry, which enables it to function as a multivibrator.

Timer's Internal Circuitry

* There is a voltage divider resistor chain consisting of resistors R_1, R_2 and R_3 each of 5 k ohms. This chain provides two internal voltage levels of 1/3 V_{CC} and 2/3 V_{CC} as indicated in the diagram. This voltage divider establishes a threshold voltage of 1/3 V_{CC} on the non-inverting (+) input of the lower comparator (2) and also a threshold voltage of 2/3 V_{CC} on the inverting (–) input of the upper comparator (1).
* The outputs of the two comparators 1 and 2 are connected respectively to the Reset and Set inputs of a flip-flop. The complement output, $\overline{Q}$, of the flip-flop is connected to the output of the discharge transistor T_1 and also to an inverting output buffer which provides the Timer's output at pin 3. We can now consider the function performed by the two comparators, which is crucial to the operation of the Timer.
* When the trigger input, pin 2, voltage of comparator 2 is less than 1/3 V_{CC}, the output of the comparator will go high, which sets the flip-flop making its $\overline{Q}$ output low and also cutting off the discharge transistor, thus creating a high resistance between pins 1 and 7. At the same time the inverting buffer output goes high.
* Now consider the upper comparator. If the input to the upper comparator. 1, rises above 2/3 V_{CC}, its output goes high, resetting the flip-flop, raising the output high and turning the transistor, T_1, on, and thereby creating a short between pins 1 and 7.

Astable Operation

Now that we have considered the basic structure of 555 Timer, its operation in the astable mode will be easy to follow.

* For the astable operation the inputs to both the comparators are connected to the timing capacitor *C*. We will now consider step-by-step what happens when power is switched on.
* As the capacitor is initially in a discharged condition, the voltage at the input, pin 2, of the lower comparator is 0, which is less than 1/3 V_{CC}. Therefore, at the commencement of the operation, the flip-flop is set, its output, $\overline{Q}$ is low and the transistor T_1 is cut off, resulting in a high resistance across the capacitor, *C*. At the same time the output at pin 3 goes up.
* From now on the capacitor progressively receives more and more charge through resistors R_A and R_B and consequently the voltage across the capacitor rises to the level of 2/3 V_{CC}.
* When this happens the output of the upper comparator goes high, thereby resetting the flip-flop, turning on the discharge transistor and lowering the output at pin 3.

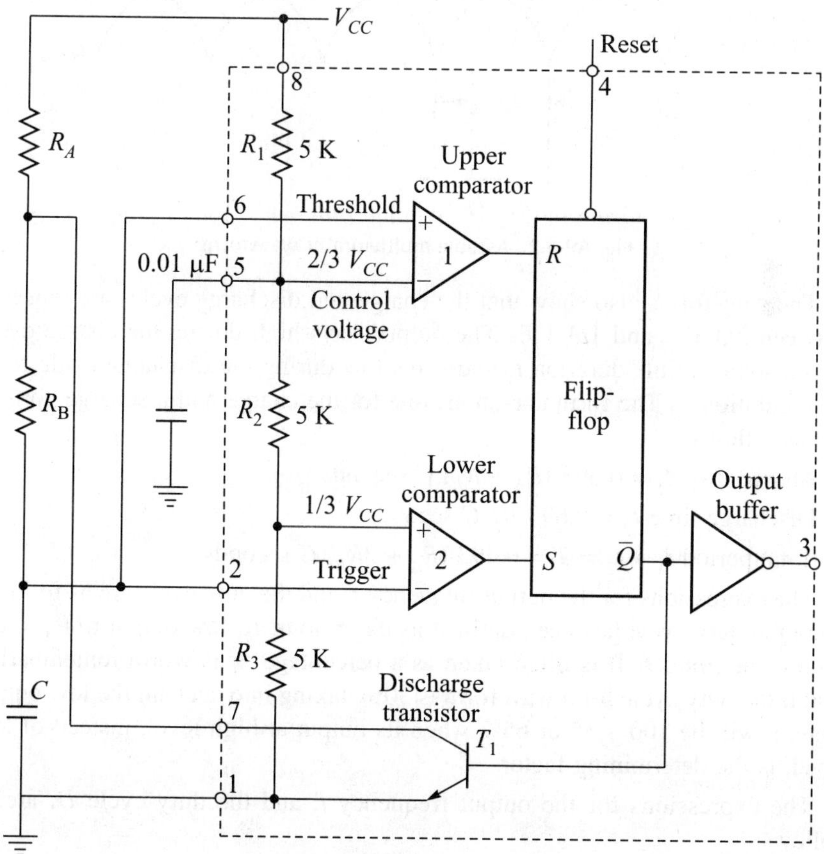

Fig. AP-5.1. Internal Structure of 555 Timer (within dotted lines).

* As the transistor is turned on it acts as a low resistance across the capacitor, thereby draining off its charge through resistor R_B.
* As the voltage across the capacitor falls below 1/3 V_{CC}, the output of the lower comparator 2 goes high, thereby setting the flip-flop, turning off the discharge transistor, and raising the output at pin 3 to a high level.
* The charge and discharge cycles are repeated as the capacitor is alternately charged to 2/3 V_{CC} and then discharged to 1/3 V_{CC}.

The charge and discharge waveforms of this astable multivibrator are shown in Fig. AP-5.2.

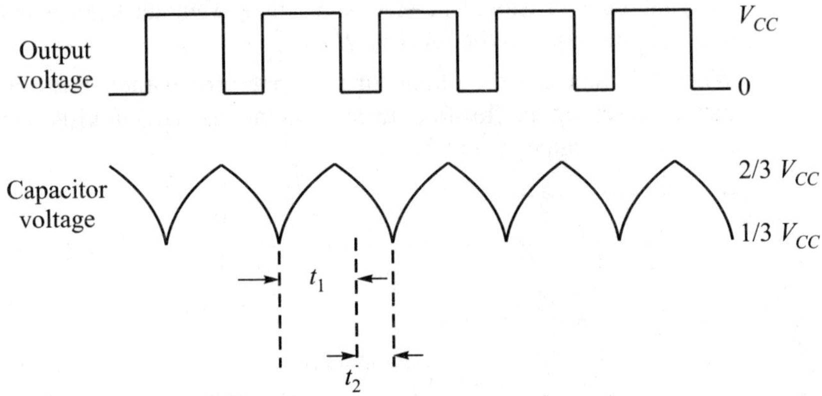

Fig. AP-5.2. Astable multivibrator waveform.

The waveforms also show that the charge and discharge cycles are repeated between 2/3 V_{CC} and 1/3 V_{CC}. The output stays high during the charge cycle which spans a time duration t_1 and stays low during the discharge cycle for a time duration t_2. The formulae applicable for the charge and discharge periods are as follows:

Charge time $t_1 = 0.693\ (R_A + R_B)\ C$ seconds

Discharge time $t_2 = 0.693\ R_B\ C$ seconds

Total period $t_1 + t_2 = T = 0.693\ (R_A + 2R_B)\ C$ seconds

The expressions for the output frequency f, and the duty cycle are as follows. Here the duty cycle has been defined as the ratio of t_2, low output time, to the total cycle time, T. It is often taken as a percentage. It is worth remembering that if the duty cycle for a waveform is 30%, taking into account the low output time, it will be 100 – 35 or 65% when its output at high level, instead of low level, is the determining factor.

The expressions for the output frequency f, and the duty cycle D, are as follows:

$$f = \frac{1}{T} = \frac{1.443}{(R_A + 2R_B)C} \qquad \text{Eq. 9.5}$$

$$D = \frac{\text{Low output time}}{\text{Total cycle time}} = \frac{t_2}{T} = \frac{R_B}{R_A + 2R_B} \qquad \text{Eq. 9.6}$$

where f is in Hz

C is in farad

and R_A and R_B are in ohm.

Since the capacitor charges through resistors R_A and R_B and discharges through resistor R_B only, the charge and discharge times are not equal. As a consequence the output is not a symmetrical square wave. This could have been possible if R_A was nearly zero, which is not possible. However, to obtain an output close to a square wave, R_A can be made much smaller than R_B and in that case the charge and discharge times will be essentially dependent on R_B and C. The frequency of the square wave will be approximately $0.693/R_B\,C$. R_A cannot be made very small as in that case a large current would flow from the supply through pin 7 when it goes low.

Monostable Operation

The monostable multivibrator has a stable and a quasi-stable state and it is also known as a 'One shot'. Its normal output is logic 0, like the reset state of a flip-flop and its $\overline{Q}$ output is normally logic 1, same as the set state of a flip-flop. The symbol for a 'One shot' is given in Fig. AP-5.3. The diagram shows the symbol for a 'One shot' which responds to a negative-going trigger pulse and, therefore, its output will change on the falling edge of the pulse as is shown in Fig. AP-5.4.

Also notice from the output waveform that the Q output which is normally 0 switches to the logic 1 state on receiving a trigger pulse and remains in that state for a predetermined time, t, of the unstable state. Subsequent trigger pulses determine its operation in the same way. The generation of a delayed pulse is a characteristic feature of one shots.

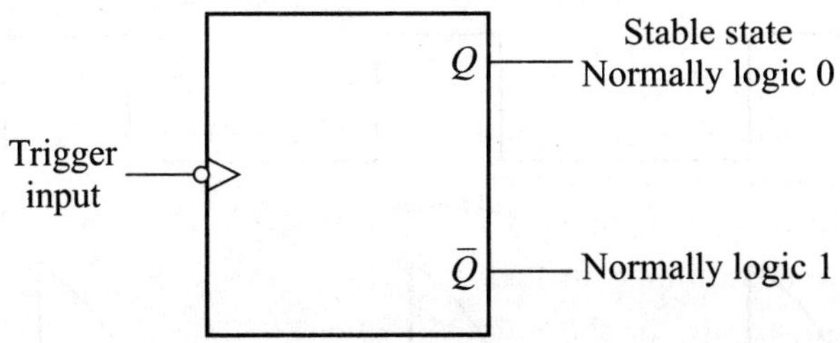

Fig. AP-5.3. Logic symbol for 'One shot'.

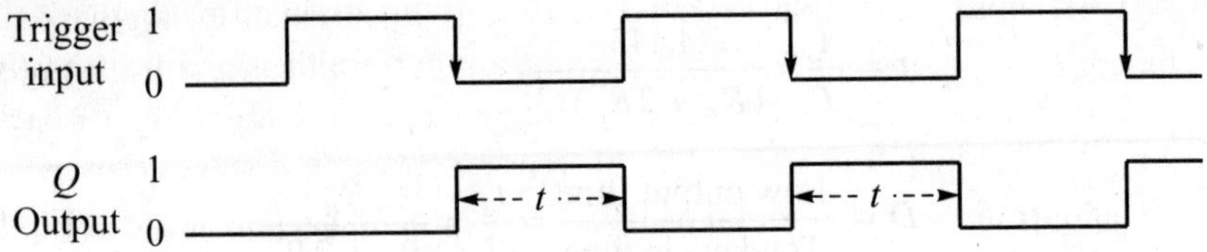

Fig. AP-5.4. Waveform.

One Shot Using 555 Timer

Figure AP-5.5 shows a 555 Timer wired for monostable operation. The waveforms generated by it are shown in Fig. AP-5.6.

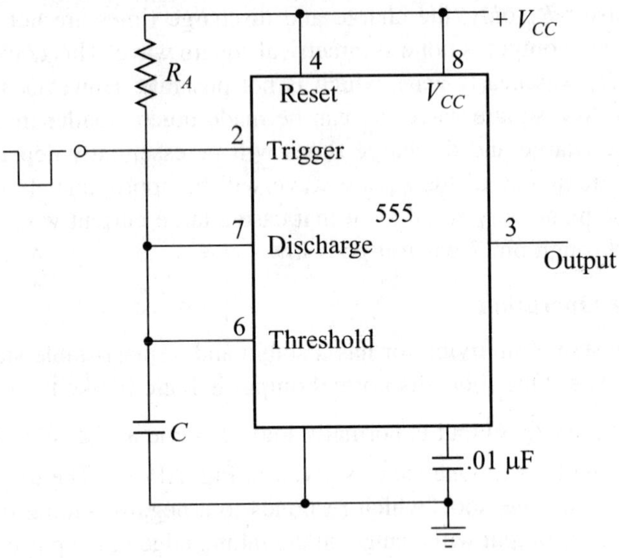

Fig. AP-5.5. Monostable multivibrator.

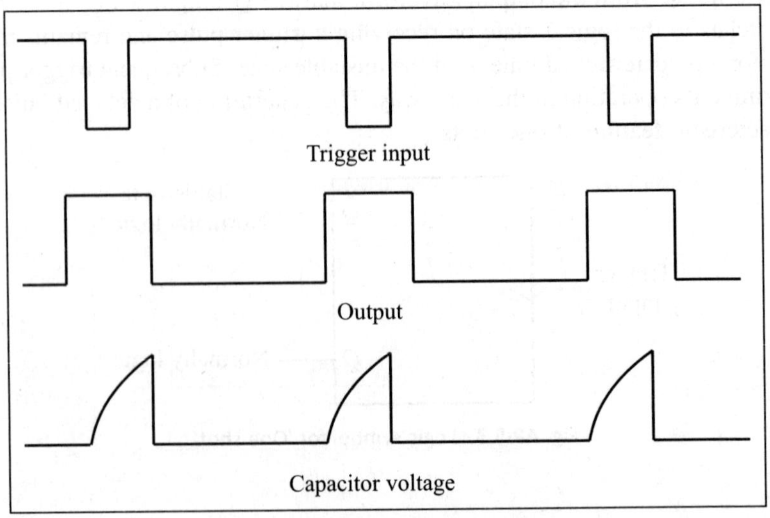

Fig. AP-5.6. Monostable waveforms.

Initially a transistor inside the Timer acts as a short across the capacitor C which is, therefore, in a discharged state. On the application of a negative-going trigger pulse at pin 2 the circuit triggers when the voltage across C reaches 1/3 V_{CC}, the flip-flop in the Timer is set, which releases the short across C and pushes the output high. At the same time the voltage across C rises exponentially with the time constant R_A C and remains in that state for a period R_A C when if it is triggered again during this interval. When the voltage across the capacitor has reached 2/3 V_{CC}, the threshold comparator resets the flip-flop which discharges C and the output is driven low. The circuit will remain in this state until the application of the next negative-going pulse at pin 2.

The pulse width can be calculated from the following equation:

$$\text{Pulse width } t = 1.1\, R_A\, C$$

where R is in ohm

C is in farad

and t is in second.

Bistable Operation

The 555 Timer can also function as a bistable flip-flop and its output is compatible with TTL devices. For bistable operation it is required to be connected as shown in Fig. AP-5.7. It is set when a negative pulse is applied to the trigger input terminal, pin 2, when the Q output goes high. A positive-going pulse when applied to the threshold terminal resets it and the Q output goes low. It can also be reset by applying a negative-going pulse to the reset terminal, pin 4. In the latter case, however, pin 6 has to be kept low.

The 555 Timer can operate on voltages from 5–15 volts. It consumes very little power and can directly drive a relay. It can sink or source a current of 200 mA.

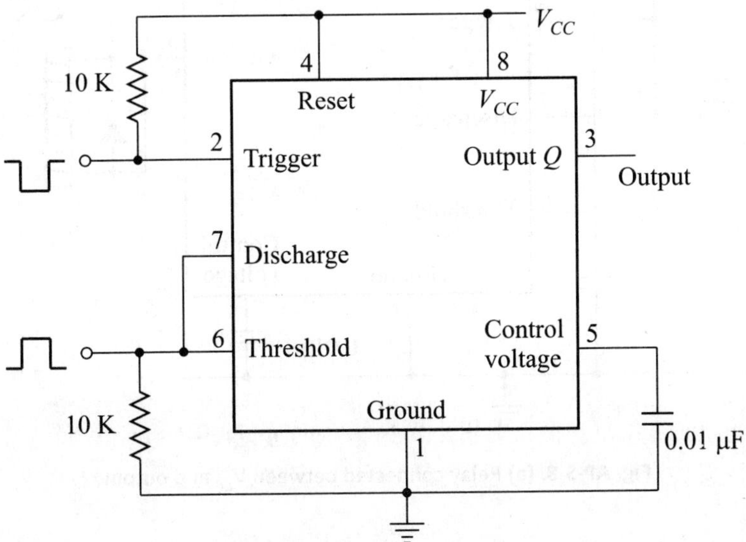

Fig. AP-5.7. Bistable operation.

Connecting Timer to Load

The output of a Timer consists of a quasi-complementary transistor circuit which is very much the same as in audio output stages. In this arrangement one end of the load can be connected to the emitter-collector junction and the other end either to V_{CC} or ground. From a Timer 555 the output can be obtained directly from pin 3.

We will first consider the circuit function when the load is connected between the V_{CC} and pin 3 of the Timer. In this arrangement when the Timer output is low a heavy current flows through the load into pin 3 of the Timer. However, when the output is high the current flowing through the load is very small.

If the load is connected between pin 3 of the Timer and ground, the output current flowing through the load is high when the Timer output is high, and it is very small when the Timer output is low.

A relay can be used as a load but in this arrangement the delay or timed interval should be more than 100 ms. The relay should not draw more than 200 mA at the operating voltage of the Timer. Figures AP-5.8 (*a*) and (*b*) show how the relay is to be connected to the Timer.

Since the relay is an inductive load, it generates a high reverse voltage when current flow is interrupted. It is, therefore, necessary to connect a diode across the relay coil to enable it to absorb the transient voltage. The diode is connected so that in normal operation it is reverse biased.

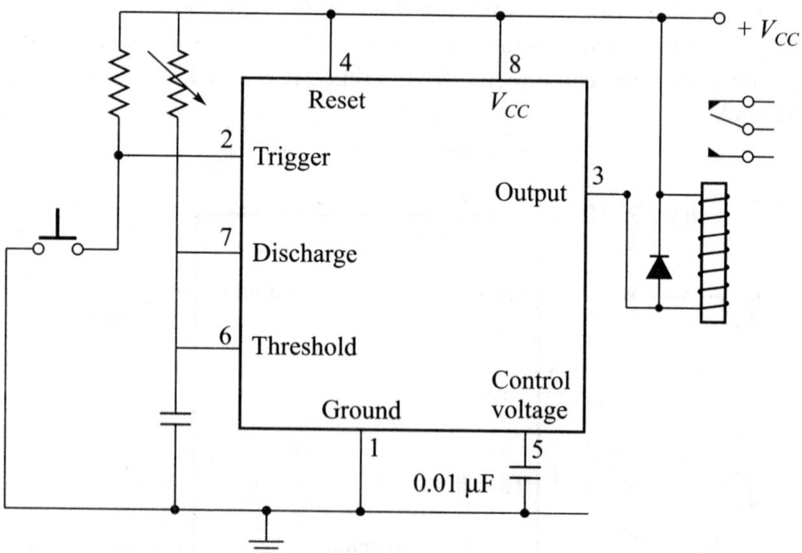

Fig. AP-5.8. (a) Relay connected between V_{CC} and output.

Figure 5.8 (*b*) shows how a relay can be connected between the output and ground. In this arrangement a diode has to be connected between the output and relay. Unless this is done the Timer may latch up because of the voltage drop across the diode.

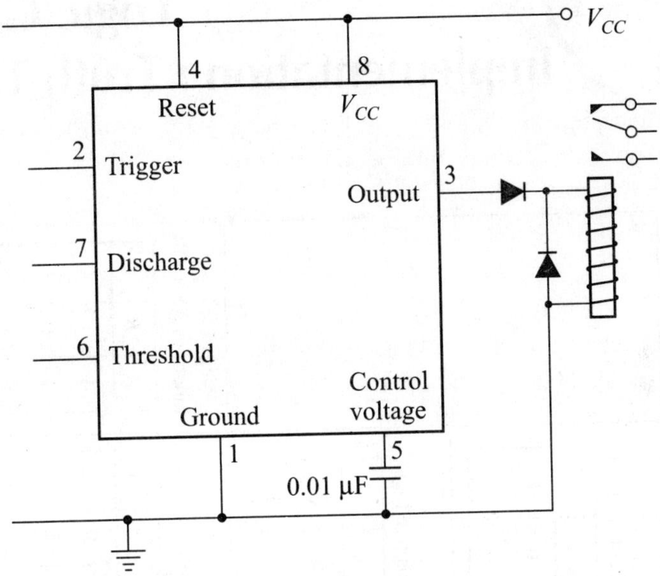

Fig. AP-5.8. (b) Relay connected between output and ground.

APPENDIX 6

Logic Gates : Implementation : Truth Tables

<table>
<tr><th>Gate symbol</th><th>Gate implementation</th><th>Truth table positive logic</th><th>Truth table negative logic</th></tr>
<tr><td>Inverter
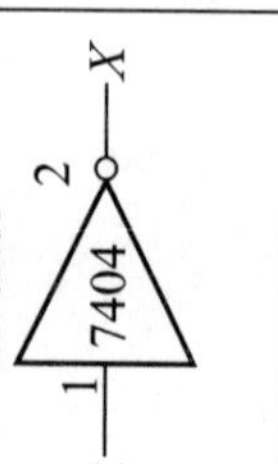</td><td>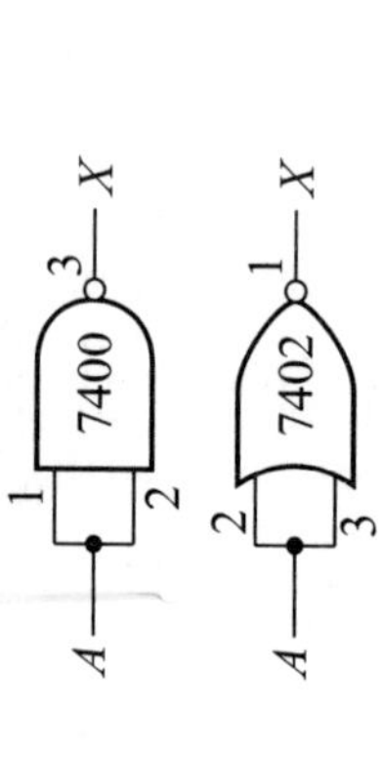</td><td>Positive logic<table><tr><th>A</th><th>X</th></tr><tr><td>0</td><td>1</td></tr><tr><td>1</td><td>0</td></tr></table></td><td rowspan="2"></td></tr>
<tr><td>Buffer
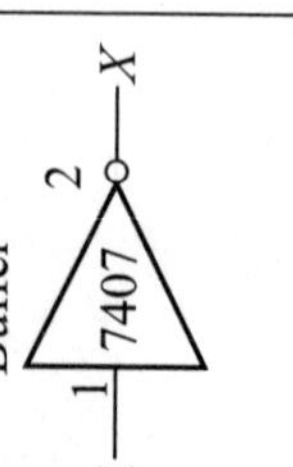</td><td>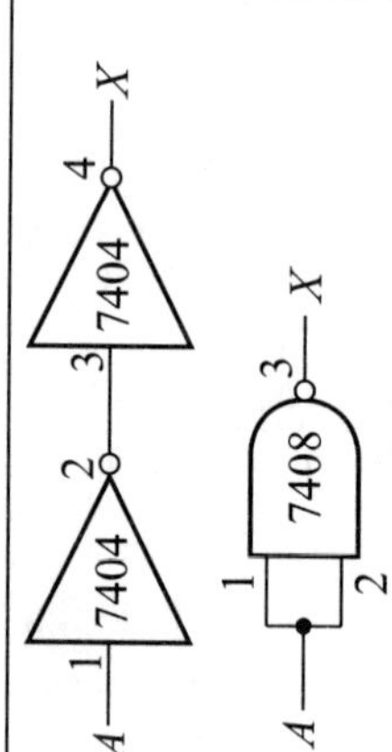</td><td>Positive logic<table><tr><th>A</th><th>X</th></tr><tr><td>0</td><td>0</td></tr><tr><td>1</td><td>1</td></tr></table></td></tr>
<tr><td>AND
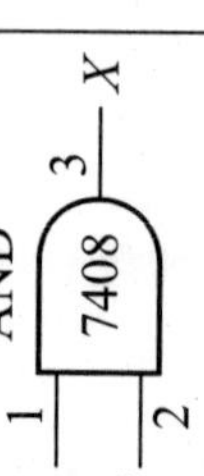
AND
Negative logic
</td><td>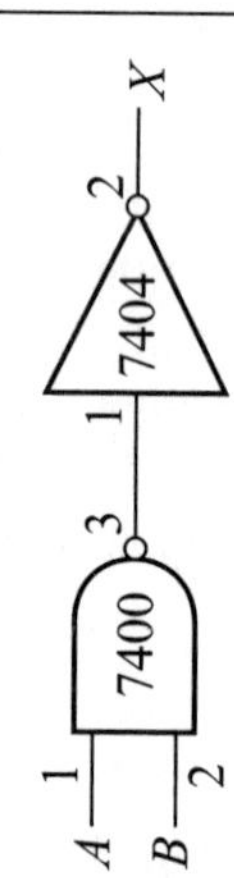
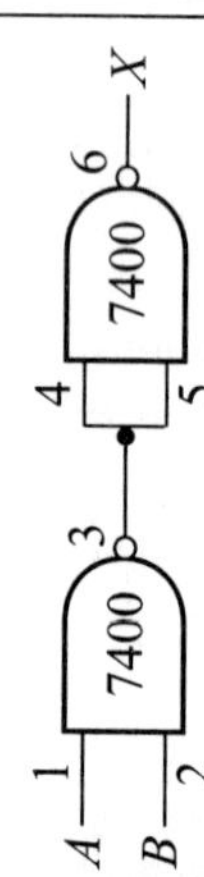</td><td>Positive logic<table><tr><th>A</th><th>B</th><th>X</th></tr><tr><td>0</td><td>0</td><td>0</td></tr><tr><td>0</td><td>1</td><td>0</td></tr><tr><td>1</td><td>0</td><td>0</td></tr><tr><td>1</td><td>1</td><td>1</td></tr></table></td><td>Negative logic functions AS + OR<table><tr><th>A</th><th>B</th><th>X</th></tr><tr><td>1</td><td>1</td><td>1</td></tr><tr><td>1</td><td>0</td><td>1</td></tr><tr><td>0</td><td>1</td><td>1</td></tr><tr><td>0</td><td>0</td><td>0</td></tr></table></td></tr>
</table>

Gate symbol	*Gate implementation*	*Truth table positive logic*	*Truth table negative logic*
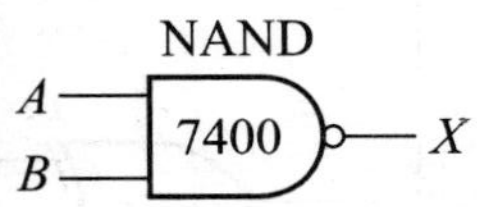	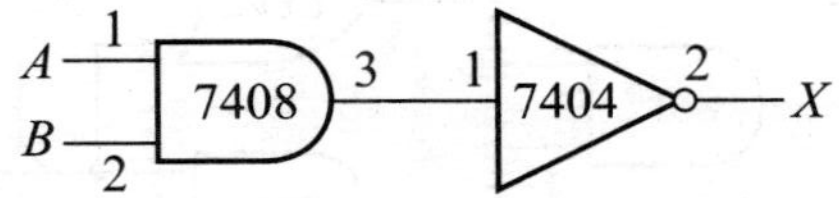	Positive logic (table 1 below)	Negative logic functions AS + NOR (table 2 below)
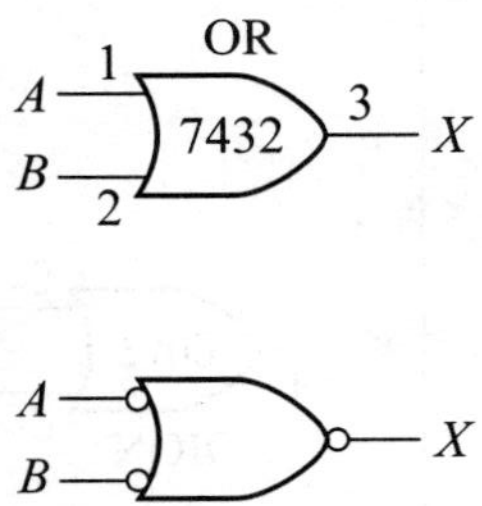	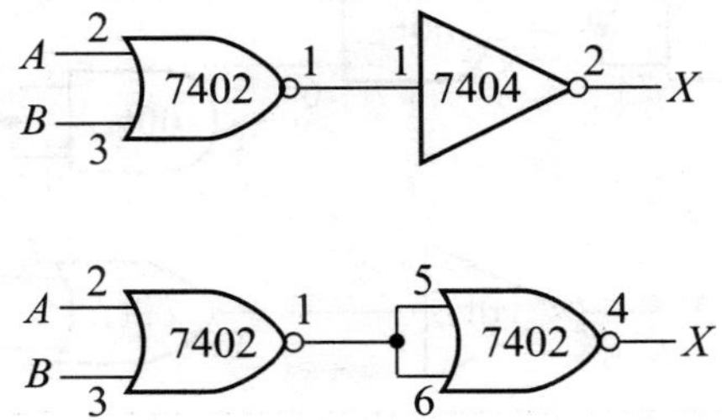	Positive logic (table 3 below)	Negative logic functions AS + AND (table 4 below)

Positive logic

A	*B*	*X*
0	0	1
0	1	1
1	0	1
1	1	0

Negative logic functions AS + NOR

A	*B*	*X*
1	1	0
1	0	0
0	1	0
0	0	1

Positive logic

A	*B*	*X*
0	0	0
0	1	1
1	0	1
1	1	1

Negative logic functions AS + AND

A	*B*	*X*
1	1	1
1	0	0
0	1	0
0	0	0

<table>
<tr><th>Gate symbol</th><th>Gate implementation</th><th>Truth table positive logic</th><th>Truth table negative logic</th></tr>
<tr>
<td>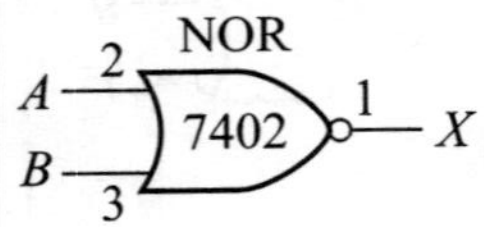</td>
<td>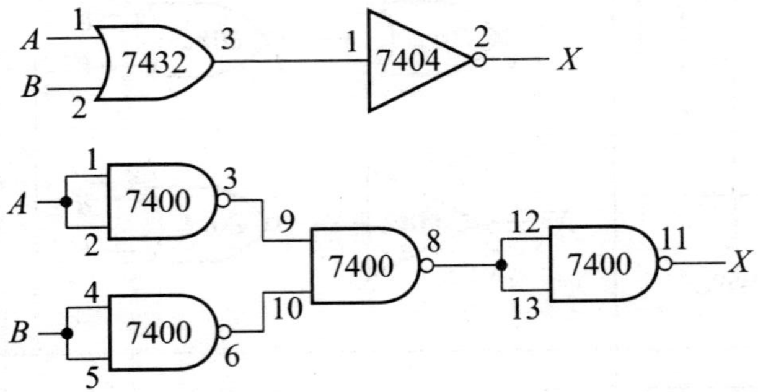</td>
<td>Positive logic
<table><tr><th>A</th><th>B</th><th>X</th></tr><tr><td>0</td><td>0</td><td>1</td></tr><tr><td>0</td><td>1</td><td>0</td></tr><tr><td>1</td><td>0</td><td>0</td></tr><tr><td>1</td><td>1</td><td>0</td></tr></table></td>
<td>Negative logic functions as + NAND
<table><tr><th>A</th><th>B</th><th>X</th></tr><tr><td>1</td><td>1</td><td>0</td></tr><tr><td>1</td><td>0</td><td>1</td></tr><tr><td>0</td><td>1</td><td>1</td></tr><tr><td>0</td><td>0</td><td>1</td></tr></table></td>
</tr>
<tr>
<td>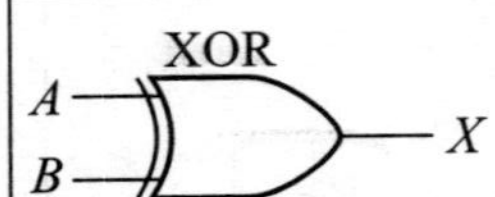</td>
<td>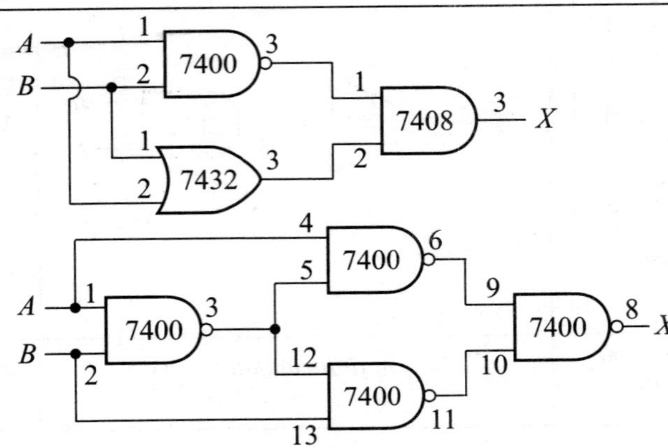</td>
<td>Positive logic
<table><tr><th>A</th><th>B</th><th>X</th></tr><tr><td>0</td><td>0</td><td>0</td></tr><tr><td>0</td><td>1</td><td>1</td></tr><tr><td>1</td><td>0</td><td>1</td></tr><tr><td>1</td><td>1</td><td>0</td></tr></table></td>
<td></td>
</tr>
</table>

APPENDIX 7

74 Series TTL Integrated Circuits

IC No.	*Function*
7400	Quad 2-point NAND Gates
7401	Quad 2-input NAND Gates (Open-Collector)
7402	Quad 2-input NOR Gates
7403	Quad 2-input NAND Gates (Open-Collector)
7404	Hex-Inverters
7405	Hex-Inverters (Open-Collector)
7406	Hex-Inverter Buffers/Drivers (Open-Collector)
7407	Hex Buffers/Drivers (Open-Collector)
7408	Quad 2-input AND Gates
7410	Triple 3-input NAND Gates
7411	Triple 3-input AND Gates
7412	Triple 3-input NAND Gates (Open-Collector)
7413	Dual 4-input NAND Schmitt Triggers
7414	Hex Schmitt Trigger Inverters
7415	Triple 3-input AND Gates (Open-Collector)
7417	Hex Buffers/Drivers (Open-Collector)
7420	Dual 4-input NAND Gates
7421	Dual 4-input AND Gates
7422	Dual 4-input NAND Gates (Open-Collector)
7425	Dual 4-input NOR Gates with Strobe
7427	Triple 3-input NOR Gates
7428	Quad 2-input NOR Buffers
7430	8-input NAND Gates
7432	Quad 2-input OR Gates
7433	Quad 2-input NOR Buffers (Open-Collector)
7437	Quad 2-input NAND Buffers
7438	Quad 2-input NAND Buffers (Open-Collector)
7439	Quad 2-input NAND Buffers

IC No.	Function
7440	Dual 4-input NAND Buffers
7441	BCD-to-Decimal Decoder/Driver (Open-Collector)
7442	BCD-to-Decimal Decoder
7443	Excess-3-to-Decimal Decoder
7444	Excess-3-Gray-to-Decimal Decoder
7445	BCD-to-Decimal Decoder/Driver (Open-Collector)
7447	BCD-to-7-Segment Decoder/Driver
7448	BCD-to-7-Segment Decoder/Driver
7449	BCD-to-7-Segment Decoder/Driver (Open-Collector)
7450	Expandable Dual 2-wide 2-input AOI Gates
7451	Dual 2-wide 2-input AOI Gates
7453	Expandable 4-wide 2-input AOI Gate
7454	4-wide 2-input AOI Gate
7460	Dual 4-input Expanders
7464	4-wide AOI Gates
7465	4-wide AOI Gates (Open-Collector)
7470	AND-Gated JK Flip-flop
7472	AND-Gated JK Flip-flop
7473	Dual JK Flip-flops
7474	Dual D Flip-flops
7475	Dual 2-bit Transparent Latches
7476	Dual JK Flip-flops
7477	Dual 2-bit Transparent Latches
7480	Gated Full-Adder
7482	2-bit Binary Adder
7483	4-bit Binary Adder
7485	4-bit Magnitude Comparator
7486	Quad EX-OR Gates
7489	16 × 4 RAM (Open-Collector)
7490	BCD Counter
7491	8-bit Serial-in Serial-out Shift-Register
7492	Divide-by-12 Counter
7493	4-bit Binary Counter
7494	4-bit Serial/parallel-in, Serial-out Shift-Register
7495	4-bit Serial/parallel-in, Parallel-out Shift-Register

IC No.	Function
7496	5-bit Serial/parallel-in/Parallel-out Serial-in/Serial-out Shift-Register
74121	Monostable Multivibrator
74122	Retriggerable Monostable Multivibrator
74123	Dual Retriggerable Monostable Multivibrator
74124	Dual Voltage Controlled Oscillators
74125	Tri-state Quad Buffers
74126	Tri-state Quad Buffers
74128	Quad 2-input NOR Buffers
74132	Quad 2-input NAND Schmitt Triggers
74133	13-input NAND Gate
74134	Tri-state 12-input NAND Gate
74135	Quad EX-OR/NOR Gates
74139	Dual 1 : 4 Demultiplexer
74141	BCD-to-Decimal Decoder/Driver (Open-Collector)
74145	BCD-to-Decimal Decoder/Driver (Open-Collector)
74147	Priority Encoder (Decimal-to-Binary)
74148	Priority Encoder (Octal-to-Binary)
74150	16 : 1 Multiplexer
74151A	8 : 1 Multiplexer
74152	8 : 1 Multiplexer
74153	Dual 4 : 1 Multiplexers
74154	1 : 16 Demultiplexer
74155	Dual 1 : 4 Demultiplexers
74156	Dual 1 : 4 Demultiplexers (Open-Collector)
74157	Quad 2 : 1 Multiplexers
74158	Quad 2 : 1 Multiplexers
74159	1 : 16 Demultiplexer (Open-Collector).
74160	Decade Up-counter
74176	Presettable BCD Counter
74177	Presettable 4-bit Binary Counter
74178	4-bit Universal Shift-Register
74179	4-bit Universal Shift-Register
74180	8-bit Parity Generator/Checker

IC No.	Function
74181	Arithmetic Logic Unit (ALU)
74182	Look-Ahead-Carry Generator
74184	BCD-to-Binary Code Converter
74185A	6-bit Binary-to-BCD Converter
74190	Decade Up/Down Counter
74191	4-bit Binary Up/Down Counter
74192	Decade Up/Down Counter
74193	4-bit Binary Up/Down Counter
74194	4-bit Bi-directional (Universal) Shift-Register
74195	4-bit Serial/Parallel-in, Parallel-out, Shift-Register
74196	Presettable BCD Counter
74197	Presettable 4-bit Binary Counter
74198	8-bit Bi-directional (Universal) Shift-Register
74199	8-bit Serial/Parallel-in, Parallel-out, Shift-Register
74206	256-bit RAM (Open-Collector)
74221	Dual Monostable Multivibrators
74240	Tri-state Octal Inverter Buffers
74241	Tri-state Octal Buffers
74244	Tri-state Octal Buffers
74246	BCD-to-7-Segment Decoder/Driver (Open-Collector)
74247	BCD-to-7-Segment Decoder/Driver (Open-Collector)
74248	BCD-to-7-Segment Decoder/Driver
74249	BCD-to-7-Segment Decoder/Driver (Open-Collector)
74260	Dual 5-input NOR Gates
74266	Quad EX-NOR Gates (Open-Collector)
74273	Octal D Flip-flops
74276	Quad JK Flip-flops
74279	Quad S.R. Latches
74280	9-bit Parity Generator/Checker
74283	4-bit Binary Adder with Fast Carry
74289	16 × 4 RAM (Open-Collector)
74290	BCD Counter
74293	4-bit Binary Counter

APPENDIX 8

Series 74C CMOS Integrated Circuits

IC No.	*Function*
74C00	Quad 2-input NAND Gates
74C02	Quad 2-input NOR Gates
74C04	Hex Inverters
74C08	Quad 2-input AND Gates
74C10	Triple 3-input NAND Gates
74C14	Hex Schmitt Trigger Inverters
74C20	Dual 4-point NAND Gates
74C30	8-input NAND Gate
74C32	Quad 2-input OR Gates
74C42	BCD-to-Decimal Decoder
74C48	BCD-to-7-Segment Decoder
74C73	Dual JK Flip-flops with Clear
74C74	Dual D Flip-flops
74C76	Dual JK Flip-flops with Clear and Preset
74C83	4-bit Binary Full Adder
74C85	4-bit Magnitude Comparator
74C86	Quad 2-input EXCLUSIVE-OR Gates
74C89	64-bit Tri-state RAM
74C90	4-bit Decade Counter
74C93	4-bit Binary Counter
74C95	4-bit Right-Shift/Left-Shift Register
74C107	Dual JK Flip-flops with Clear
74C154	4-line to 16-line Decoder/Demultiplexer
74C157	Quad 2-input Multiplexers
74C160	Decade Counter with Asynchronous Clear
74C16I	Binary Counter with Asynchronous Clear

IC No.	*Function*
74C162	Decade Counter with Synchronous Clear
74C163	Binary Counter with Synchronous Clear
74C164	8-bit Serial-in, Parallel-out Shift-Register
74C174	Hex D Flip-flops
74C175	Quad D Flip-flops
74C192	Synchronous 4-bit Up/Down Decade Counter
74C193	Synchronous 4-bit Up/Down Binary Counter
74C195	4-bit Register
74C221	Dual Monostable Multivibrators

APPENDIX 9

Handling Precautions for MOS Devices

Because MOS devices have very high input resistance, they are susceptible to damage when exposed to extremely high static electrical charges. To avoid possible damage to devices during handling, testing or actual operation the following precautions should be taken:

1. MOS devices which are not in use should be retained in their original packages.
2. When this is not convenient, store MOS integrated circuits in conductive foam or in aluminium foil. Avoid storing them in white styrofoam, which is an excellent static electricity generator.
3. P.C. boards containing ICs should be stored in conductive plastic or metallic envelopes.
4. All tools in use which may come in contact with MOS device pins should be connected to a good ground.
5. The soldering iron tip should be invariably connected to a reliable ground.
6. Wear a shorting strip around your wrist to discharge static electricity.
7. Avoid synthetic material clothing while handling these devices.
8. Connect to ground your metal bench and the chassis of all test equipment in use.
9. Preferably work on an antistatic treated work bench or alternatively use an antistatic mat.
10. The supply voltage to these devices should never exceed the prescribed limit.
11. Great care should be taken to ensure that the power supply connections are not reversed.
12. All connections should be checked before power is switched on.
13. Changes in circuit connections should not be made without first disconnecting the power supply.
14. Power should be turned off before removing ICs from their sockets or inserting them in their sockets.

15. The input signal should always be less than the supply voltage. It is, therefore, important that the power supply should not be disconnected while the signal source is still connected.
16. Before any signal is connected to a device input, power should be on. All unused inputs on ICs should be connected to V_{CC} or V_{DD} whichever of these provides the proper logic state at the input.
17. It is worthwhile to connect a 10 K ohm resistor in series with the signal source as it may afford some protection from excessive input voltages.
18. After MOS units have been mounted on circuit boards, proper handling precautions should still be observed. Until these sub-assemblies are inserted into a complete system in which the proper voltages are applied, the board is no more than an extension of of the leads of the devices mounted on the board.
19. It is a good practice to put conductive clips or conductive tape on the circuit board terminals. This precaution prevents static charges from being transmitted through the board wiring to the devices mounted on the board.

APPENDIX 10

Boolean Algebra

George Simon Boole, an Englishman, will always be remembered as the father of Boolean Algebra. Although the Boolean algebra has some similarities with conventional algebra, but in many respects it has considerable differences. Boolean algebra is now indispensable in the design of digital systems. It is invaluable in expressing logic circuit functions and in analyzing their design. It provides a sound mathematical foundation for a proper understanding and evaluation of digital circuits.

Boolean expressions contain variables which can only have either of two values, that is 0 or 1. The final result of a Boolean function can only be either of these two values 0 or 1. For a thorough understanding of digital circuits, their design, function and operation it is essential that the student must acquire an indepth knowledge of the application of Boolean algebra.

Here we will highlight some of the basic postulates and laws of Boolean algebra. The postulates have their origin in the AND, OR and Invert operations. These nine postulates, summed up below, have a single variable and as they have their origin in the three logic gates referred to above, they can be proved by the application of the truth tables of these logic gates.

Postulates which follow from the AND operation	*Postulates which follow from the OR operation*
$A \cdot 1 = A$	$A + 1 = 1$
$A \cdot 0 = 0$	$A + 0 = A$
$A \cdot A = A$	$A + A = A$
$A \cdot \overline{A} = 0$	$A + \overline{A} = 1$

The postulates which follow from the NOT operation are as follows:

If $A = 1$ $\quad \overline{A} = 0$

and $\quad \overline{\overline{A}} = 1$

It also follows that $\overline{1} = 0$

and $\quad \overline{0} = 1$

Normally there are more than a single variable in Boolean expressions. The laws which govern multivariable identities have been given in Table AP-10.1 and are found very useful in understanding and simplifying Boolean functions.

DeMorgan's theorem referred to as Laws 21 and 22 in Table AP-10.1 will be of considerable interest and require some explanation.

This theorem states as follows:

"For a given Boolean expression, if all + (OR) are changed to '.' (AND), and vice-versa, and each variable is replaced by its complement (replace A with $\overline{A}$ and $\overline{A}$ with A etc. for all variables), the result will be the complement of the original expression." Summarized briefly, it can be stated as follows in equation form.

$$\overline{A} \cdot \overline{B} = \overline{A + B} \qquad \text{(NOR expression)}$$

$$\overline{A} + \overline{B} = \overline{A \cdot B} \qquad \text{(NAND expression)}$$

Using this theorem we can convert an expression using AND relationship to one which uses OR relationship.

Table AP-10.1

Summary of Boolean Algebra Laws

Classification	*Algebraic definition*	*Law*
1. Laws of Intersection	$A \cdot 1 = A$	1
	$A \cdot 0 = 0$	2
2. Laws of Union	$A + 1 = 1$	3
	$A + 0 = A$	4
3. Laws of Tautology	$A \cdot A = A$	5
	$A + A = A$	6
4. Laws of Complements	$A \cdot \overline{A} = 0$	7
	$A + \overline{A} = 1$	8
5. Law of Double Negation	$\overline{\overline{A}} = A$	9
6. Laws of Commutation	$A \cdot B = B \cdot A$	10
	$A + B = B + A$	11
7. Laws of Association	$(A \cdot B)C = A \cdot (B \cdot C)$	12
	$(A + B) + C = A + (B + C)$	13
8. Laws of Distribution	$A \cdot B + A \cdot C = A \cdot (B + C)$	14
	$(A + B)(A + C) = A + (B \cdot C)$	15
9. Laws of Absorption	$A(A + B) = A$	16
	$A + A \cdot B = A$	17
	$A(\overline{A} + B) = AB$	18
	$A \cdot B + \overline{B} = A + \overline{B}$	19
	$A \cdot \overline{B} + B = A + B$	20
10. DeMorgan's Theorem	$\overline{A + B} = \overline{A} \cdot \overline{B}$	21
	$\overline{A \cdot B} = \overline{A} + \overline{B}$	22

APPENDIX 11

Operational Amplifier

Operational amplifiers are very high gain linear amplifiers. They can be used to amplify both a.c. and d.c. signals and they are also useful for carrying out mathematical operations. The basic circuit of an operational amplifier is a differential amplifier and we will devote some time to see what function it performs and how it accomplishes it.

Differential Amplifier

The basic circuit of a differential amplifier is given in Fig. AP-11.1.

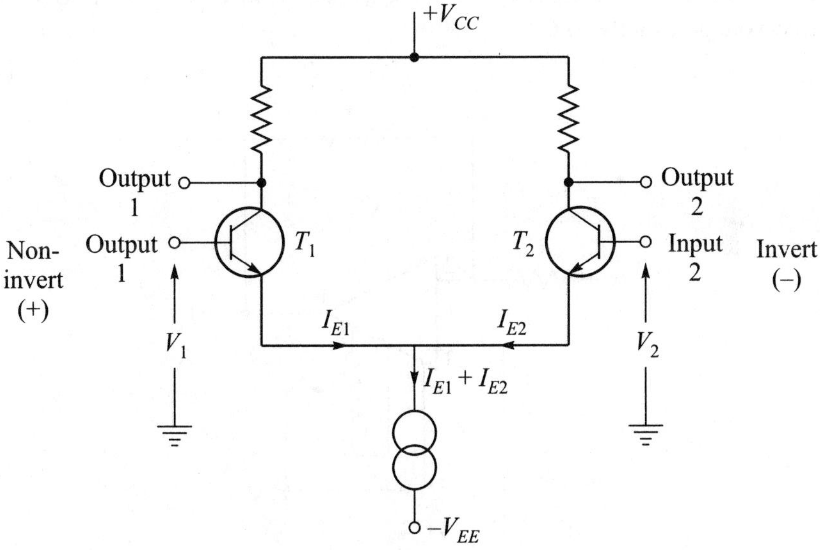

Fig. AP-11.1. Differential amplifier.

Basically a differential amplifier consists of two identical common-emitter stages with their emitters connected to a constant-current source. When we apply a positive signal V_1 to the base of T_1 its collector current will increase. At the same time the collector current of T_2 will decrease by the same amount, since both T_1 and T_2 are connected to a constant current source. If the gain of T_1 and T_2 in the common-emitter configuration is A, a voltage AV_1 will appear at output 2. It is important to note that a positive signal at the input of T_1 produces a positive output. This input is, therefore, a non-inverting input.

For the same reasons a positive signal V_2 at the input of T_2 will produce an output of AV_2 at output 2. The negative sign shows that input 2 is an inverting input. If inputs V_1 and V_2 are applied simultaneously the output at T_2 will be $A(V_1 - V_2)$. In conclusion we can say that a differential amplifier amplifies the difference of the two input signals.

Input D.C. Offset Voltage

Operational amplifiers are carefully designed to ensure that with zero input the output is also zero. Even a very slight mismatch between internal components may still result in a small d.c. output voltage even if both the inputs are grounded. The larger the gain of the Op-amp the greater will be the magnitude of this voltage. The offset voltage presents problem when the Op-amp is used for amplification of d.c. signals. This problem does not upset the operation with a.c. signals as the d.c. voltage can be blocked from the output by a capacitor. Some Op-amps like 741 have pins available where an external potentiometer can be connected as shown in Fig. AP-11.2 to set the output voltage exactly to 0.

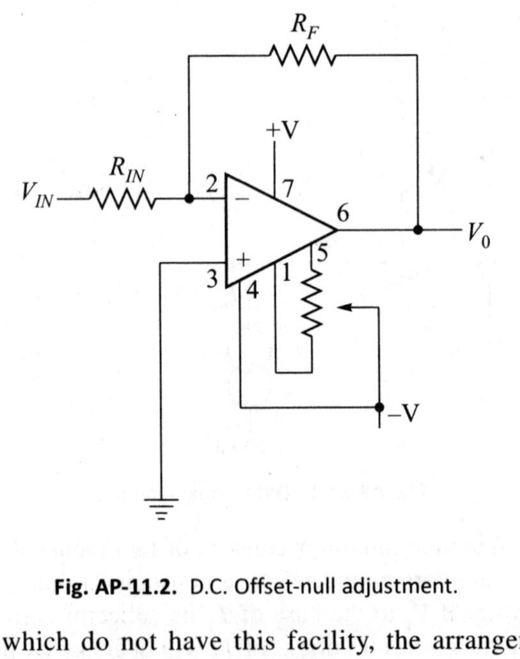

Fig. AP-11.2. D.C. Offset-null adjustment.

In Op-amps which do not have this facility, the arrangement shown in Fig. AP-11.3 may be used for null adjustment.

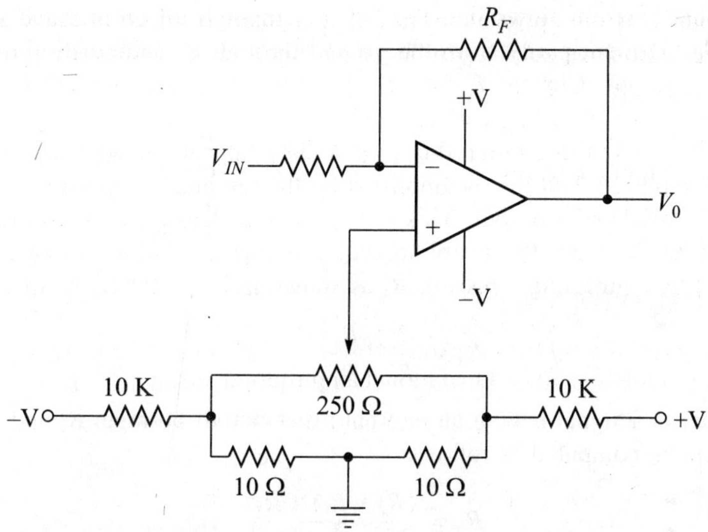

Fig. AP-11.3. Alternative set up for D.C. Offset-null adjustment.

Input Bias Current

The input bias current also contributes to the offset error. The Op-amp input requires a bias current for proper operation; but this gives rise to a d.c. output voltage even in the absence of an input signal.

To understand this problem let us consider the inverting amplifier in Fig. AP-11.4.

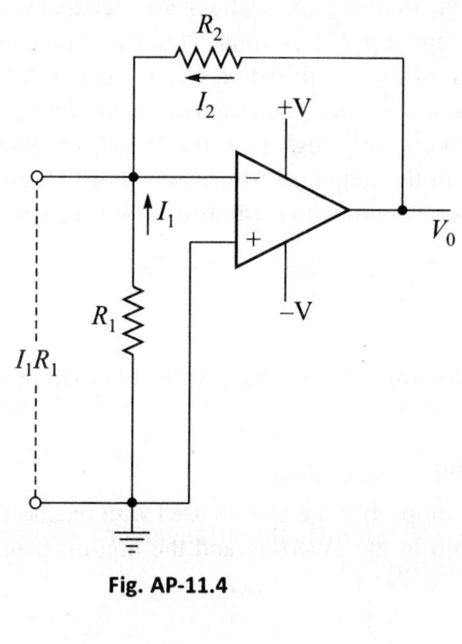

Fig. AP-11.4

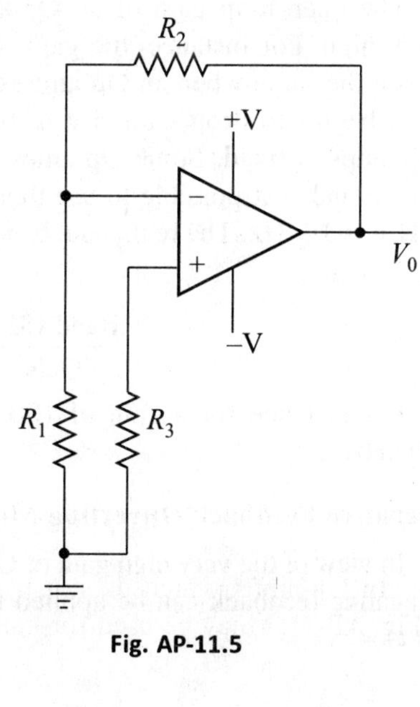

Fig. AP-11.5

The bias current flows into the N.I. (+) input from ground and into the inverting (–) input partly from the ground through R_1 and partly through R_2 from the output. Current I_1 flowing into the inverting input from ground causes a voltage drop I_1R_1 between the inverting input and ground. As the N.I. input is connected to ground, this voltage drop appears between the inverting and N.I. inputs. When this is amplified by the Op-amp, it appears as a much larger voltage at the output. This voltage is equal to $I_2 R_2$. If the feedback resistor, R_2 is large, the problem really requires a solution. This can be achieved by connecting a resistor R_3 as shown in Fig. AP-11.5 so that

$$R_3 = \frac{R_1 \times R_2}{R_1 + R_2}$$

If there is a signal having an internal resistance R_s between R_1 and ground, R_3 should be computed as follows:

$$R_3 = \frac{(R_s + R_1) \times R_2}{R_s + R_1 + R_2}$$

In a non-inverting amplifier it is difficult to compensate for the effect of the bias current. The normal practice in such cases is to use a signal source having an internal resistance R_s so that

$$R_s = \frac{R_1 \times R_2}{R_1 + R_2}$$

Gain-Bandwidth Product

The open-loop gain of an Op-amp, that is gain without any feedback is very high. For instance, the gain of Op-amp 741 is about 200,000. The gain is not the same when an Op-amp is used for amplification of a.c. signals. The gain begins to drop with rise in frequency. The gain-bandwidth product for Op-amps is fixed. Some Op-amps have a very high gain-bandwidth product, which makes it possible to use them in the region of Megahertz. For Op-amp 741 it is 1 MHz. The realizable bandwidth upto which amplification is possible is

$$\frac{\text{Bandwidth}}{\text{Gain}}$$

For instance for a gain of 100 Op-amp 741 the bandwidth obtainable is 10 kHz.

Negative Feedback (Inverting Mode)

In view of the very high gain of Op-amps they are always used with feedback. Negative feedback can be applied both in the inverting and the non-inverting modes.

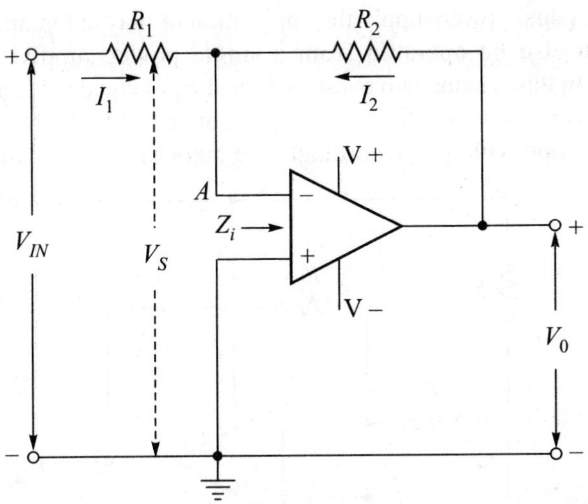

Fig. AP-11.6. Op-amp with negative feedback in inverting mode.

An Op-amp amplifier using negative feedback is given in Fig. AP-11.6. In this arrangement, the input voltage V_{in} forces a current I_1 through R_2 into the inverting input terminal of the Op-amp. However, this current is negligible because the input impedance of the Op-amp is very high. The input current has a tendency to raise the voltage V_s which causes a decrease in the output voltage V_0. When V_0 falls, it pulls the voltage V_{in} down to oppose the change in V_0. This is a property of circuits using negative feedback. It tends to oppose the change in the output resulting from a change in the input.

In this arrangement it is worth observing that, as no current flows into the Op-amp, the current flowing through R_1 from the input must be the same as the current flowing through R_2 towards the inverting input of the Op-amp. These two currents meeting at point *A* have a tendency to cancel each other. Point *A* is, therefore at zero potential. This phenomenon is called 'virtual earth'.

We can now calculate as follows, the voltage gain of this amplifier

$$I_1 = \frac{V_{in}}{R_1}$$

and

$$V_0 = -I_1R_2 = \frac{V_{in}}{R_1} \times R_2$$

Therefore,

$$A_v = \frac{V_0}{V_{in}} = \frac{R_2}{R_1}$$

Op-amp Power Supply Requirements

Op-amps are designed to operate from a dual symmetrical power supply, one delivering positive voltage and the other negative voltage with respect to ground. With a dual power supply the application of bias to Op-amps is simple. Op-amps can also be operated from a single power supply as shown in Fig. AP-11.7. In this scheme two resistors R_3 and R_4 of equal value are connected across the power supply and the centre tap is connected to the N.I. input which is at half the supply voltage. Accordingly the output is also at half the voltage.

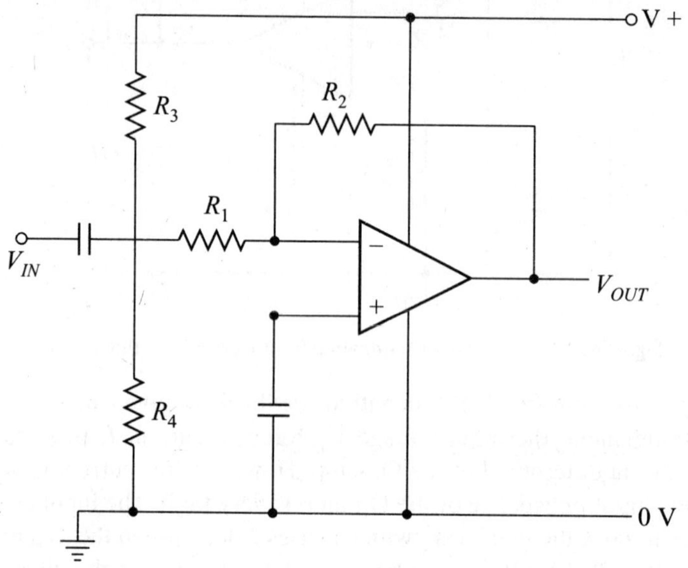

Fig. AP-11.7. Inverting A.C. amplifier on a single rail supply.

The voltage gain of this amplifier is R_2/R_1 and its input impedance is R_1.

A non-inverting A.C. amplifier with negative feedback operating on a dual voltage supply can be implemented as shown in Fig. AP-11.8. The voltage gain of this amplifier, as before, will be R_2/R_1 and its input impedance will be R_3.

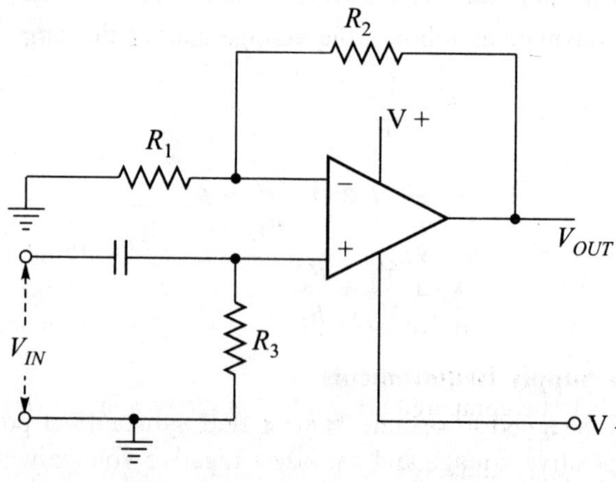

Fig. AP-11.8. Non-inverting A.C. amplifier on a dual power supply.

Op-amp Current Summer

Op-amps can be used for summation of input currents. A circuit for this purpose is given in Fig. AP-11.9.

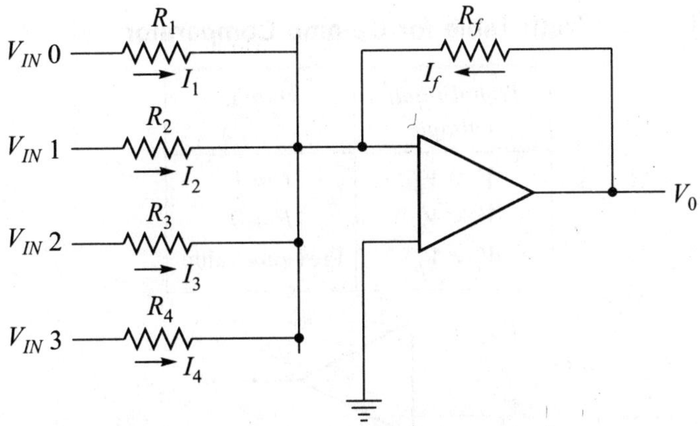

Fig. AP-11.9. Op-amp current summer.

Since feedback current I_f has to be equal to the sum of the input currents, we can equate the two.

Therefore,

$$I_f = I_1 + I_2 + I_3 + I_4$$

$$-\frac{V_0}{R_f} = \frac{V_{in}0}{R_1} + \frac{V_{in}1}{R_2} + \frac{V_{in}2}{R_3} + \frac{V_{in}3}{R_4}$$

and
$$-V_0 = \frac{R_f}{R_1}V_{in}0 + \frac{R_f}{R_2}V_{in}1 + \frac{R_f}{R_3}V_{in}2 + \frac{R_f}{R_4}V_{in}3$$

This shows that the output voltage reflects the weighted sum of the input voltages. The factor by which it is weighted is a function of the ratio of the feedback resistor to the input resistor in each case. The output is inverted because the inputs are connected to the inverting input of the Op-amp.

Comparator

A comparator is used to compare two signals to determine which of the two is greater in magnitude. When Op-amps are used without feedback, a small input signal causes the Op-amps output to switch between its two saturation limits. This property of Op-amps is made use of in making voltage comparators.

The two signals to be compared are applied at the two inputs of an Op-amp. When the non-inverting input is larger than the voltage at the inverting input, the output voltage goes high. When the inverting voltage is larger than the non-inverting voltage, the output goes low. The truth table for its operation as a comparator is given in Table AP-11.1.

Table AP-11.1

Truth Table for Op-amp Comparator

Signal input voltage	*Output*
$V_a > V_b$	$F = 1$
$V_a < V_b$	$F = 0$
$V_a = V_b$	Previous value

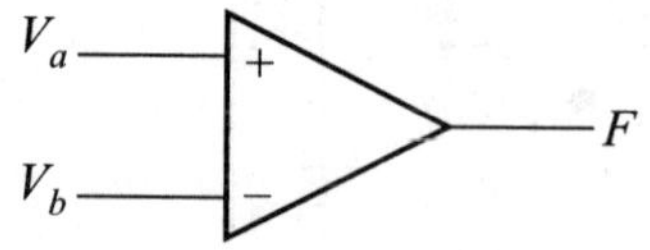

Fig. AP-11.10. Op-amp voltage comparator.

We will discuss the comparator shown in Fig. AP-11.11. It is run on a dual power supply and its inverting input terminal is grounded. A voltage V_{in} is applied at the non-inverting input. Since negative feedback is not used the Op-amp will be saturated even by a small increase in the input voltage. The saturation will be in the positive direction and the output voltage will jump to $+ V_{CC}$. If V_{in} is even slightly negative with respect to ground the Op-amp will be saturated in the negative direction and the output voltage will fall down to $- V_{sat}$ as shown in Fig. AP-11.11. The voltage at which the switch over occurs is called the trip point. In this case the trip point is 0 V. When V_{in} is larger than the trip point the output voltage is high and when it is less than the trip point, the output voltage is low.

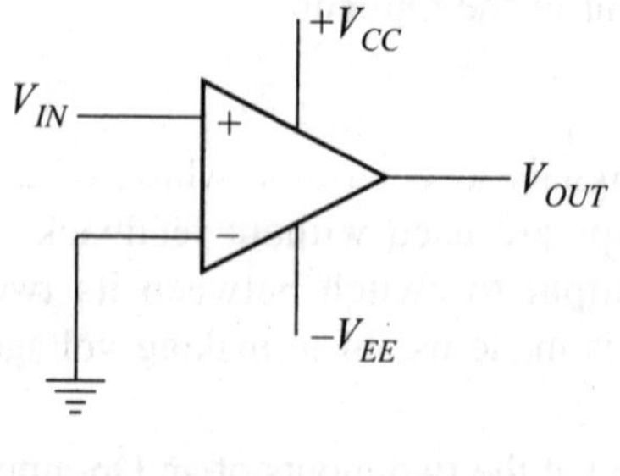

Fig. AP-11.11. (a) Op-amp Comparator.

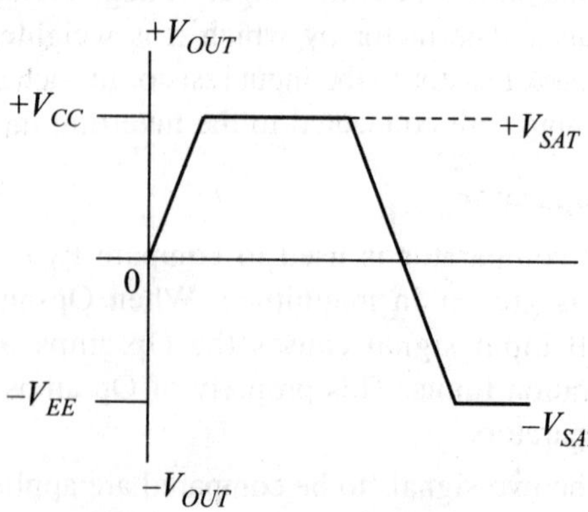

Fig. AP-11.11. (b) Comparator characteristics.

This Op-amp configuration is often referred to as a zero-crossing detector. It is more convenient to use a single power supply as shown in Fig. AP-11.12 (*a*). The V_{EE} input is now grounded and voltage is applied to the inverting output from V_{CC} through a voltage divider. The output can now swing between V_{CC} and $V_{CC} - V_r$, where V_r is the voltage at the inverting input. As long as V_{in} is greater than V_r, the output will be high and when it is smaller than V_r, the output will be low; but it will be positive in both the cases as shown in Fig. AP-11.12 (*b*).

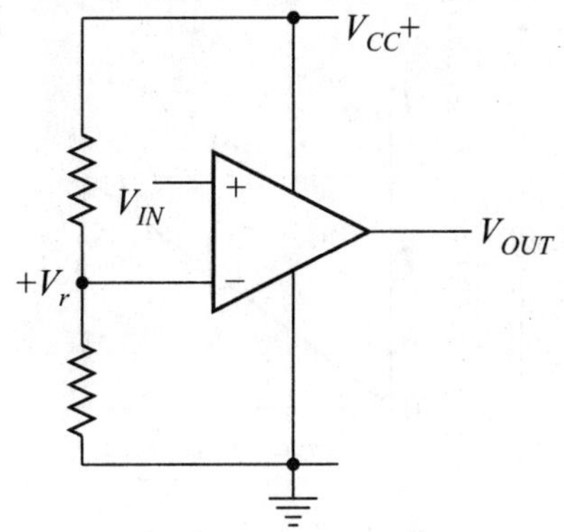

Fig. AP-11.12. (a) Comparator using single power supply.

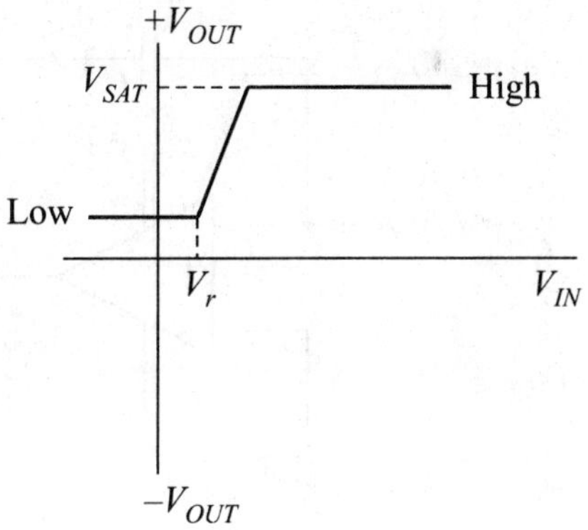

Fig. AP-11.12. (b) Comparator characteristics.

Voltage comparators are used in digital multimeters for the measurement of voltage. In this application the voltage to be measured is applied at the N.I. input of the Op-amp and it is compared with a staircase (ramp) voltage, which is generated internally. Figure AP-11.13 shows a rising ramp voltage which is applied at the inverting input of the Op-amp. As long as the voltage to be measured is greater than the ramp voltage, the output voltage is high. When the ramp voltage equals or is larger than the input voltage, the output goes low.

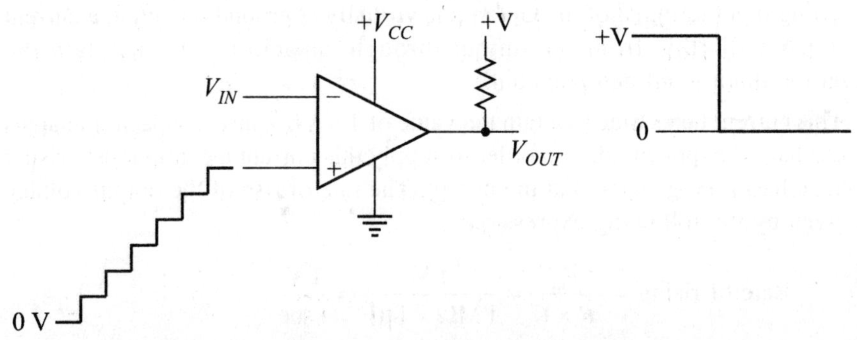

Fig. AP-11.13. Comparison of ramp voltage with the input voltage.

The ramp voltage takes a finite time, which serves as a measure to determine the magnitude of the input voltage. If the input voltage is large, the ramp voltage will take more time to come to the level of the input voltage.

Op-amps as Integrators

Integrators find application in digital voltmeters for producing rising and falling waveforms. Here we will consider how Op-amps can be used to generate rising waveforms. Figure AP-11.14 (*a*) shows how an Op-amp can be used to generate a rising waveform.

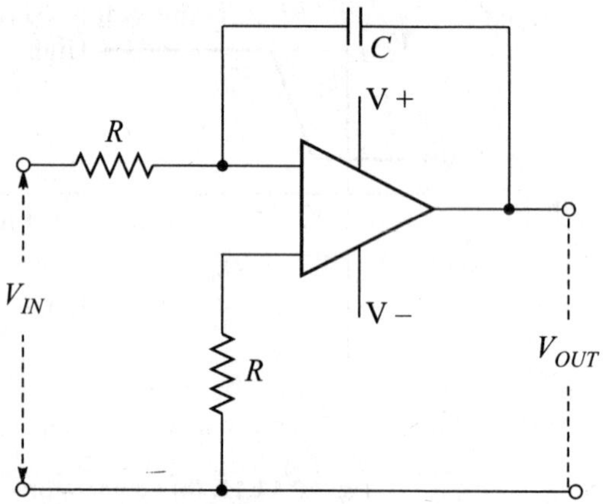

Fig. AP-11.14. (a) Integrator.

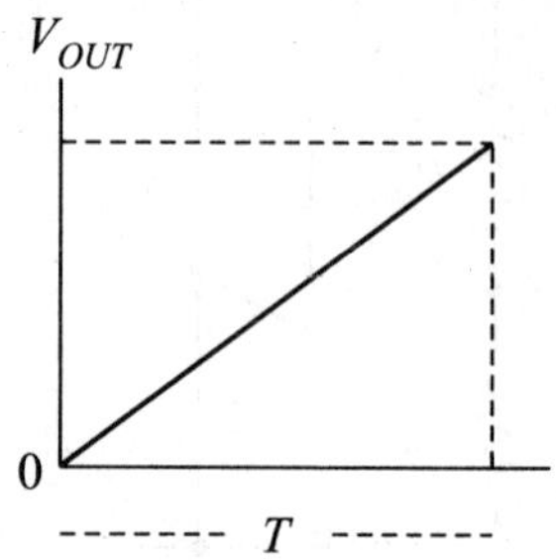

Fig. AP-11.14. (b) Integrator waveform.

Input to the inverting amplifier shown in Fig. AP-11.14 (*a*) is applied through a resistor *R* and feedback is applied by a capacitor *C*, which enables the Op-amp to function as an integrator. A resistor *R* of the same value is connected between ground and the N.I. input to reduce drift due to bias currents.

If a voltage of – 1 V is applied at the input and the input resistor is 1 M ohm, a current of 1 μA will flow through this resistor. Because the inverting input terminal of the Op-amp is virtually at ground potential, a current of 1 μA will flow from the output through capacitor *C* to maintain the inverting input at ground potential.

This current takes time to attain the value of 1 μA because a capacitor charges or discharges exponentially. In order to supply this current a continuously rising voltage has to be generated at the output. The rate of rise of the output voltage is given by the following expression:

$$\text{Rate of rising} = \frac{V_{in}}{R \times C} = \frac{1\text{ V}}{1\text{ M}\Omega \times 1\,\mu\text{F}} = \frac{1\text{ V}}{1\text{ sec}}$$

With the values we have chosen, the rate of rise of the output voltage is 1 V per second. The output voltage will continue to rise at this rate until the input voltage is changed or the Op-amp reaches saturation. For instance, with these values of V_{in}, R and C, the output voltage will be 1 V after a lapse of 1 second, if the input voltage remains at – 1 V.

The output voltage of the circuit which we have considered can be calculated from the following equation:

$$V_{out} = -\frac{1}{CR}\int_0^t V_{in}\,dt$$

You will notice from this expression that if V_{in} is positive, V_{out} is negative and *vice-versa*. If the output is integrated over a time t, the output voltage will be given by

$$V_{out} = \frac{1}{CR} \times V_{in}\,t \quad \text{or} \quad \frac{t}{CR} \times V_{in}$$

This shows that if the period of integration t, is equal to the time constant, CR, the output voltage will be equal to the input voltage; but it will be opposite in sign. In other words in one time constant the output voltage equals the input voltage.

If you refer to Fig. AP-11.14 (*b*), which shows the rise of output voltage with time, you will notice that the output voltage rises linearly with time. That is, the slope of the rise of voltage with time is constant. This is important because we cannot build a voltmeter if the rise of output voltage with time is not linear.

APPENDIX 12

Logic Trainer Circuits

12.1 LOGIC TRAINER LAYOUT

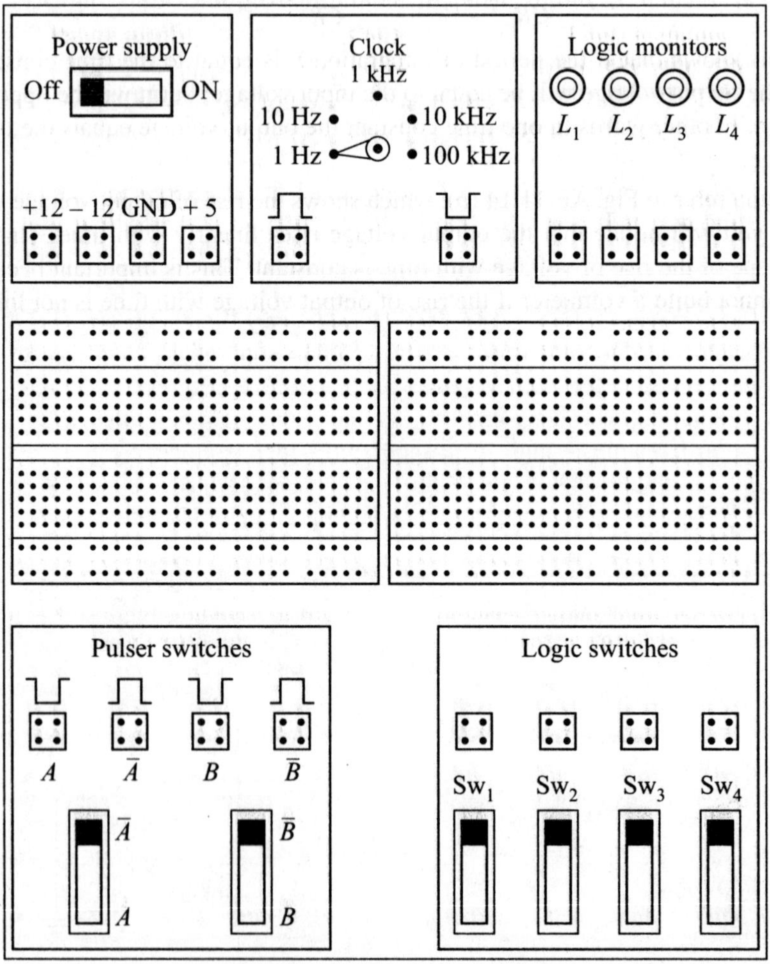

Fig. AP-12.1. Logic trainer layout.

12.2 LOGIC MONITOR

Figure AP-12.2 gives a circuit diagram for a bank of four logic monitors. The LEDs in this logic monitor begin to glow faintly as the voltage begins to rise above 2 V and glow brightly at 2.4 V. They do not glow if the input voltage is 2 V or under.

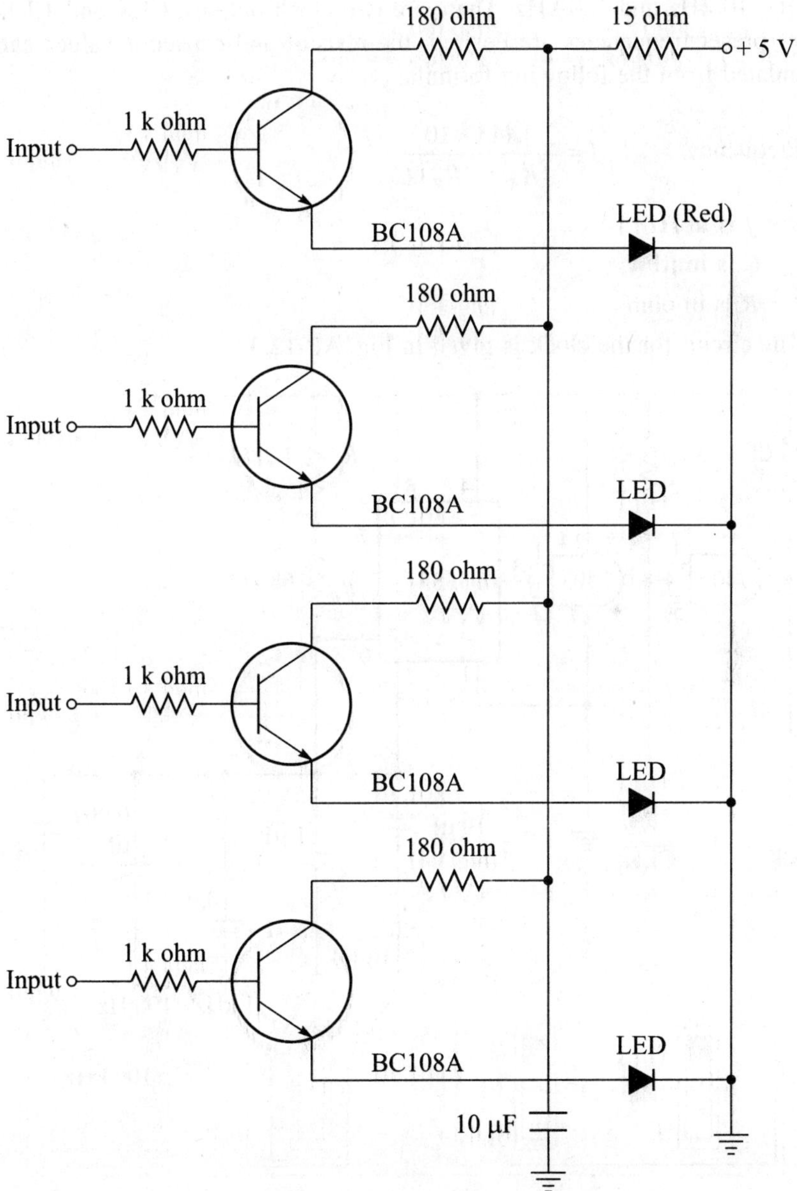

Fig. AP-12.2. Logic monitor.

LOGIC TRAINER CIRCUITS
CLOCK

12.3 CLOCK

The clock circuit described here has preset frequencies of 1 Hz, 10 Hz, 1 kHz, 10 kHz and 100 kHz. There are two clock outputs, CLK and $\overline{\text{CLK}}$. If other preset frequencies are desired, the resistor and capacitor values can be calculated from the following formula:

$$\text{Frequency,} \qquad f = \frac{1.443 \times 10^6}{(R_A + 2R_B)\,C}$$

where f is in Hz

C is in µF

and R is in ohm

The circuit for the clock is given in Fig. AP-12.3.

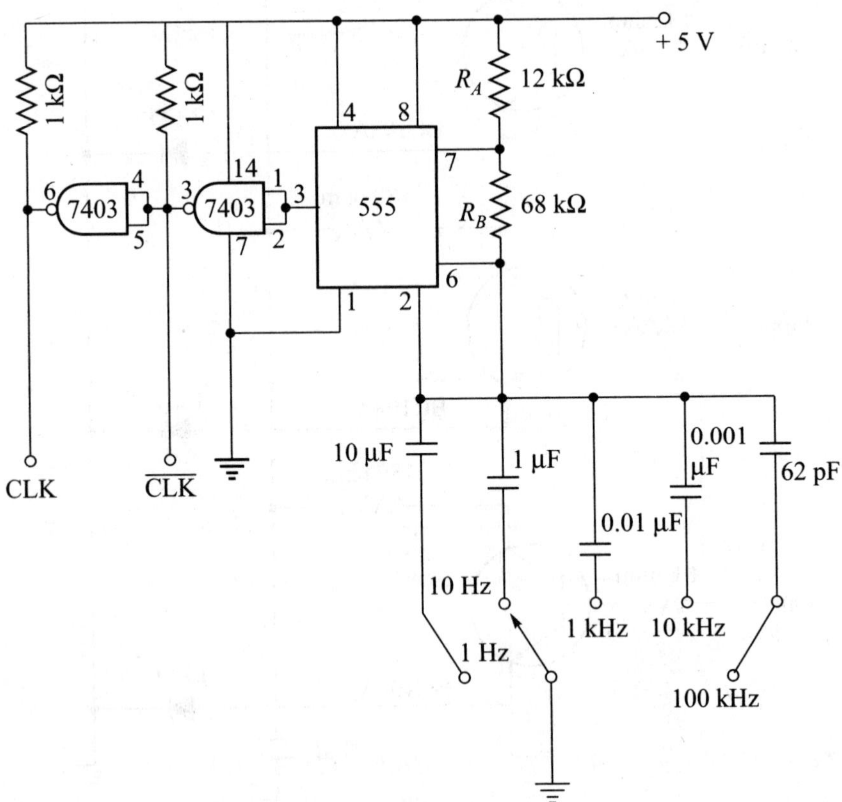

Fig. AP-12.3. Clock.

LOGIC TRAINER CIRCUITS
PULSER SWITCHES

12.4 PULSER SWITCHES

A circuit diagram for pulser switches is given in Fig. AP-12.4. These pulser switches provide pulses with complementary outputs.

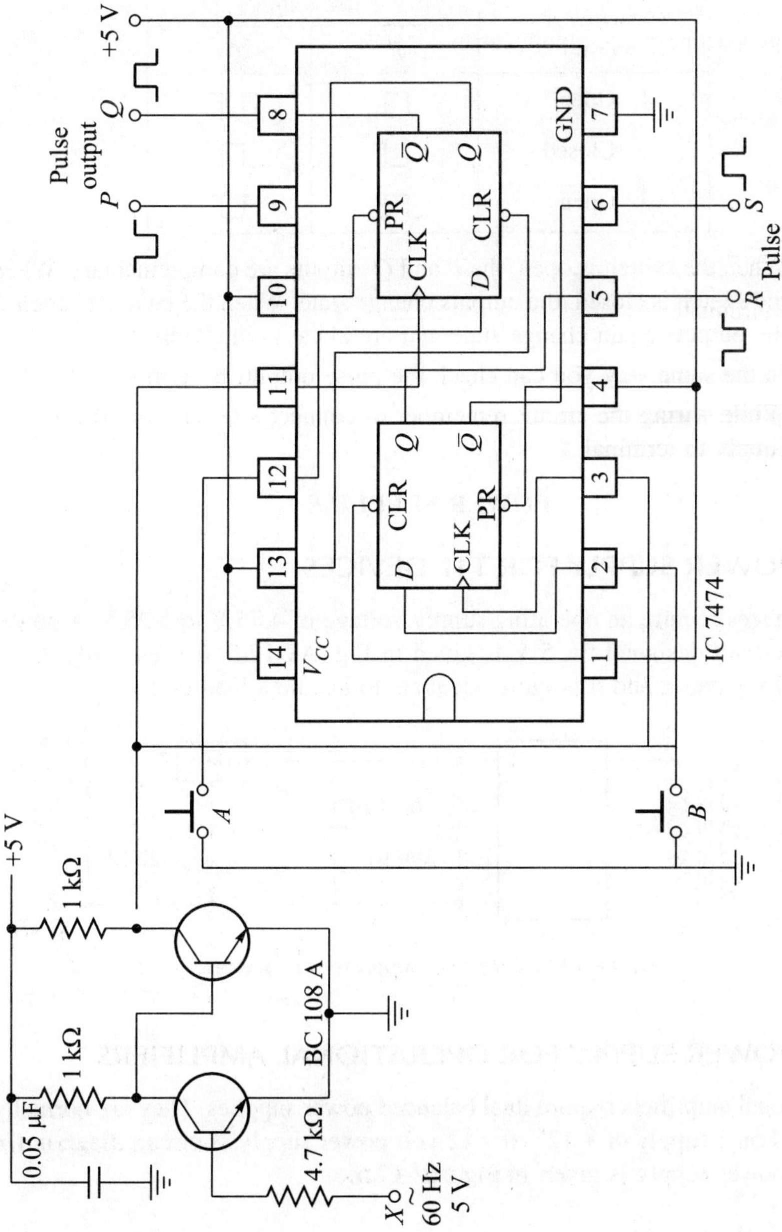

Fig. AP-12.4. Pulser switch.

After you have assembled the pulser switch, you may, if you like, test it as follows:

1. Connect pulse output *P*, *Q*, *R* and *S* to logic monitors and operate switch *A* as indicated in Table AP-12.1.

Table AP-12.1

Switch A	*Pulse outputs*	
	P	*Q*
1. Open		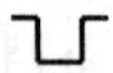
2. Closed		
3. Open		

2. When the switch is open, the *P* and *Q* outputs are complementary. When the switch is closed, the outputs change state. When the switch is opened the outputs again change state and are always complementary.
3. In the same way you can check the pulse outputs at terminals *R* and *S*.
4. While wiring the circuit remember to connect a 60 Hz (or 50 Hz) 5 V supply to terminal *X*.

POWER SUPPLIES

12.5 POWER SUPPLY FOR TTL DEVICES

TTL devices require an operating supply voltage of 4.75 V to 5.25 V. A power supply circuit designed for 5 V is given in Fig. AP-12.5. It uses 7805, three-terminal regulator, and it is quite adequate to handle a load of 1 A.

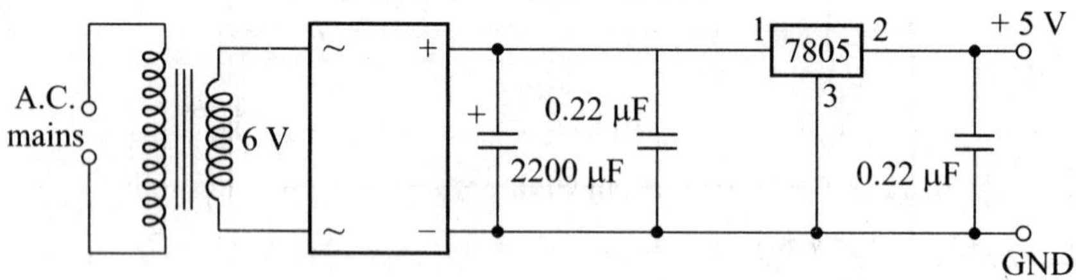

Fig. AP-12.5. 5 V power supply for TTL devices.

12.6 POWER SUPPLY FOR OPERATIONAL AMPLIFIERS

Operational amplifiers require dual balanced power supplies. They are normally operated on a supply of + 12 – 0 – 12 volt power supply. A circuit diagram for such a power supply is given in Fig. AP-12.6.

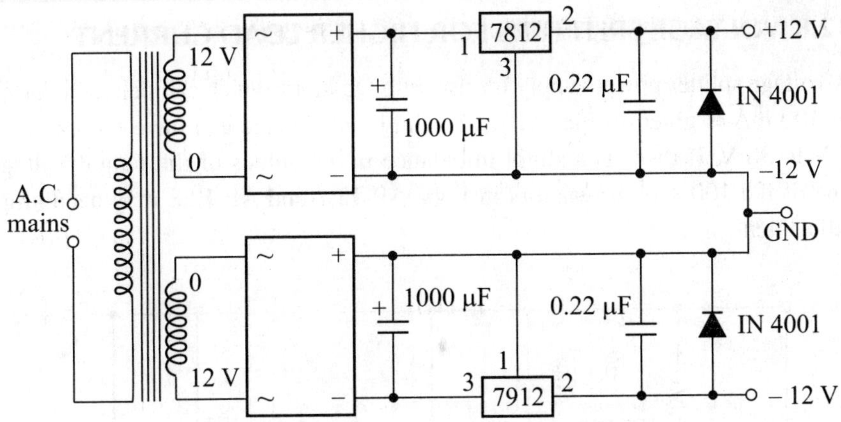

Fig. AP-12.6. Dual balanced power supply.

12.7 VOLTAGE SPLITTERS FOR OP-AMP POWER SUPPLIES

In the absence of dual balance power supplies, or when dual power supplies do not have the required voltage, voltage splitters are used to convert a single rail supply to a dual balanced power supply. A circuit for this purpose based on IC 741 is given in Fig. AP-12.7.

This splitter is used when the current requirement does not exceed 25 mA, which is the maximum current rating for IC 741. The input voltage can have a range from 5 V to 30 V.

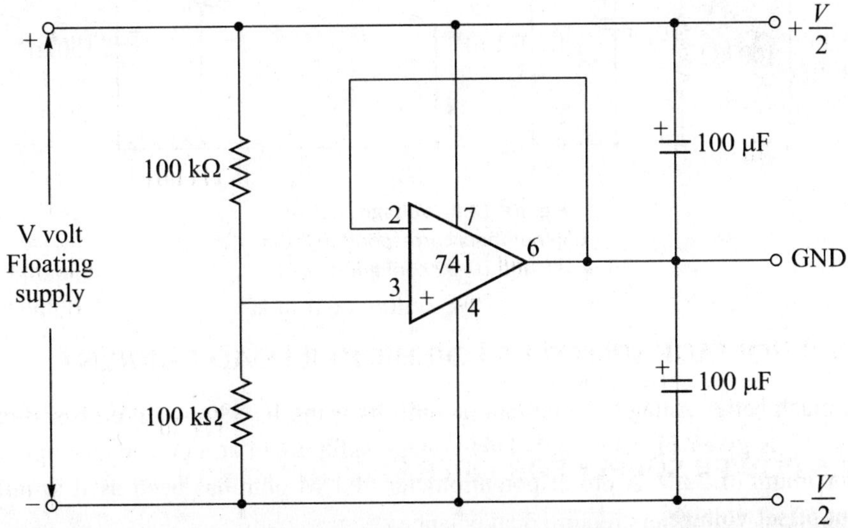

Fig. AP-12.7. Voltage splitter power supply. (Pin numbers correspond to 8-pin DIP and TO-99 configurations.)

12.8 VOLTAGE SPLITTERS FOR HIGHER LOAD CURRENT

A voltage splitter power supply for use with Op-amps which can deliver a current of 100 mA is given in Fig. AP-12.8. The input voltage can have a range from 5 V to 30 V. If there is a slight imbalance in the values of the output voltage, one of the 100 k ohm resistors in Figs. AP-12.7 and AP-12.8 may need slight adjustment.

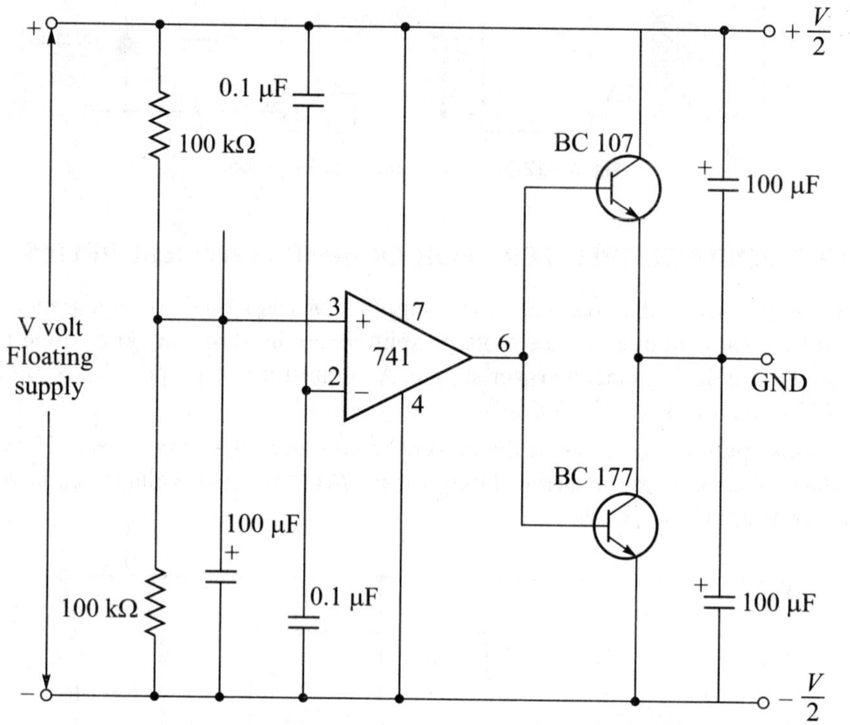

Fig. AP-12.8. Voltage splitter.
(Pin numbers correspond to 8-pin
DIP and TO-99 configurations.)

12.9 VOLTAGE SPLITTERS FOR HIGHER LOAD CURRENT

A much better voltage splitter can be built by using IC 380. A circuit based on this IC is given Fig. AP-12.9. This voltage splitter can accept votlages upto a maximum of 22 V. A preset potentiometer of 1 M ohm has been used to null out offset voltage.

This voltage splitter can be used where load currents do not exceed 300 mA provided an adequate heat sink is used.

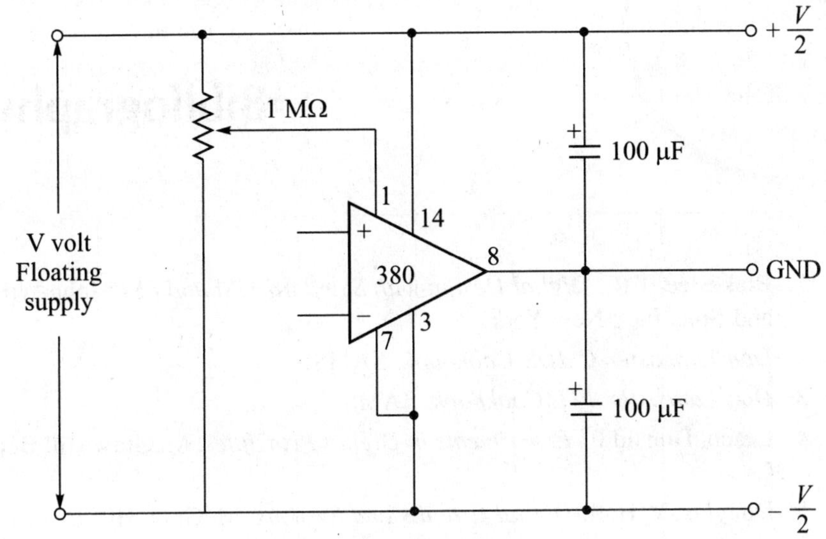

Fig. AP-12.9. Voltage splitter.
(Pin numbers correspond to 14-pin DIP configuration.)

Bibliography

1. Blakeslee, T.R.; *Digital Design with Standard MSI and LSI* : John Wiley and Sons Inc., New York.
2. *Don Lancaster CMOS Cookbook,* SAMS.
3. *Don Lancaster TTL Cookbook,* SAMS.
4. Leach, Donald P.; *Experiments in Digital Principles*: McGraw Hill Book Co.
5. Douglas, V. Hall; *Digital Circuits and Systems*: McGraw Hill Book Co.
6. Hnatek, E.R.; *A User's Handbook of Semiconductor Memories*: John Wiley and Sons Inc., New York.
7. Hnatek, E.R.; *A User's Handbook of Integrated Circuits*: John Wiley and Sons Inc., New York.
8. Kumar, V.; *Digital Technology*: *Principles and Practice*, New Age International (P) Limited, New Delhi.
9. Nashelsky, L.; *Digital Technology*: John Wiley and Sons Inc., New York.
10. *The TTL Data Book for Design Engineers*; Texas Instruments.
11. Texas Instruments Staff; *Designing with TTL Integrated Circuits*: McGraw Hill Book Co.
12. *TTL Logic Data Book, 1988* : SDLD001A : Texas Instruments.
13. *Overview of IEEE Standard 91-1984 SDYZ001* (Condensed overview of the symbolic language for digital logic circuits); Texas Instruments.
14. Tokheim, R.L.; *Digital Principles*; McGraw Hill Book Co.

Index